Constructing Insurable Risk Portfolios

Constructing Insurable Risk Portfolios offers a data-driven approach to devising risk retention programs that safeguard firms from a multitude of risks. Because firms face many risks, including fire damage to their buildings, liability from management misconduct, and external threats like cyberattacks, this book treats these potential liabilities as a "portfolio." Drawing inspiration from Markowitz portfolio theory, the text leverages techniques from probability, statistics, and optimization to build algorithms that construct optimal risk insurable portfolios under budget constraints.

Features

- Through engaging case studies and supporting statistical (R) code, readers will learn how to build optimal insurable risk portfolios.
- This book illustrates a frontier that depicts the trade-off between the uncertainty of a portfolio and the cost of risk transfer. This visual representation, mirroring familiar Markowitz investment tools, enables informed decision-making and easy adoption by risk advisors.
- This book lays the mathematical groundwork for constructing optimal insurable risk portfolios in an effective and aesthetically pleasing manner.
- For those interested in the detailed mathematical aspects of insurable risk portfolio optimization, comprehensive proofs and derivations are available in an online supplement.

This book equips students, academics, and practitioners with quantitative tools to analyze real-world risk portfolios. Additionally, it empowers financial analysts to provide data-driven insights that enhance their advisory roles for risk managers.

Edward W. Frees is an emeritus professor affiliated with the University of Wisconsin-Madison where he served as the Hickman Larson Chair of Actuarial Science. Until recently, he enjoyed a fractional research appointment with the Australian National University. He earned his PhD in mathematical statistics from the University of North Carolina at Chapel Hill. He works at the intersection of data science and actuarial studies; he is a fellow of the American Statistical Association and a fellow of the Society of Actuaries (SOA) (the only fellow of both organizations).

Prof. Frees has provided extensive service to the profession, including serving as the founding chairperson of the SOA Education and Research Section, member of the SOA Board of Directors, trustee of the Actuarial Foundation, editor of the *North American Actuarial Journal*, and actuarial representative to the Social Security Advisory Board's Technical Panel on Methods and Assumptions. He has written three books, edited a two-volume series on predictive modeling applications in actuarial science, and is editing an online, open source book, *Loss Data Analytics*.

Regarding his research, Prof. Frees has published extensively and won several awards for his work. He has won the Society of Actuaries' Annual Prize for the best paper published by the Society, the SOA's Ed Lew Award for research in modeling, the Casualty Actuarial Society's Hachmeister award, and the Halmstad Prize for the best paper published in the actuarial literature (four times).

Constructing Insurable Risk Portfolios

Edward W. Frees

CRC Press
Taylor & Francis Group
Boca Raton London New York

CRC Press is an imprint of the
Taylor & Francis Group, an **informa** business
A CHAPMAN & HALL BOOK

First edition published 2025
by CRC Press
2385 NW Executive Center Drive, Suite 320, Boca Raton FL 33431, U.S.A.

and by CRC Press
4 Park Square, Milton Park, Abingdon, Oxon, OX14 4RN

CRC Press is an imprint of Taylor & Francis Group, LLC

ISBN: 9781032745046 (hbk)
ISBN: 9781032745107 (pbk)
ISBN: 9781003469568 (ebk)

DOI: 10.1201/9781003469568

Typeset in CMR10
by codeMantra

To my brother Joe and his wife Mary

In honor of their courageous stand against ALS. Their steadfastness and inspiring spirit in the face of adversity have touched the hearts of all who know them.

Contents

Preface

Date: 20 December 2024

Firms face many risks, such as damage to their buildings due to fire, liability resulting from management misbehavior, and threats external to the organization like cyber attacks. Because the financial impact of these risks can be mitigated by using risk transfer tools like insurance, we think of them as *insurable* risks. Firms are financially responsible for these risks, essentially "owning" them. This naturally leads to the idea of managing them as a *portfolio*. This book shows how risk managers can manage a risk portfolio cost effectively by adapting financial concepts of diversification to insurable risk management.

Who is this book for? It is written for financial analysts who advise risk managers, such as commercial insurers, risk brokers, and reinsurers, who are responsible for risk retention and risk portfolio management. Risk managers are accustomed to using their knowledge of industry practices and experience in the field to make retention recommendations. This book introduces techniques that provide an objective framework to supplement, and corroborate, an advisor's thoughts. While no prior industry knowledge is required, familiarity with the applications will enhance the reading experience. As such, it is also suitable for students and other academics seeking an entry point into industry applications as well as industry analysts who would like an introduction to new developments. The book equips students, academics, and practitioners with quantitative tools for analyzing real-world risk portfolios. Further, it empowers financial analysts to advise risk managers with data-driven insights to enhance their advisory roles.

What is this book about? It introduces a novel approach for constructing data-driven insurable risk portfolios. Drawing inspiration from Markowitz portfolio theory, it leverages techniques from probability, statistics, and optimization to build algorithms that create optimal risk insurance portfolios under budget constraints. As such, outputs of the algorithm include a frontier that portrays the uncertainty of a portfolio versus a cost of transferring risks. A visual display of this frontier, mirroring familiar Markowitz investment tools, allows for informed decision-making and seamless adoption by risk advisors.

The Table of Contents provides an overview of the topics covered, organized into four parts. The first four chapters comprise the foundations.

- The first chapter sets the stage for the problem, emphasizing its importance, available resources such as data, and illustrating potential approaches through minimal scenarios.
- The second chapter summarizes different contractual forms commonly used for decomposing a risk into that portion retained and that transferred. This chapter also introduces tools for summarizing the uncertainty of the risk retaining and the cost of that transferred.

- The third chapter dives more deeply into the trade-off between the uncertainty of the risk retained and the cost of transferring risk. Different approaches historically used are summarized; the approach based on constrained optimization tools is emphasized. This approach will be used in the rest of the book.
- To simplify the development, the first three chapters emphasize a single risk (as does most of the academic literature) whereas the fourth introduces multivariate risks. This is done in the context of reinsurance (where the risk owner is an insurance company), introducing copulas to capture dependence among risks, and describing numerical optimization techniques for selecting the best risk retention strategy subject to budget constraints. This chapter also sets the problem of "retain-transfer" risks in the more general context of risk exchanges.

The second part, the fifth chapter through the eighth, constitutes the core material of the book. Specifically:

- The fifth chapter delves into the details of managing two risks where notions of portfolio uncertainty and risk transfer costs, with their sensitivities, are examined with a fair bit of mathematical rigor.
- The sixth chapter steps away from the mathematical development and seeks to provide context as to how quantitative liability portfolio tools could be applied. Risk management fundamentals are reviewed, together with case studies introduced in the first chapter and more recent developments such as robo-advising and peer-to-peer risk exchanges.
- Extending the fifth chapter, the seventh chapter examines the multivariate risk retention problem using simulation techniques. This approach generates an optimal insurable risk portfolio – **it provides the basic strategy advocated in the book.**
- To see how the seventh chapter strategy works with real applications, the eighth chapter practically applies the chapter seven strategy to the two case studies that were introduced in the introduction and developed further in the sixth chapter.

The third part, the ninth chapter through the twelfth, examines the properties and characteristics of insurable risk portfolio strategies.

- The ninth chapter focuses on the reliability of the optimal insurable risk portfolio results from three perspectives. First, reliability can be gauged by examining results under differing scenarios known as stress testing. Second, one can examine the "sensitivity" of optimization results to small changes in the input assumptions. Third, one can gauge the "robustness" of results by planning for the worst possible outcome within a broad region of an assumption and then try to optimize results.
- The tenth chapter further develops the notion of sensitivity by developing gradients that quantify the changes in the optimal decision variables as a result of a change in one of the inputs. Further, it uses this novel notion to quantify changes in optimal risk portfolio coefficients due to changes in the risk distribution parameters.
- The eleventh chapter establishes conditions that must hold at the optimal strategy. This framework is used to establish conditions to achieve (1) a binding budget constraint, (2) a balance among retention parameters at the optimum, and (3) boundary constraints.

- The twelfth chapter describes the impact of dependence on the composition of insurable risk portfolios.

The fourth part provides various materials that supplement the main body of the text.

Book Features

Practice Features

This book aims to develop the mathematical underpinnings for constructing optimal insurable risk portfolios in a manner that is both effective and aesthetically pleasing. Another key objective is the development of a suite of algorithmic tools that is potentially appealing to industry analysts.

To this end, this book dives deep into the Markowitz investment asset allocation problem as this approach is widely known to financial risk managers and will help them interpret results of liability portfolio risk retention. The case studies developed throughout the book provide further evidence as to how these tools might be used in practice. The Australian National University (ANU) case represents applications commonly confronted by risk management brokers and others who advise medium and large corporations on risk retention strategies. The Wisconsin property fund case represents applications commonly encountered by multi-line general insurers.

Important additional practice features include the accompanying statistical code and data, as follows.

Statistical Code and Data

With this text, you will be able to **construct optimal insurable risk portfolios**. Although mathematical justifications are important for understanding the "why" for risk portfolio applications, it is also crucial to be versed with the "how".

You can do so by downloading the sample data of this text and practicing with illustrative statistical code. Many sets of code are available in an online supplement to the book. In the beginning chapters, I remind readers of their availability using a note such as:

Code. See the *Online Supplement.*

In later chapters, I assume interested readers will be used to looking for the code and thus do not include explicit reminders. The findings and results are highlighted in the text, while the methods to achieve them are provided in the code. Shorter versions of the code are also available in an html supplement using convenient "show/hide" buttons. Even for analysts more interested in the mathematical development of insurable risk portfolio strategies, I believe that there are insights to be gained by working through the code. Certainly for analysts interested in business applications, the code provided here should be a useful springboard for additional investigations.

The data (and html versions of the code) are available at an online version of the book, Frees (2025a). The online supplement is available in Frees (2025b).

A Variety of Mathematical Techniques

This book addresses real-world insurable risks with a flexible toolbox of mathematical methods. It is not a book that develops a set of techniques and then demonstrates their relevance on a variety of applications. Rather, this is a book focused on a set of problems and then brings to bear on them a variety of mathematical techniques including constrained optimization, simulation, probabilistic arguments, and empirical analytics. (As with the author, ...) It is likely that one or more of these techniques will be new to readers and so you will learn about them in the context of insurable risks. For most sets of techniques, explanations and references are provided to get you started on a new area. Hopefully, you will see the opportunity to learn about a new technique, e.g., constrained optimization, as a benefit of this book.

The mathematical framework and supporting code are designed to allow practicing analysts and others beginning their research journeys to be able to quickly determine optimal insurable risk portfolios that can be utilized in business settings. This mathematical framework was not created in a vacuum but rather carefully selected from an extensive academic literature. The chapter sections on "Further Resources and Readings" will help readers appreciate the connections between the underpinnings of this framework and the literature. Readers will find that much of this academic literature is based on advanced mathematical constructs, which, although yielding deep insights, are not directed toward practical considerations. This book is meant to be a conduit between practical considerations and the academic literature; it also may help to re-direct the academic literature to place greater emphasis on practice features.

Technical Supplements

Curious about the mathematical intricacies behind insurable risk portfolio optimization? Detailed mathematical proofs and derivations are given in an online supplement to the book, indicated by the icon,

Under the Hood. See the *Online Supplement.*

For readers with the necessary mathematical background, the supplements provide additional material that can be useful when communicating ideas to technical audiences. Think of an analogy of driving a car. Most users do not need to understand the inner workings of an auto in order to drive a car. However, to advance the development of automobiles, engineers (a type of quantitative analyst) need to understand how a car engine works. To capture this perspective, an old saying is that one needs to understand "what is going on **under the hood**."

Also available is an html supplement that employs convenient "show/hide" buttons. These technical supplements reinforce and extend the results in the main body of the text by giving a more formal, mathematical treatment of the material. This treatment is actually a supplement because the primary results and findings are described in the main body of the text. While readers may skip over these technical supplements during their first read-through, they can serve as valuable resources for understanding the inner workings of the mathematical models and answer more detailed questions.

Suggested Courses. This book provides a versatile framework for advanced courses in actuarial and financial analytics. For all audiences, the main emphasis is on five key chapters: 1, 4, 6, 7, and 8. This basic treatment should include considerable attention to the statistical implementation of the algorithms and resulting frontiers. Additionally:

- **Undergraduates and entry-level graduate students.** Apart from the five key chapters, the foundational materials in Chapters 2, 3, and 5 should also be carefully explored, including the mathematical developments.
- **Advanced graduate students**. This audience is likely to be interested in mathematical developments and should review the theoretical underpinnings provided in the supplements. In particular, time should be devoted to the mathematical developments in Chapters 5 and 9-11. However, the material in Chapters 2 and 3 can be treated as review and covered lightly.

- **Industry analysts.** Alongside the basics, one could include a light introduction to ideas and notation from Chapters 2, 3, 5, and 9. Following Chapter 8, this audience will likely be interested in spending time exploring applications in industry.

Acknowledgments

I began writing this book six years ago, following my retirement from the University of Wisconsin-Madison. Six years is a long time for a book project; there are many who have helped me along the way who I wish to acknowledge. The ideas underpinning this book were conceived when I was writing an earlier paper, Frees (2017), and assisted Gee Lee with his doctoral thesis, Lee (2017). Gee (now at Michigan State) was not only there at the beginning but was also one of the prime collaborators in the development of the Wisconsin Property Fund data featured in this text. He also provided very helpful reviews of earlier drafts.

Soon after my retirement from the University of Wisconsin-Madison, I took on a part-time position at the Australian National University (ANU) and am grateful for their support of my work. In particular, I thank colleagues Adam Butt and Tim Higgins for their collaboration in developing the ANU data featured in this text and facilitating my presentation of a short course in which I was able to sharpen my thoughts on quantifying insurable risk portfolios.

Over the years, I have had opportunities to present this stream of work at numerous conferences and university seminars. In addition, I am grateful for the guidance and critical feedback from many reviewers, including:

- Vali Asimit (Bayes Business School)
- Catalina Bolancé Losilla (Universitat de Barcelona)
- Yiqing Chen (Drake University)
- Tolulope (Tolu) Fadina (University of Illinois Champaign - Urbana)
- Fei Huang (University of New South Wales)
- Shimeng Huang (University of Wisconsin-Madison)
- Hirokazu (Iwahiro) Iwasawa (Waseda University, Tokyo University, and the Institute of Actuaries of Japan)
- Jacie Liu (Australian National University)
- Emiliano Valdez (University of Connecticut)
- Ping Wang (Saint John's University)
- Shaun Shuxun Wang (Southern University of Science and Technology)
- Chengguo Weng (University of Waterloo)
- Armando Zarruk Rivera (Universidad Nacional de Colombia)
- Jinggong Zhang (Nanyang Technological University)
- Wenjun Zhu (Nanyang Technological University)

In addition, I am pleased to acknowledge the contributions from my colleague Peng Shi (University of Wisconsin-Madison) and for his permission to include our joint work on sensitivity that is featured in Chapters 9 and 10. Peng and I have collaborated on insurance analytics for many years and much of this book indirectly reflects his thoughts. I thank him for his generosity in sharing his insights with me and the world.

Saving the most important for last, I thank my family for their support. Ten thousand thanks to my brothers Randy, Guy, and Joe; my wife Deirdre; our sons Nathan and Adam; and Adam's wife, Claire, and their children, Frankie and Felicity.

1

Introduction

Chapter Preview. Organizations face many risks, some of which are readily quantifiable whereas others are not. To set the stage for the rest of the book, Section 1.1 describes some of the many risks that concern organizations. These risks can be transferred to other entities, such as insurers, for a price and this section emphasizes the importance of the economic activity of the insurance sector.

Instead of only abstract theories, Section 1.2 introduces two case studies that not only motivate the problem of transferring risks but also provide real examples where quantitative tools for managing multiple risks would be of value. These case studies will be referred to throughout the book as bases for pilot studies when new risk management tools are introduced.

Methods for managing portfolios of risks faced by firms are similar to, yet differ from, methods that have been developed for investment portfolios. Drawing parallels from the world of finance, Section 1.3 introduces asset portfolio management strategies. Motivated by the pioneering work of Markowitz (1952), managers of financial assets examine trade-offs between the expected gains of a portfolio of assets relative to its uncertainty, typically calibrated using the variance or standard deviation of returns. In the same way, the collection of an organization's non-asset risks can also be thought of as a portfolio and, as argued in this book, can be managed using methods analogous to asset portfolio techniques.

To appreciate the nature of the tools being developed here, Section 1.4 analyzes some "minimal scenarios," situations where we strip away almost all features that could make a tool practicable. Minimally complex by design, these scenarios are sufficient to demonstrate the resolution of a problem without any additional complexity that could make the resolution harder. Through these scenarios, we are able to demonstrate the importance of dependence among risks, introduce budget constraints as a cost of transferring risks, and provide advice on alternative forms of risk transfer.

Section 1.5 closes the chapter with an overview of the rest of the book.

1.1 Firms Face Multiple Risks

Organizations face a multitude of risks. The specific risks vary by industry, size of the firm, location of the business and many other factors. The risks can vary from broad concerns such as "economic slowdown" that affect virtually every organization to risks such as "regulatory changes" that are of concern to firms in a specific industry and regulatory jurisdiction. Some risks are readily quantifiable such as "property damage" whereas others such as "reputation damage" are less so.

One way to get insights into the breadth of risk that a firm faces is to look to concerns expressed by risk managers. To this end, I cite a survey of global risk managers conducted by Aon in 2019, Aon Risk Solutions (2019) (for another viewpoint, see a Deloitte survey, Insights (2019)). From the survey (Page 13) "Each year we offer respondents the chance to

DOI: 10.1201/9781003469568-1

assess their future risk landscape and project the top 5 risks that their organizations will face in three years' time." The survey identifies 69 risks, some which are insurable, some partially insurable, and some uninsurable. For example, the top five risks are:

1. Economic slowdown/ slow recovery - uninsurable
2. Damage to reputation/brand - partially insurable
3. Accelerated rates of change in market factors - uninsurable
4. Business interruption - insurable
5. Increasing competition - uninsurable

This survey was conducted in mid-2019 before the onslaught of the COVID pandemic. Interestingly, global risk managers only identified "Pandemic risk/ health crises" as rank number 60 risk factor. Unlike pandemic risks, we will analyze risks that a manager might transfer to another organization and so focus on insurable risks. From the survey, the top insurable risks (with their rank in the survey) are given in Table 1.1.

Although the techniques developed in this book can be applied to personal insurance and reinsurance, our focus is on transferring risks of companies. Even though insurance is only one mechanism for doing so, commercial insurance by itself plays an important role in the global economy. According to the Insurance Information Institute, commercial lines accounted for 50.8% of non-life insurance premiums in 2022. From the OECD (2022), non-life premiums account for about 51.7% of insurance premiums (with life insurance taking up the balance). And finally, also from the OECD, insurance premiums are approximately 9.5% of world GDP (gross domestic product). Taken together, one can argue that commercial premiums are approximately 2.3% of world GDP, a significant portion of the world economy.

To summarize general (non-life) insurance, Table 1.2 employs the classification used by the *Organization for Economic Co-operation and Development (OECD).* The publication OECD (2020) provides many statistics, including premiums written by class of insurance business. Table 1.2 gives a sense of relative activity by summarizing gross premiums paid in each class of insurance; from OECD (2020), "Gross premiums are total premiums written, excluding any premium taxes or other charges, but before deduction of commission or reinsurance outwards. It is acknowledged that the inclusion of reinsurance will mean that there is some element of double-counting in the figures provided."

In this book, a risk is said to be *insurable* if it potentially can be transferred to another party for a fee. Insurance is the most important mechanism for doing so but others such as self-insurance pools, peer to peer risk exchanges, and so forth are also available. Descriptions of these alternatives is a topic taken up in Chapter 6.

TABLE 1.1
Top Insurable Risks Facing Firms.

Risk	Risk
4. Business interruption	44. Directors and Officers personal liability
19. Counter-party credit risk	47. Fraud
21. Property damage	52. Theft
22. Environmental risk	55. Terrorism/sabotage
23. Weather natural disasters	56. Safety and Pharmacovigilance
24. Third party liability	58. Embezzlement
28. Injury to workers	61. Harassment/discrimination
40. Product recall	66. Kidnap and ransom

Source: Aon Risk Solutions (2019)

TABLE 1.2
World Premiums by Classes of Non-Life Insurance.

Classification	Premiums	Classification Components
Motor Vehicle Insurance	589,572	*Land Vehicles* (other than railway rolling stock) covering all damage to or loss of: land motor vehicles, and land vehicles other than motor vehicles
Marine Aviation and Other Transport Insurance	32,783	*Railway rolling stock and other transport* covering all damage to or loss of railway rolling stock *Aircraft* covering all damage to or loss of aircraft *Ships* (sea, lake, and river and canal vessels) covering all damage to or loss of: river and canal vessels, lake vessels, and sea vessels *Aircraft Liability* covering all liability arising out of the use of aircraft (including carrier's liability) *Liability for Ships* (sea, lake, and river and canal vessels) covering all liability arising out of the use of ships, vessels or boats on the sea, lakes, rivers or canals (including carrier's liability)
Freight Insurance	35,132	*Goods In Transit* including merchandise, baggage and all other goods consisting of all damage to or loss of goods in transit or baggage, irrespective of the form of transport
Fire and other Property Damage Insurance	377,572	*Fire and Natural Forces* covering all damage or loss of property (other than land vehicles, railway rolling stock, aircraft, ships and goods in transit) due to: fire, explosion, storm, natural forces other than storm, nuclear energy, and land subsidence *Other Damage to Property* covering all damage to or loss of property (other than land vehicles, railway rolling stock, aircraft, ships and goods in transit) due to hail or frost, and any event such as theft, other than those mentioned under Fire and Natural Forces

(*Continued*)

TABLE 1.2 ***(Continued)***
World Premiums by Classes of Non-Life Insurance.

Classification	Premiums	Classification Components
Pecuniary Loss Insurance	64,620	*Credit* including insolvency (general), export credit, installment credit, mortgages, and agricultural credit
		Suretyship
		Miscellaneous Financial Loss including employment risk, insufficiency of income (general), bad weather, loss of benefits, continuing general expenses, unforeseen trading expenses, loss of market value, loss of rent or revenue, indirect trading losses other than those mentioned above, other financial loss (non-trading), and other forms of financial loss
General Liability Insurance	234,993	*General Liability* covering all liability other than motor vehicle liability, aircraft liability and liability for ships
Accident and Health Insurance	1,073,109	*Accident* including industrial injury and occupational diseases, consisting of fixed pecuniary benefits, benefits in the nature of indemnity, combinations of the two, and injury to passengers
		Sickness consisting of fixed pecuniary benefits and benefits in the nature of indemnity
Other Non-Life Insurance	102,875	*Legal Expenses and Costs of Litigation, Assistance, Miscellaneous*
Treaty Reinsurance	5,665	*Treaty Reinsurance*
Total	2,516,320	

[a] *Note:* 2018 Premiums are in millions of U.S. dollars.
[b] *Source:* OECD, https://stats.oecd.org/Index.aspx?DatasetCode=INSIND

1.2 Illustrative Case Studies

To motivate the conceptual development, the text features two illustrative case studies. The Australian National University (ANU) case describes a situation comparable to that commonly encountered by risk managers at medium and large corporations. The Wisconsin Property Fund case describes an insurance risk pool that has many of the features of a typical general insurance company that offers multiple coverages, or lines, of business.

1.2.1 Australian National University

Motivated by the importance of insurance to the global economy, we are interested in risk transfers of companies. To be concrete, the "company" that we consider in detail is the Australian National University (ANU). This university is a large, complex organization that faces many risks. To illustrate its size, according to its annual report ANU (2020), in 2020 it enjoyed an asset base of about 4.6 billion AUD (which is approximately 3.4 billion USD). To illustrate its complexity, its financial statements summarize many risks including property damage, general liability, cyber security, and so forth. There are 15 risks in total that are further described in Section 6.3.2; these are the "usual suspects," property and liability risks that are typically encountered by many firms. In our quantitative work, we represent losses from these risks as $(X_1, \ldots, X_{15})$. In absence of risk transfer agreements, ANU has responsibility for the aggregate sum of risks,

$$S = X_1 + \cdots + X_{15}.$$

Although the university is responsible for the aggregate sum of losses, it follows common practice and tracks each risk type separately. Different risk types exhibit different distributions; combining them without tracking each risk type can result in poor management decisions.

As a subject of a case study, ANU is ideal in part because we have access to detailed information about its risk management processes. In addition, in 2020 it suffered a major property loss (250 million AUD) meaning that its risk appetite has been subject to renewed scrutiny, see Net (2020). That is, for this year the realized value of X_1 was 250 million, suggesting that we will need to employ models that allow for heavy-tail distributions that permit very high values of outcomes with not insignificant probabilities.

1.2.2 Wisconsin Property Fund

We also demonstrate the conceptual development of insurable risk portfolios using data from the Wisconsin Local Government Property Insurance Fund. The fund was established to provide property insurance for local government entities that include counties, cities, towns, villages, school districts, fire departments, and other miscellaneous entities. Properties covered under this fund include government buildings, vehicles, and equipment. For example, a county may need coverage for its snow plowing trucks, in addition to its buildings and their contents; Frees and Lee (2015) and Frees et al. (2016). More details about the project may be found at the Wisconsin Property Fund Site.

Risk Types. The data consist of six coverage groups; building and contents (BC), contractor's equipment (IM), and four types of motor vehicle coverages. For the motor vehicle types, there are collision and comprehensive coverages, subdivided into new and old vehicles. Hence, an organization can potentially have collision coverage for new vehicles (CN), collision coverage for old vehicles (CO), comprehensive coverage for new vehicles (PN),

TABLE 1.3
Property Fund Risk Types.

Coverage	Segments	Description
Building and Contents		This coverage provides insurance for buildings and the properties within. In case the policyholder has purchased a rider, claims in this group may reflect additional amounts covered under endorsements.
Contractors Equipment		IM, an abbreviation for inland marine, is used as a coverage code for equipment coverage, which originally belong to contractors.
Collision		This provides coverage for impact of a vehicle with an object, impact of vehicle with an attached vehicle, or overturn of a vehicle.
Comprehensive		Direct and accidental loss or damage to motor vehicle, including breakage of glass, loss caused by missiles, falling objects, fire, theft, explosion, earthquake, windstorm, hail, water, flood, malicious mischief or vandalism, riot or civil common, or colliding with a bird or animal.
	N - New	This code is used as an indication that the coverage is for vehicles of current model year, or 1 to 2 years prior to the current model year.
	O - Old	This code is used as an indication that the coverage is for vehicles three or more years prior to the current model year.

and comprehensive coverage for old vehicles (PO). These six coverage groups are further described in Table 1.3.

In modeling work, one may label the six risk types as $(X_1, \ldots, X_6)$ or use the more informative labels $(X_{BC}, X_{IM}, X_{CN}, X_{CO}, X_{PN}, X_{PO})$. Further, as will be seen in Section 6.3.3, we will consider 1,098 members of the fund. If one uses the label i to identify each member, then we might be interested in aggregate claims for that member. Without risk transfers, this is

$$S_i = X_{i,BC} + X_{i,IM} + X_{i,CN} + X_{i,CO} + X_{i,PN} + X_{i,PO}.$$

Moreover, we will be interested in risk transfer agreements at the fund level. In absence of any risk transfer agreement, the fund is responsible for the aggregate losses, $S = \sum_{i=1}^{1098} S_i$.

1.3 Investment Portfolio Strategies

This book approaches diversifying risky portfolios similar to the strategy pioneered by Markowitz (1952) in his development of asset portfolio theory. In this approach, the investor considers p asset risks and wishes to determine c_j, the proportion of wealth allocated to the jth risk. The goal is to minimize the uncertainty of the portfolio subject to a budget constraint based on the portfolio earning a required expected return level. Example 1.1

provides a concrete example of risks typically encountered in the investment allocation problem.

Example 1.1. Portfolio of Insurance Stock Returns. I consider a portfolio of $p = 8$ asset risks from the insurance sector. For illustrative purposes, the largest companies (by market capitalization) in each of eight countries were chosen. With ticker symbols in parens (), they are:

- United Health (UNH) - *USA*
- AIA Group Limited (1299.HK) - *Hong Kong*
- Ping An Insurance (Group) Company of China, Ltd. (2318.HK) - *China*
- Chubb (CB) - *Switzerland*
- Allianz (ALV.DE) - *Germany*
- Aon (AON) - *UK*
- AXA (CS.PA) - *France*
- Tokio Marine (8766.T) - *Japan.*

Daily stock prices were extracted from Yahoo Finance for the period 5 January 2015 to 14 September 2023 and are displayed in Figure 1.1. Over the entire period, United Health had the largest price increase.

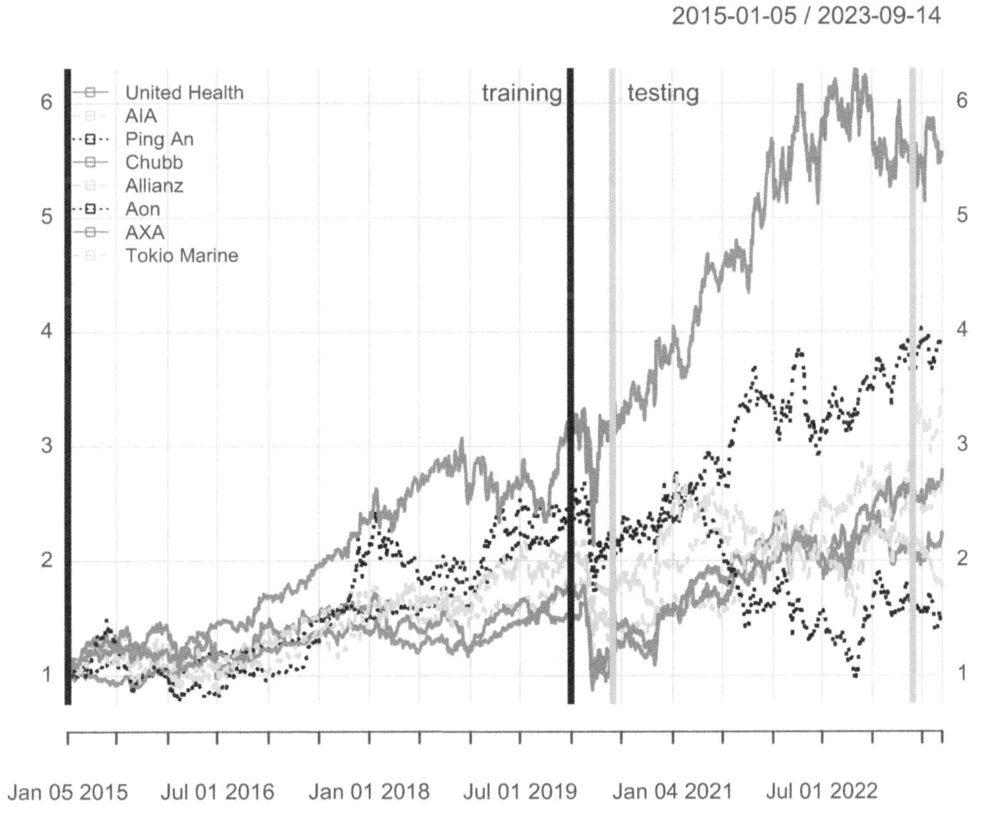

FIGURE 1.1
Normalized Stock Prices of Eight Insurers.

TABLE 1.4
Insurance Sector Performance Measures.

	Train Ann Return	Train Std Dev	Train Sharpe (Rf=0)	Test Ann Return	Test Std Dev	Test Sharpe (Rf=0)
United Health	0.254	0.213	1.193	0.206	0.220	0.938
AIA	0.147	0.239	0.613	0.022	0.311	0.070
Ping An	0.189	0.264	0.714	-0.108	0.377	-0.286
Chubb	0.088	0.160	0.546	0.162	0.237	0.684
Allianz	0.150	0.193	0.779	0.092	0.228	0.403
Aon	0.182	0.170	1.072	0.183	0.235	0.776
AXA	0.116	0.241	0.480	0.211	0.245	0.862
Tokio Marine	0.119	0.267	0.447	0.295	0.248	1.190

As is common in finance, insurance, and other areas that employ data science, I split the data into two components. The first is a five-year *training* period, 5 January 2015 to 31 December 2019, that will be used to develop alternative portfolios. The second is a three-year *test* period, 1 June 2020 to 31 May 2023, that will be used to assess the performance of the portfolios. From Figure 1.1, one can see that during the period omitted (in between the training and test periods), most stock prices dropped dramatically due to the global effect of the Covid pandemic. There are various ways of modeling such outlying features - for this analysis, I choose to omit this aberrant behavior. Table 1.4 summarizes several performance measures over training and testing periods.

R *Code.* The *Online Supplement* provides Example 1.1 illustrative statistical code.

For notation, let $R_1, \ldots, R_8$ be the returns from the eight risks. With investment allocations c_j that sum to one, one can form the portfolio return

$$PR = c_1 R_1 + \cdots + c_8 R_8.$$

The goal is to determine the investment allocations c_j in a way that minimizes the uncertainty of this portfolio return subject to a budget constraint.

Liability Risk Portfolio Strategy

For diversifying a portfolio of insurable risks, one can also start with a set of p risks. But, instead of these risks being assets that might be purchased, they are obligations of a person or firm. The risk owner may transfer a portion of each insurable risk to another organization (e.g., an insurance pool or an insurer) using one of several potential mechanisms available (e.g., an upper limit, deductible, coinsurance, and so on), each having parameters θ_j that calibrate the extent of the transfer. The uncertainty of the retained risk portfolio is to be minimized. The minimization is subject to a budget constraint based on restricting the amount that a risk owner is able to pay for the risk transfer. Because the goal is to minimize the uncertainty through risk transfer and retention, think of this too as optimal diversification of the portfolio.

Comparing Asset Investment and Liability Risk Portfolio Strategies

Despite investments and risk management arising from very different application areas, their overall premise is similar, minimizing the amount of uncertainty from a portfolio of risks subject to costs associated with acquiring the risks.

However, in their applications, they differ in one fundamental aspect; asset allocations have traditionally focused on the proportion of wealth to be invested in each risk as the decision variable. This is a *linear* allocation. In contrast, contracts for managing insurable risks routinely utilized non-linear mechanisms such deductibles and upper limits. As will be seen, the linearity in asset allocations means that investment strategies can be analyzed using linear and convex optimization methods; these methods are readily scalable to large numbers of risks (e.g., in the thousands). In contrast, the non-linearities in managing insurable risks means that, in many important problems, convexity is not available. This means that numerical optimization methods provide local, not global, solutions and can typically be implemented based on far fewer decision variables (e.g., twenty will be large, not in the thousands).

There are other less prominent, but nonetheless important, distinctions between applications in investment and risk management.

1. Asset allocations have traditionally used the variance and standard deviation as measures of uncertainty. Even for assets, they are not good practical measures of risk since they penalize both the unwanted high losses and the desired high gains in the same way. Further, these measures emphasize deviations from central tendencies. In contrast, alternative measures for risk (such as the value at risk) focus on extreme outcomes that can be far from the center of the distribution.
2. Another distinction is the importance of *boundary cases*. In investments, an allocation of $c_j = 0$ means no investment in the jth asset and similarly, in many cases a retention parameter $\theta_j = 0$ means no (or full) retention of the jth risk. In investments, many analysts also permit short sales (a negative investment essentially achieved by borrowing money) so that c_j is allowed to become negative and meaning that $c_j = 0$ is not even a boundary case. In contrast, in insurance the concept of "negative insurance" is typically viewed as contrary to public policy (it becomes a gamble) and so is not permissible. Thus, there is a large insurance literature on boundary cases that attempts to identify when a risk (typically, a person) is fully or only partially insured.
3. Yet another distinction is the availability of information about the distribution of a risk. In investments, the typical model assumes that risk outcomes are observable on a public market and that all parties have knowledge of its distribution. In insurance applications, information can be more private. As a consequence, pricing the transfer of risk is more commonly realized as a transaction between two parties, not a price that is observed on a public market. Therefore, care is needed when discussing each party's perspective on a transfer and in particular when discussing budget constraints involving the "cost of acquiring" an insurable risk.

1.4 Risk Transfers and Effects of Dependence

Management techniques for handling the many risks that organizations face is further discussed in Sections 6.1 and 6.2. As already suggested by the economic activity summarized in Table 1.2, risk transfer techniques such as insurance play a prominent role. Further, it is not surprising that these multiple risks naturally depend upon one another. This section gives our first introduction on how one can quantify the effects of dependence when transferring risks.

Retained Risk. To emphasize the impact of the dependence, this section uses a *minimal scenario* starting with only two risks. Specifically, suppose that a company already has a suite of risks, denoted by the random variable X_1, and is considering transferring either all or a portion of a second risk, X_2. The suite in X_1 are sometimes known to as *background* risks in that they are not part of the risks being transferred but, as will be seen, can influence risk management decisions. Companies want to transfer risks that are potentially harmful and so agreements specifying that the firm retains $\min(X_2, u)$ and transfers an amount $X_2 - \min(X_2, u)$ are common. Here, the fixed constant u is known as an *upper limit* of risk that the company will retain. Mildenhall and Major (2022) refer to upper limit contracts as the "basic building block of insurance" and so these contracts provide an appropriate starting point for our investigations. With this, the total risk retained by the company is

$$S(u) = X_1 + \min(X_2, u).$$

There can be other expenses in the total retained risks including claim adjustment costs for retained risk, as well as external (brokers, consultants and other vendors) and internal (staff and related) risk management costs. However, in keeping with our minimal scenario, for now let us restrict attention to $S(u)$ as a measure of total retained risk.

Additional Motivation. To provide more context, suppose that X_2 represents a company's workers' compensation claims. The upper limit u might be set as part of a so-called *large deductible program* with an insurer. For this type of an agreement, the insurer is the first payer on all losses and it will bill the company (the insured) for the amount that falls less than u. If the company does not or no longer can pay its claims, then the insurer is responsible for paying those losses. To make sure that the company will honor its obligation to cover claims less than u, insurers often require an amount to be set aside as collateral, typically as a line of credit, cash, or bonds. This collateral depends on the distribution of X_2 but may also be determined accounting for the company's ability to meet all of its obligations through the distribution of $S(u)$.

Risk Transfer Cost. In general, the risk amount $X_2 - \min(X_2, u)$ is transferred (to an insurer or a pool) and naturally there is a cost for this transfer. In this minimal scenario, we use the *risk transfer cost*

$$RTC(u) = \mathrm{E}\left\{X_2 - \min(X_2, u)\right\}, \tag{1.1}$$

that is, the expected, or "fair," cost. Other cost structures will be considered in subsequent sections and we will see that typically the cost of insurance depends on the market supply of insurance, among other factors. The key feature of this structure is that the cost decreases as the amount retained increases. For example, in equation (1.1) $RTC(u)$ decreases with the upper limit u.

In keeping with the principle of constructing a minimal scenario devoid of unnecessary complications, assume that X_1 and X_2 are both distributed Uniform(0,1). With this, it is easy to check that $RTC(u) = \frac{1}{2}(1 + u^2) - u$. However, these two risks may depend on one another and so we consider their joint distribution function. In particular, the joint distribution function of two uniform (0,1) variables is known as a *copula*. For this initial example, we utilize Frank's copula as it is commonly used in insurance studies. To keep the book self-contained, an introduction to copulas is presented in Section 4.2 where Frank's copula is formally defined.

1.4.1 Minimal Scenario 1. Effects of Dependence

There are many ways to measure the uncertainty of $S(u)$, the total risk retained by the firm after the risk transfer; an overview of these measures will appear in Section 2.2.3. All measures are based on the distribution of $S(u)$ and so we begin with its distribution function.

Using a Distribution Function to Measure Uncertainty. In particular, if one evaluates the distribution function at a particular level, say Q, then in risk management applications one can interpret Q as the "capital" required by the firm for the retained risk. For example, Q might represent the collateral in a large deductible program. That is, if a company holds assets Q to cover retained risks, then $\Pr(S(u) \leq Q)$ gives the probability that retained risks will be less than these assets. A company wishes to set aside the appropriate amount of assets to cover risks. Too low a value of Q means that assets are insufficient to cover retained risks meaning that the company is under insured. Too high a value of Q means that capital is unnecessarily deployed on risk transfer, at the expense of other business enabling activities.

Using a Quantile to Measure Uncertainty. To complement the distribution function, we also look to the value at risk, denoted as VaR. The value at risk is simply a quantile or an inverse distribution function. For example, we use $VaR_{0.95}$ to represent the 0.95 confidence level of the VaR. This is the amount so that there is a 95 percent probability that retained risks are less than or equal to VaR. Risk managers regularly employ value at risk concepts and they will take on a prominent role in our development. Section 2.2.3 provides additional background on VaR.

By assumption, the risk transfer cost defined in equation (1.1) decreases as the amount of retained risk increases. It is not surprising that the probability of having sufficient capital

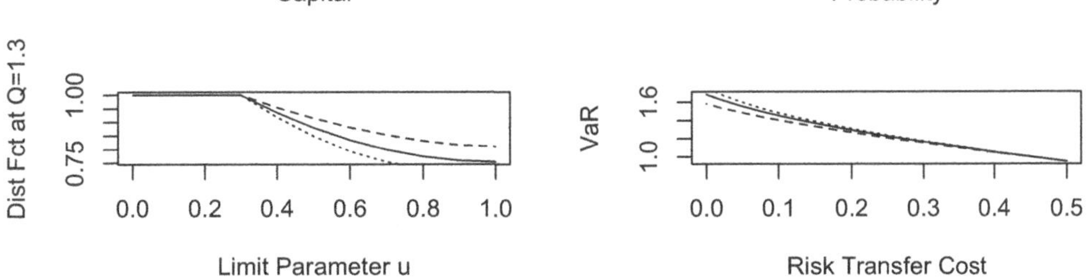

FIGURE 1.2
Portfolio Uncertainty in a Minimal Scenario. The solid black line corresponds to the case of independence. The dotted line is for a positive Spearman correlation 0.3, the dashed line is for a negative correlation -0.3.

decreases as the amount of retained risk increases. This example shows that this probability is a function of the dependence between the two risks. To support these observations, Figure 1.2 summarizes several important points.

- The upper left-hand panel shows the distribution function $\Pr(S(u) \leq Q)$ as a function of the capital amount Q. For this demonstration, I use $u = 0.6$ to limit the retention for the second risk. As anticipated, the distribution function increases as the capital Q increases. For each Q, the probability is smaller for positive dependence, the dotted line that corresponds to a Spearman correlation of 0.3, and larger for negative dependence, the dashed line that corresponds to a Spearman correlation of -0.3.
- The upper right-hand panel gives the same information in terms of the value at risk $VaR_{0.95}$. As the upper limit for the first retained risk is 1 and is 0.6 for the second risk, it is not possible to exceed 1.6. For any probability level, the value at risk for positively associated risks is greater than the independence case, and is lower for negatively associated risks. This is consistent with the information presented in the upper left-hand panel and conforms to our intuition.
- The lower left-hand panel presents the distribution function, fixing the capital level at $Q = 1.3$, but allowing the retention limit u to vary. If $u \leq 0.3$, then the combined risk S must be below $Q = 1.3$ and so the probability equals one. If $u > 0.3$, then the probability decreases as u increases because the amount of retained risks increases.
- The lower right-hand panel presents the value at risk $VaR_{0.95}$ compared to the risk transfer cost, $RTC(u)$, over different choices of u. This is the trade-off the portfolio manager faces, smaller risk transfer costs mean a greater need for retained capital. Note that the risk transfer cost does not "depend on the dependence" in contrast to the value at risk.

Figure 1.3 emphasizes the effects of dependence.

- The left-hand panel presents the distribution function at $Q = 1.3$ and $u = 0.6$ but allows the dependence, as quantified through Spearman's correlation, to vary. For

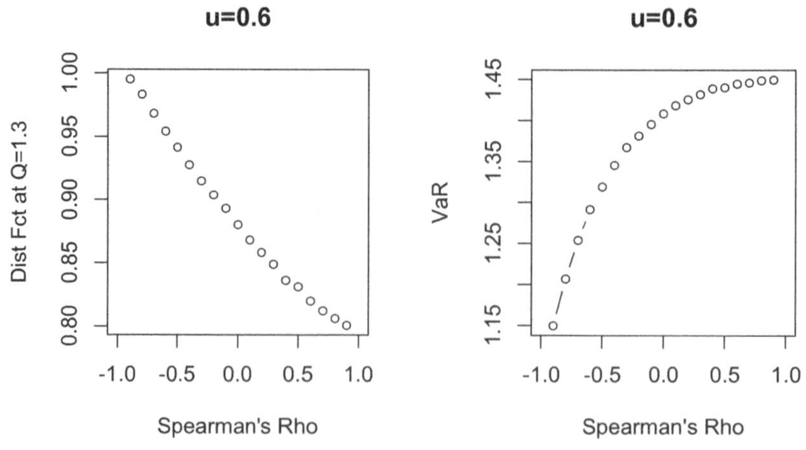

FIGURE 1.3
Effects of Dependence on Measures of Portfolio Uncertainty.

a given retention level, the probability of retained risks being less than or equal to a given level of capital decreases as the level of association increases.

- The right-hand panel presents the value at risk $VaR_{0.95}$ at $u = 0.6$ but allows the dependence, as quantified through Spearman's correlation, to vary. The required capital increases as the level of association increases.

Code. The *Online Supplement* provides Section 1.4.1 illustrative statistical code.

1.4.2 Minimal Scenario 2. Form of Risk Transfer

There are many alternative methods, or forms, of transferring risks; Section 6.1 will provide a broad discussion on whether or not to transfer risk and ways of doing so. At this stage, we wish to emphasize by means of an example that the financial consequences of the choice of the risk transfer form and show how it "depends on the dependence" among risks.

Risk Transfer Forms. Again, consider the minimal scenario in which a manager with background risks in X_1 would like to transfer a portion of a second risk X_2. The prior Section 1.4.1 considered the case where the manager retains $\min(X_2, u)$. As an alternative, the manager may wish to retain a simple proportion of the second risk, say cX_2, where c is a fixed constant. Later, we will call this the *coinsurance* case. As another alternative, the manager may wish to retain $X_2 - \min(X_2, d)$ where we think of d as a *deductible.*

As before, we assume that both X_1 and X_2 have uniform distributions and that risk transfer costs are fair and determined via taking expectations. We now denote the risk transfer cost of the upper limit policy as $RTC_u(u) = \frac{1}{2}(1 + u^2) - u$. In the same way, it is easy to check that risk transfer costs for the alternatives are

$$RTC_c(c) = \frac{1}{2}(1 - c) \quad \text{and} \quad RTC_d(d) = d - \frac{d^2}{2}.$$

For a meaningful comparison, assume that these alternatives produce the same risk transfer cost, say RTC_0. Then, one can compute the value of the upper limit u needed to attain this cost through the relation $RTC_u(u) = RTC_0$. This turns out to result in

$$u = 1 - \sqrt{2 \times RTC_0}.$$

In the same way, we can determine the coinsurance and deductible parameters that result in the risk transfer cost RTC_0. They are

$$c = 1 - 2 \times RTC_0 \quad \text{and} \quad d = 1 - \sqrt{1 - 2 \times RTC_0}.$$

Figure 1.4 shows, for each type of risk transfer, that the probability of being less than or equal to a required capital amount ($\Pr[S(u) \leq Q]$, $Q = 1.1$ for this demonstration) is largest for the case of negatively dependent risks and smallest for positively dependent risks, as before.

To emphasize the importance of the risk transfer type, Figure 1.5 switches the presentation. Here, the middle and right-hand panels of Figure 1.5 represent the case of independence and positive dependence (correlation = 0.3), respectively. They show that the upper limit is the preferred form of risk transfer agreement in the sense that it has the largest probability of being less than or equal to a required capital amount. The coinsurance form is close, particularly for the case of independence. The left-hand panel representing negative dependence (correlation = −0.3) suggests that there is no clear winner among the three risk transfer types.

Code. The *Online Supplement* provides Section 1.4.2 illustrative statistical code.

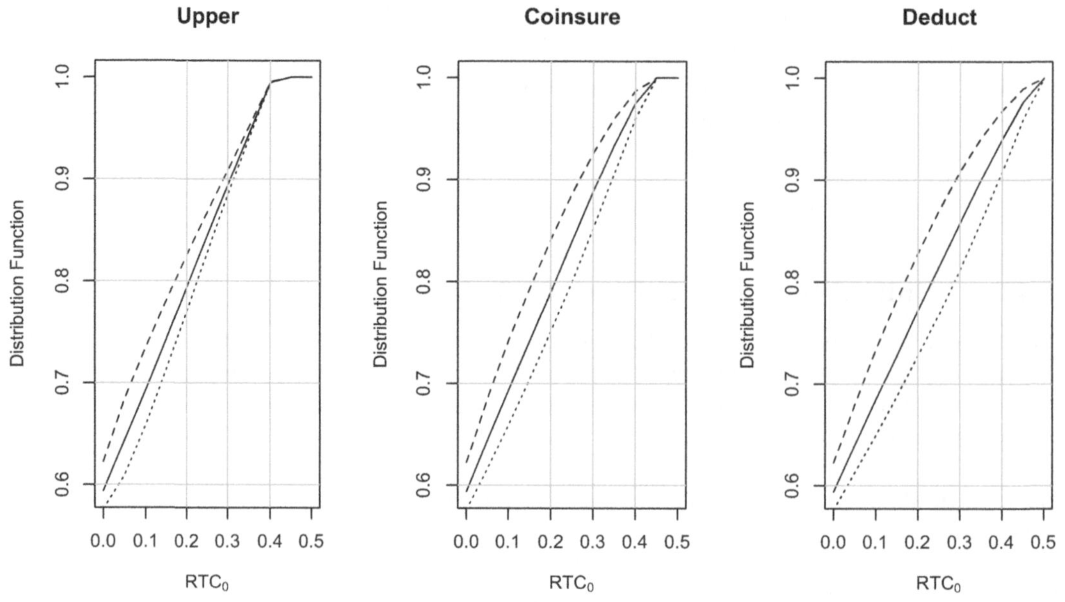

FIGURE 1.4
Distribution Function versus the Risk Transfer Cost for a Risk, by Risk Transfer Type. Within each risk transfer type, the solid black line corresponds to the case of independence. The dotted line is for a positive Spearman correlation 0.3, the dashed line is for a negative correlation −0.3.

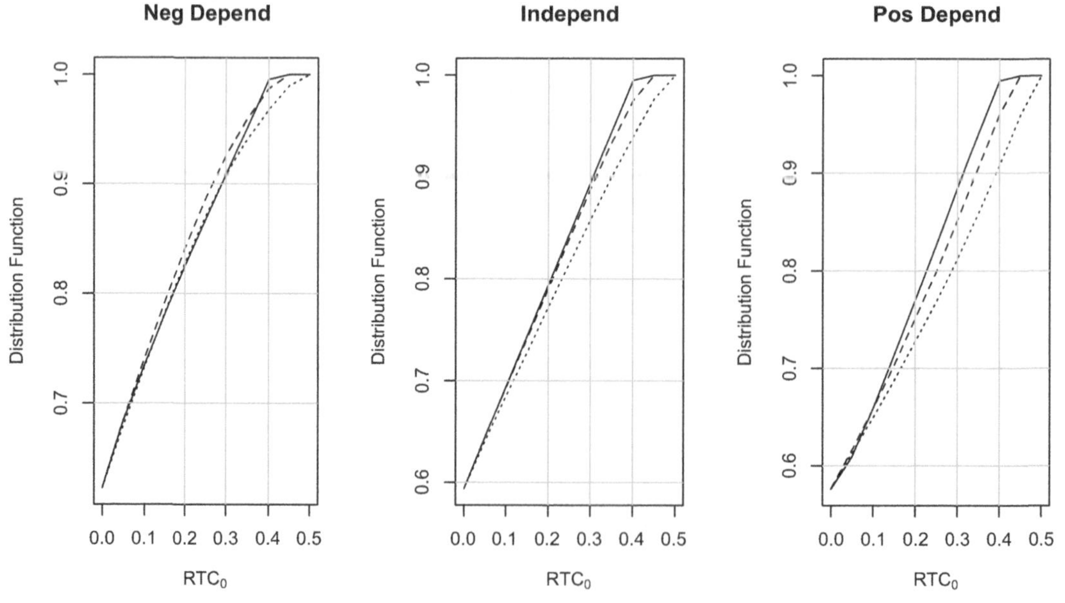

FIGURE 1.5
Distribution Function versus the Risk Transfer Cost for a Risk, by Dependence Type. Within each dependence type, the solid black line corresponds to the case of upper limit. The dotted line is for a deductible, the dashed line is for coinsurance.

1.5 Overview of the Book

In the stripped-down examples of Section 1.4, we saw in action some of the key elements that drive liability portfolio risk strategies: (1) different contracts for dividing risks into portions transferred and retained, (2) summarizing the uncertainty of the portion retained and the cost of the portion transferred, and (3) the impact of dependence among risks on the uncertainty of retained risks.

As argued in this book, the risk owner is concerned about the set of all risks in the portfolio as well as the cost of transferring portions of each risk so that the retained amount of risk is acceptable. The Section 1.4 examples emphasize that the retained risk portfolio uncertainty depends on the association, or dependence, among the components. In contrast, transfer costs are oftentimes exogenous, given by external market forces (supply and demand of availability), that typically do not depend on other risks in the portfolio. So, the risk portfolio manager has room to maneuver and can make selections in the risk owner's best interests. This book develops techniques to take advantage of these opportunities and is organized as follows.

Part I. Foundations. The first part, consisting of four chapters, sets the stage.

Chapter 1 introduces the problem of multiple risks faced by organizations and the importance of risk transfers through the insurance marketplace. It describes case studies that further motivate the problem and introduces minimal scenarios to underscore the role of uncertainty and costs for mitigating this uncertainty, as well as the importance of dependence among risks.

Chapter 2 on *Risk Retention Functions* describes fundamental quantitative tools for managing risks, separating the overall set of risks into those retained by the firm and those that can be offloaded, or transferred, to another organization. For a risk retention policy, measures are introduced to quantify the uncertainty of the risk retained by the firm and the costs of offloading risks.

Chapter 3 on *Balancing Retained Risk and Risk Transfer Cost* provides a deeper discussion of quantifying the balance between the risk retained and the costs of offloading risks. Three approaches for quantifying this balance are introduced: a ratio that provides an intuitive measure that summarizes this balance, a formal total cost approach that can be optimized, and a more flexible constrained optimization approach.

Chapter 4 on *Transferring Multiple Risks including Reinsurance* extends the Chapter 3 approaches to include dependence and the reinsurance market yet still restricting consideration to the classic case of variance as a measure of risk uncertainty. Even with this classic measure, the nonlinearities introduced by retention policies motivate the copula as a natural vehicle for modeling dependence. Reinsurance applications suggest the introduction of a host of complex retention policies that expand the scope of applications. Reinsurance also provides a natural platform for introducing general theoretical structures for exchanging risks, the subject of Section 4.5.

Part II. Core Material. The second part, Chapters 5-8, provides the core material for developing optimal insurable risk portfolio strategies.

Chapter 5 is on *Excess of Loss for Two Risks.* The first four chapters emphasize the classical case of (1) variance as a risk measure, (2) independent risks, and (3) proportional sharing of risks. In contrast, this chapter considers excess of loss, a non-proportional risk sharing mechanism, for dependent risks. It develops the distribution, quantiles, and associated derivatives for the retained risk, allowing us to quantify general (quantile) risk measures. Due to computational challenges, this chapter restricts considerations to two risks.

Chapter 6 is on *Managing Portfolios of Insurable Risks* emphasizes the *management* of insurable risks and describes methods used in practice. This chapter focuses on risks faced by general commercial firms (not just insurers) but also discusses peer-to-peer exchanges that are more geared for individuals.

Chapter 7 is on *Constructing Multivariate Portfolios* describes how simulation can be used to optimize insurable risk portfolios. The chapter describes how to (i) simulate a portfolio of dependent risks, (ii) split the portfolio into a portion retained by the risk owner and another portion to be transferred to one or more other parties, and (iii) optimize the uncertainty of retained risks subject to budget constraints. The detailed illustrations focus on a multivariate version of an excess of loss transfer arrangement with considerations of alternative ways of measuring the uncertainty of retained risks. The strategy underpinning these illustrations can be applied to general risk retention arrangements that are also outlined in this chapter.

Chapter 8 is on *Case Studies* and features two illustrative examples: the Wisconsin Property Fund case, based on an insurance risk pool with many features found in a typical multi-line general insurance company, and the Australian National University (ANU) case, which considers a portfolio of risks commonly encountered by risk managers at medium and large corporations. The ANU case emphasizes stress testing due to the relative paucity of data available for some risk types.

Part III. Properties and Characteristics. The third part, Chapters 9 - 12, develops several properties and characteristics of optimal insurable risk portfolios that apply broadly. These properties and characteristics provide insights for developing interpretation and more robust risk retentions strategies.

Chapter 9 is on *Stress Testing, Sensitivity, and Robustness.* The preceding chapters showed how to use constrained optimization methods to suggest desirable insurable risk portfolios. However, like any mathematical representation, constrained optimization results are based on precise model assumptions. So, when interpreting findings, it is natural to ask questions about the reliability of the results, something that risk managers are keen to understand. Stress testing, examining results under different scenarios, is an approach already used in prior chapters and is emphasized as a basic approach to examine results reliability. This chapter also introduces a differential approach using a tool from economics known as the Envelope theorem to quantify sensitivities. Here, *sensitivities* refer to changes in the optimization results when one changes any one of the input assumptions by a small amount. One can also entertain alternative optimization procedures that explicitly recognize parameter uncertainty. This chapter describes *robust optimization* in the context of investment portfolios to give readers a feel for this approach.

Chapter 10 continues the development of sensitivities and provides linkages to *data uncertainty.* Specifically, this chapter describes a novel extension of the Envelope theorem that provides explicit gradients for decision variables based on changes in an input assumption, referred to here as a *perturbation sensitivity.* After showing how this general result can be used in a risk retention context, applications to collection of input parameters are considered. The most natural such collection is a set of risk parameters and this approach allows the introduction of *stochastic sensitivities* which can be used to gauge data uncertainty.

Chapter 11 is on *Risk Retention Conditions.* When using the method of constrained optimization to suggest a desirable insurable risk portfolio, it can be useful to understand what conditions might be suitable or desirable to transfer a risk. This chapter introduces the Karush-Kuhn-Tucker, or *KKT*, conditions that must hold for an optimal risk transfer solution. These conditions provide a convenient framework to summarize work done in

Chapter 3 for a single risk. The conditions can be used to verify classical conditions on balance among risks for multivariate excess of loss and extend them to the case of dependence, as well as achieving a binding budget and establishing that parameters are on the boundary. These conditions also help us learn about the equivalence of different optimization techniques when using simulation.

Chapter 12 is on *The Role of Dependence in Managing Insurable Risks*. Managers form portfolios to rely on benefits of diversification and these benefits can be critically impacted by dependence. This chapter provides a structure for thinking about the ways in which dependence can make important differences. This structure is based on the number of risks in the portfolio and the purposes of modeling such as determining risk transfer costs, the uncertainty of a portfolio, and the risk retention parameters obtained when constructing an optimal insurable risk portfolio. In addition to sensitivity analysis, extending the work in Chapter 9, this chapter introduces models of *contagion* where there is an association or spread of loss from one entity to another within a portfolio.

Part IV. Supporting Materials. The fourth part provides various materials that support and supplement the main body of the text.

Outline of the Investment Portfolio Problem. Many readers will likely be interested in the development of asset investment portfolio problem due to Markowitz, sometimes called the asset allocation problem in this book. This alternative name is employed to make sure that we utilize the intuition from this well-developed area of investments but do not accept conventions learned from this other industry without verifying them. To encourage readers to develop expertise is this sister industry, here is an outline of our development of the asset allocation problem.

- The problem has already been introduced in Section 1.3.
- The solution of the investment portfolio problem using constrained optimization is given in Section 3.4, complete with illustrative code and data. This is the traditional approach based on minimizing the variance.
- The minimization based on expected shortfall is given in Section 4.4.1 where it is identified as a convex problem. Again, sample code is available to encourage readers to construct their own optimal portfolios.
- Section 9.4 summarizes asset allocation sensitivity and introduces another approach for managing parameter uncertainty, asset allocation robustness.
- Chapter 10 discusses the data uncertainty in the asset allocation problem. Here, "data uncertainty" refers to the observation that small changes in the risk distribution parameters can lead to large material changes in optimal risk portfolio coefficients. Data uncertainty is a major limitation of the Markowitz strategy which, fortunately, turns out to be less severe in liability risk portfolio applications.

1.6 Supplemental Materials

1.6.1 Further Resources and Readings

For this chapter and the remainder of the book, the intent is that practicing analysts, as well as other readers with experience, will be able to read the main body of the text and get the big ideas. Some readers may wish to access code and details of mathematical justifications (or the longer derivations in the appendices) that provide a deeper experience but these can be skipped without loss of continuity.

Students, or those with less experience in the field, are encouraged to take deeper dives into the code, the theory, or preferably both. Moreover, you may wish to acquaint yourself with some related fields that provide the foundations for the development here.

- First and foremost is a fundamental understanding of quantifying losses. Readers wanting a quick refresher to some basic statistical and actuarial concepts may refer to the online book Loss Data Analytics written by the Actuarial Community (2020). In particular, Chapter 3 describes calculation of risk transfer costs with basic retention limitations and Chapter 14 provides an introduction to copulas. Other notable introductions that provide comparable material include Tse (2009), Gray and Pitts (2012), and Klugman et al. (2012).
- This book accounts for dependence among risks using a copula framework. See Hofert et al. (2018) for a modern book-length introduction of copulas; Genest and Mackay (1986) and Frees and Valdez (1998) provide classic introductory articles. Joe (2014) provides excellent in-depth background to statistical aspects of copulas at a higher mathematical level.
- The goal of this book is to develop a suite of quantitative techniques for risk managers and those who advise them. As such, it follows the spirit of the pioneering book aptly entitled *Quantitative Risk Management* by Embrechts et al. (2005).
- The aim here is to develop a set of tools that can be applied to commercial insurance, as well as personal insurance, reinsurance, and other risk financing techniques. Many of the problems that we address can be motivated by the reinsurance field. An excellent quantitative introduction to this area can be found in Albrecher et al. (2017).
- Technique-wise, this book employs constrained optimization extensively. In the developing this book, we used Boyd and Vandenberghe (2004) and Simon and Blume (1994) although there are many good resources available in this area.

On Risk and Uncertainty. This book interprets a *risk* to be a random outcome (a random variable), typically having financial implications. The *uncertainty* is the driver of the randomness and we can measure the level of uncertainty through risk measures. This is consistent with the practice of applied insurance risk modeling, see for example, Mildenhall and Major (2022) (Chapter 3) and Gutterman (2017). In contrast, some readers are likely to be familiar with terminology due to Knight (1921) where a distinction is made between a risk, described by an objective probability distribution, and uncertainty, which is not related to any precise probability distribution. Alternatively, many economists now utilize terms such as *ambiguity* to describe this type of uncertainty, cf. Machina (2013). The goal of this book is to provide quantitative modeling tools for managing collections of risks and these tools depend on the language of probability theory. The probabilities may be objective (with grounding in empirical data), including frequentist and Bayesian, or subjective (reflecting individual judgment or degree of belief); many applications utilize all three types. The adopted interpretations of risk and uncertainty are consistent with the goal of developing a suite of tools founded on our best available information about probabilities of financial outcomes.

1.6.2 Exercises

See the *Online Supplement* for selected exercise solutions.

Exercise 1.1. Trade-off between Price and Uncertainty, including the Effects of Dependence. This exercise mimics and extends the development of the first minimal scenario in Section 1.4.1. The extension is to move beyond uniform (0,1) outcomes to distributions that are more representative of potential liabilities that a firm might face.

Specifically, in this exercise, assume:

- That X_1 has a gamma distribution with mean 10,000 and standard deviation 7,071. With these assumptions, the 95th percentile of X_1 is 23,719.
- That X_2 has a Pareto distribution with mean 1000 and standard deviation 1,732. With these assumptions, the 95th percentile of X_2 is 3,429.
- The risks X_1, X_2 may be dependent, assuming Frank copula to construct the dependence.

For this exercise, you should:

a. Examine the statistical code in Section 1.4.1 and outline a strategy for making changes to the code. (Subsequent chapters will develop the explicit tools for doing so. At this stage, just think about the strategy.)

b. Figure 1.6 summarizes results from these new distributional assumptions, comparable to Figure 1.2. Interpret the upper left-hand panel, emphasizing the role of the dependence.

c. Interpret the upper right-hand panel, emphasizing the role of the dependence.

d. Interpret the lower left-hand panel, emphasizing the role of the dependence.

e. Interpret the lower right-hand panel, emphasizing the role of the dependence.

Exercise 1.2. Using the assumptions of Exercise 1.1, we now mimic the minimal scenario 1 results presented in Figure 1.3. For a fixed upper limit, Figure 1.7 shows values of the distribution function at a point and the value at risk over different values of the association parameter.

a. Interpret the left-hand panel.

b. Interpret the right-hand panel.

1.6.3 On Your Own

This text does *not* emphasize procedures for fitting models of claims from empirical data. This is an important step and readers will find it useful to get a sense of the uncertainties that arise from this step. As will be described in later chapters (see in particular Chapter 9 on sensitivity and robustness), these uncertainties play a role in the optimal insurable risk portfolios that we develop.

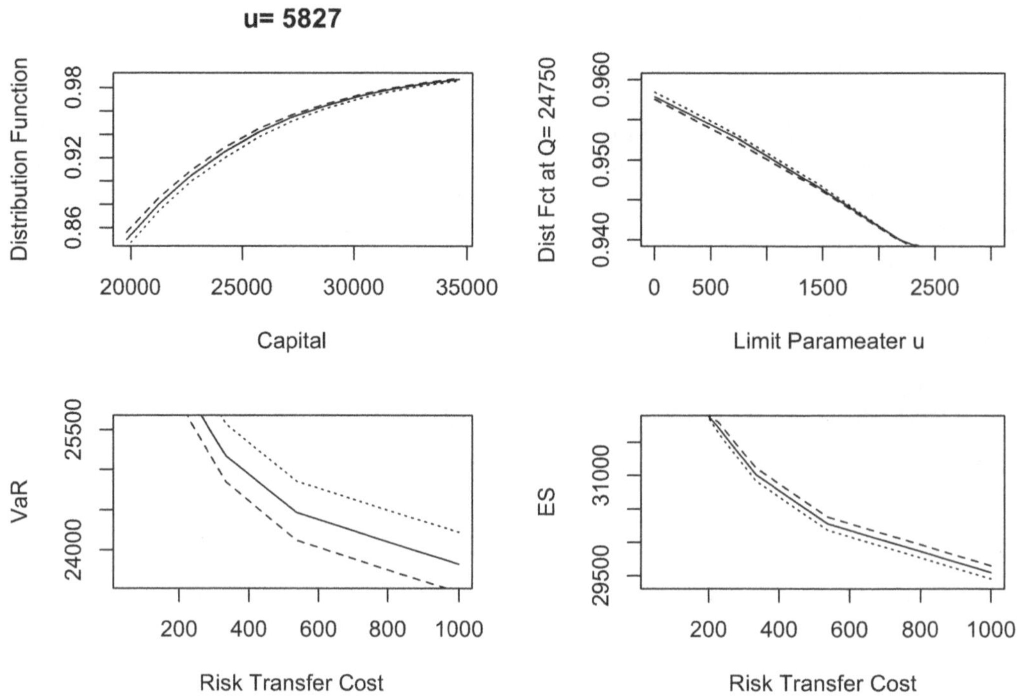

FIGURE 1.6
Single Parameter Problem. The solid black line corresponds to the case of independence. The dotted line is for a positive Spearman correlation 0.3, the dashed line is for a negative correlation -0.3.

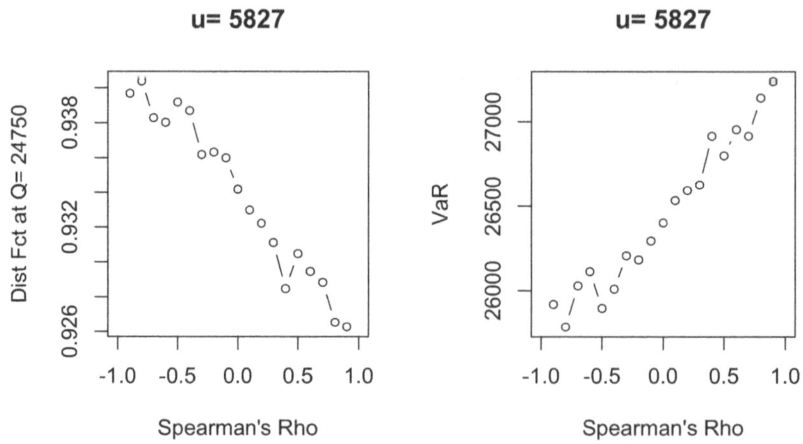

FIGURE 1.7
Effects of Dependence on Measures of Portfolio Uncertainty.

On your own, you may wish to review model fitting resources for the two case studies. These resources not only provide access to sample data but also illustrative `R` code for model fitting. These are:

- Frees and Butt (2022) for a detailed description of models of ANU risks summarized in Section 1.2.1
- Frees et al. (2016) for a detailed description of models of the Wisconsin property fund summarized in Section 1.2.2.

As another option, many readers will wish to explore the financial asset data as introduced in Section 1.3. Use the illustrative code provided in this section to download your own sample of data, mimic the data summaries provided there, and you will be in a position to form your own stock portfolios in later chapters.

2

Risk Retention Functions

Chapter Preview. In Chapter 1, risk transference emerges as a vital managerial tool. Risk transfer agreements split risk: one part remains with the owner, while the rest shifts to others at a cost. A *risk retention function* determines this division of risk. When insurance is used for risk transference, the standard retention functions are sometimes referred to as "modifications" of a base risk. Classic examples of insurance coverage modifications include imposing a deductible where small risks are not covered, a coinsurance where only a fraction of a risk is covered, and an upper limit that restricts the impact of a large risk. Deductibles, coinsurances, and upper limits are examples of *risk retention parameters* upon which a risk retention function is based.

To quantify the impact of risk transfer, this chapter lays the foundations by describing commonly used measures to quantify risk uncertainty. Risk owners want to understand the uncertainty of risks, both before and after a transfer. Risk owners are also concerned with the cost of offloading risks; for simplicity, this chapter emphasizes a "fair" cost based on expectations.

To lay the groundwork, this chapter restricts attention to a single risk that one can interpret to be the risk owner's portfolio. Subsequent chapters will develop a structure to determine optimal choices of risk retention parameters. To provide the groundwork for these developments, this chapter quantifies **changes** in the measures of uncertainty and risk transfer costs based on changes in the risk retention parameters. It is convenient to use a differential approach when summarizing the effects of small changes in parameters; these effects are known as *sensitivities.*

2.1 Measures of Uncertainty

To evaluate alternative risk transfer options, the risk owner's manager needs principles for determining amounts of risk before and after the transfer, as well as the cost of a risk transfer. Of course, the exact reduction in obligations only becomes known once all financial obligations are settled at the contract's conclusion. During the decision-making process, only a distribution of potential risk outcomes is available.

As with any random variable, tools such as the distribution function provide analysts with an understanding of the potential outcomes' distribution. Risk managers are especially concerned with the **uncertainty** of the potential outcomes for which there are several commonly used measures. Because of its simplicity, historically the most commonly used measure has been the variance, $\mathrm{Var}(X) = \mathrm{E}[X - \mathrm{E}(X)]^2$, and its positive square root, the standard deviation.

In a broad context, a risk measure is a mapping of a loss random variable to a numerical value that indicates its level of uncertainty. Managers utilize risk measures as a guide to determine the necessary asset reserves for covering contingent losses stemming from risks.

 DOI: 10.1201/9781003469568-2

For an introductory overview of risk measures, refer to the online book Actuarial Community (2020), Chapter 10.

In our development we focus on risk measures that are based on quantiles of a random variable. As indicated in Section 1.4.1, the primary measure is the value at risk (VaR) which represents the quantile of X. In other words, for a specific confidence level α, the VaR is the smallest value of x for which the associated distribution function is greater than or equal to α. Mathematically, we can write this as

$$VaR_{\alpha}[X] = \inf\{x : F_X(x) \geq \alpha\}, \tag{2.1}$$

for $0 \leq \alpha \leq 1$. Here, *inf* is the infimum, or greatest lower bound, operator. To emphasize the quantile nature of the VaR, we sometimes use the notation $VaR_{\alpha}[X] = F_{\alpha}^{-1}$ and occasionally write $F^{-1}(\alpha)$ (for complex expressions in the place of the argument α). *Note on Notation:* To distinguish the value of risk (VaR) from the variance (Var), we follow a common practice in the literature and capitalize the **R** in the value at **R**isk. You will also see some sources that use $V@R$ to promote this distinction.

Another commonly used quantile-based risk measure is the **Expected Shortfall**, ES, that can be interpreted as an "average" value at risk. Mathematically, we can express this as

$$ES_{\alpha}(X) = \frac{1}{1-\alpha} \int_{\alpha}^{1} VaR_a(X) da. \tag{2.2}$$

Under some mild assumptions on the continuity of the distribution function, there are a few other forms of the ES that will be useful to us. As described in Section 2.5.1, one can show that

$$\begin{array}{rll} ES_{\alpha}(X) & = \frac{1}{1-\alpha} \int_{\alpha}^{1} VaR_a(X) da & \text{Expected Shortfall} \\ & = F_{\alpha}^{-1} + \frac{1}{1-\alpha}\left\{\mathrm{E}X - \mathrm{E}[X \wedge F_{\alpha}^{-1}]\right\} & \text{Tail VaR} \\ & = \mathrm{E}(X|X > F_{\alpha}^{-1}) & \text{Conditional VaR.} \end{array} \tag{2.3}$$

From the third expression, we see that ES can also be interpreted to be the expected amount given that the loss exceeds the VaR_{α}. There is a bit of divergence in the literature, these closely related measures are also known as *tail conditional expectation*, *conditional tail expectation*, and *expected tail loss*, among others.

Another Note on Notation: The expected shortfall is denoted with capital italic letters as ES. In contrast, the expectation operator is the capital roman letter $\mathrm{E}(\cdot)$. These both represent common practice in the literature but, because notationally they are so similar, it is easy to confuse these two very different concepts.

Under the Hood. See the *Online Supplement* for the Section 2.1 ES development.

Example 2.1. Property Fund Claims Distribution. To illustrate, we consider commercial property claims from the Wisconsin Property Fund, introduced in Section 1.2.2. These data can be accessed from the online book, Actuarial Community (2020), Chapter 22. This example is based on 1,377 claims from 2010 for damages to state government properties and their building contents.

These data were used to estimate an empirical distribution function, without reference to a parametric model. The empirical distribution function was then used to estimate VaR and ES over several confidence levels. Figure 2.1 summarizes these results. From the figure, we see that the two summary measures of uncertainty provide qualitatively similar results in that they increase in the same way as the confidence level α increases. In addition, the right-hand panel compares the two measures - this comparison reminds us that, for any

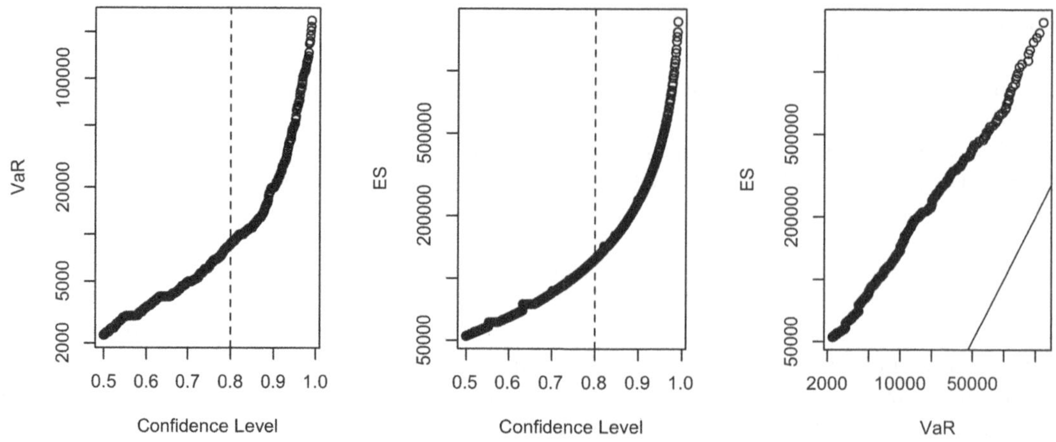

FIGURE 2.1

Property Fund *VaR* and *ES* Plots. The left-hand panel shows the value at risk VaR for several confidence levels and the middle panel gives similar information for the expected shortfall (ES). The confidence level $\alpha = 0.80$ is marked with a dashed vertical line. Note that the vertical axes ranges differ. This is emphasized by direct comparison in the right-hand panel where the 45-degree solid line falls below the empirical values.

given level of confidence, that the ES measure typically exceeds the VaR. (See the middle expression in Display (2.3).)

Code. The *Online Supplement* provides Example 2.1 illustrative statistical code.

A risk measure that encompasses both the value at risk and the expected shortfall is the **range value at risk**, as presented in Embrechts et al. (2018). This is defined as

$$RVaR_{(\alpha,\beta)}(X) = \begin{cases} VaR_{\alpha}(X) & \text{if } \beta = 0 \\ \frac{1}{\beta}\int_{\alpha}^{\alpha+\beta} VaR_{a}(X)da & \text{if } 0 < \beta \le 1-\alpha \end{cases}. \tag{2.4}$$

To see the connection, first note that as $\beta \to 0$, $RVaR_{(\alpha,\beta)}(X) \to VaR_{\alpha}(X)$. Second, by choosing $\beta = 1-\alpha$ and using equation (2.3), we have $RVaR_{(\alpha,1-\alpha)}(X) = ES_{\alpha}(X)$. From this perspective, the VaR, ES, and $RVaR$ can each be represented as an "average" quantile and so we refer to them as *quantile-based* risk measures.

Example 2.2. Property Fund and the Pareto Distribution. To show how these uncertainty measures might work with parametric distributions, we consider the Pareto distribution because it is commonly used in insurance applications. This distribution function can be expressed as

$$F_X(x) = 1 - \left(\frac{\gamma}{x+\gamma}\right)^{\eta}, \tag{2.5}$$

for $x > 0$, where γ is the scale parameter and η is the shape parameter. Using a Pareto distribution, the property fund data was fit using maximum likelihood. The estimated parameters turn out to be $\hat{\gamma}_P = 2,282$ and $\hat{\eta}_P = 0.9991$. With these parameter values, the value at risk can be estimated using the expression

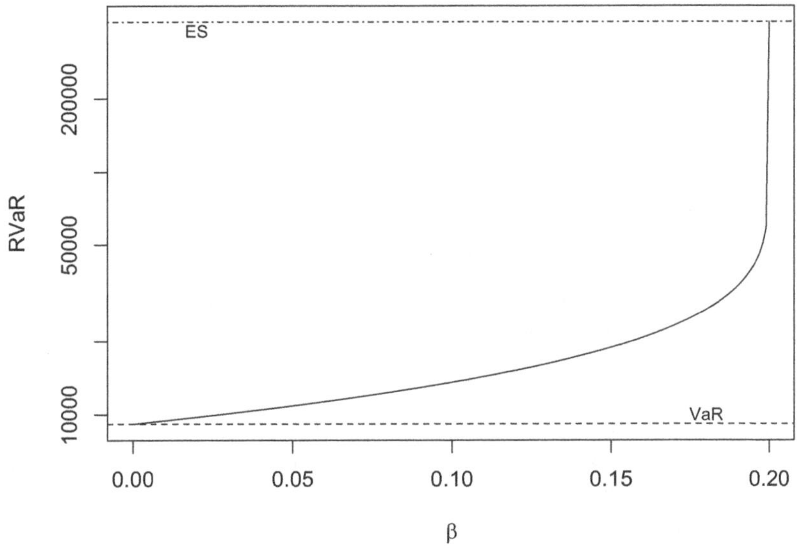

FIGURE 2.2
Property Fund *RVaR* Plot. For benchmarking, recall that VaR and ES represent lower and upper values of $RVaR$.

$$\widehat{VaR}_\alpha = \hat{\gamma}_P \left[(1-\alpha)^{-1/\hat{\eta}_P} - 1\right].$$

This gives the estimated value for $RVaR_{(\alpha,0)}$. In addition, for $\beta > 0$, easy calculus shows that

$$\begin{aligned}\widehat{RVaR}_{(\alpha,\beta)}(X) &= \frac{1}{\beta}\int_\alpha^{\alpha+\beta} \hat{\gamma}_P \left[(1-\gamma)^{-1/\hat{\eta}_P-1} - 1\right] d\gamma \\ &= \frac{\hat{\gamma}_P}{\beta}\left\{\frac{(1-\alpha)^{1-1/\hat{\eta}_P} - (1-\alpha-\beta)^{1-1/\hat{\eta}_P}}{1-1/\hat{\eta}_P}\right\} - \hat{\gamma}_P.\end{aligned}$$

Using a level of $\alpha = 0.80$, Figure 2.2 summarizes the fitted value of $RVaR$ over different values of β. For reference, this level of α yields estimated values of $\widehat{VaR}_{0.80} = 9{,}145$ and $\widehat{ES}_{0.80} = 413{,}540$. These values are marked in Figure 2.2 with the horizontal lines. They represent lower and upper values of $RVaR$.

Code. The *Online Supplement* provides Example 2.2 illustrative statistical code.

2.2 Risk Retention Function

A *risk retention* function is a mapping that indicates what portion of a risk is retained by the risk owner after a risk transfer is in place. For notation, let $g(\cdot)$ denote a generic risk retention function. As in Section 1.4.1, one might use $g(x) = \min(x, u)$ where u is the upper limit of the loss to be retained. In general, the loss retained is $g(X)$ and the loss transferred, or *ceded*, is $X - g(X)$. Both are random variables and each has their own distribution. Unless specified otherwise, you may assume that $0 \leq g(x) \leq x$. The lower bound means that negative retained losses are not possible and the upper bound means that one cannot have a retained loss that exceeds the loss.

2.2.1 A Risk Retention Function

For an important special case that covers many situations of interest, consider the following expression.

Risk Retention Function

$$\boxed{g(x;d,c,u) = \begin{cases} 0 & \text{if } x < d \\ c(x-d) & \text{if } d \le x < u \\ c(u-d) & \text{if } x \ge u \end{cases}.} \tag{2.6}$$

Here, d is the deductible, c is the coinsurance amount, and u is the upper limit of coverage. This function can also be expressed as

$$g(x;d,c,u) = c\left\{(x \wedge u) - (x \wedge d)\right\}.$$

The wedge symbol $\wedge$ is used for the minimum operator and is defined as $x \wedge z = \min(x,z)$. The variable $X \wedge u$ is called a *limited loss variable* and it is often used for limited expected value calculations in actuarial contexts.

The risk retention parameters may be collected into a vector $\boldsymbol{\theta} = (d,c,u)$. With this notation, one can use integration by parts and express the *mean retained loss* as

$$\begin{aligned} \mathrm{E}[g(X;\boldsymbol{\theta})] &= \int_d^u c(x-d)dF(x) + c(u-d)[1-F(u)] \\ &= -c(x-d)[1-F(x)]\Big|_d^u + c\int_d^u [1-F(x)]dx + c(u-d)[1-F(u)] \\ &= c\int_d^u [1-F(x)]dx. \end{aligned} \tag{2.7}$$

This development assumes that d is finite. Although not needed for our applications, the case of $d = -\infty$ can also be handled in a straightforward fashion. We can also write

$$\mathrm{E}[g(X;\boldsymbol{\theta})] = c\left\{\mathrm{E}(X \wedge u) - \mathrm{E}(X \wedge d)\right\},$$

as the limited expected value expression $\mathrm{E}(X \wedge x) = \int_0^x [1-F(z)]dz$ is routinely tabulated in software.

Consistent with the motivating examples in Section 1.4, we use summary measures of the amount ceded, $X - g(X)$, such as mean or a quantile, to represent the **risk transfer cost** (RTC) . In the baseline case of using means, we can calculate the risk transfer cost as the overall mean minus the mean retained loss, that is, $RTC(\boldsymbol{\theta}) = \mathrm{E}(X) - \mathrm{E}[g(X;\boldsymbol{\theta})]$.

The risk retention function introduced in equation (2.6) is motivated by insurance risk transfers and covers many applications of interest. If the risk owner is a company, one can imagine small losses being handled by a service contractor and large losses being covered by an insurer. If the risk owner is an insurance company, small losses may be the responsibility of a policyholder with large losses being covered by a reinsurer (see Chapter 4). Other interpretations are possible. Moreover, the risk retention function provides the groundwork for extensions that will be taken up in subsequent chapters covering risk exchanges among multiple parties in Section 4.5 and multiple risks in Chapter 7.

2.2.2 Distribution Function

With the vector of risk retention parameters, $\boldsymbol{\theta} = (d,c,u)$, the distribution function of the retained risk $g(X;\boldsymbol{\theta})$ can be expressed as

$$F_{g(X;\boldsymbol{\theta})}(z) = \begin{cases} F(d) & \text{if } z = 0 \\ F\left(\frac{z}{c} + d\right) & \text{if } z < c(u-d) \\ 1 & \text{if } z \ge c(u-d) \end{cases}. \tag{2.8}$$

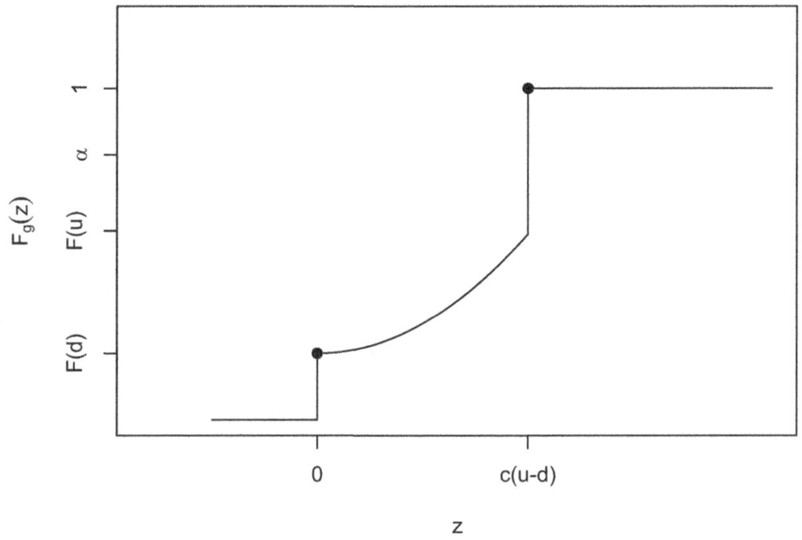

FIGURE 2.3
Distribution Function of Retained Risk. Even if the loss X has a continuous distribution, there can be discontinuities induced in the distribution of retained loss through the introduction of deductibles d and upper limits u.

There is additional discreteness due to the parameters d and u. Figure 2.3 allows one to visualize the distribution function of the retained loss with discreteness induced by the insurance parameters.

2.2.3 Risk Measures

Recall that $F_\alpha^{-1} = VaR_\alpha(X)$ is a quantile for a random variable X. Quantiles have several attractive properties. For example, they preserve location translations in the sense that $VaR_\alpha(X - a_0) = VaR_\alpha(X) - a_0$ for a constant a_0. Further, they preserve scale translations in the sense that $VaR_\alpha(a_1 X) = a_1 VaR_\alpha(X)$ for a positive a_1. Summarizing, we have

$$VaR_\alpha(a_1 X - a_0) = a_1 \; VaR_\alpha(X) - a_0, \quad \text{for } a_1 > 0. \tag{2.9}$$

From this expression and equations (2.3) and (2.4), we see that the expected shortfall and the range value at risk also preserve location and scale translations.

Now use Figure 2.3 to help determine the quantile of the retained loss distribution, $VaR_\alpha[g(X;\boldsymbol{\theta})]$. From the figure, if $F(d) < \alpha < F(u)$, then one finds $VaR_\alpha[g(X;\boldsymbol{\theta})]$ has the solution z of the equation $\alpha = F\left(\frac{z}{c} + d\right)$. Straight-forward algebra shows this to be $VaR_\alpha[g(X;\boldsymbol{\theta})] = c\left(F_\alpha^{-1} - d\right)$. From the figure, if $\alpha \geq F(u)$, then $VaR_\alpha[g(X;\boldsymbol{\theta})] = c\,(u - d)$. Summarizing, we have

$$VaR_\alpha[g(X;\boldsymbol{\theta})] = \begin{cases} 0 & \text{if } \alpha < F(d) \\ c(F_\alpha^{-1} - d) & \text{if } F(d) \leq \alpha < F(u) \\ c(u-d) & \text{if } F(u) \leq \alpha \end{cases} . \tag{2.10}$$

In the same way, we have

$$ES_{\alpha}[g(X;\boldsymbol{\theta})] = \begin{cases} \frac{c}{1-\alpha}\left\{\mathrm{E}(X\wedge u) - \mathrm{E}(X\wedge d)\right\} & \text{if } \alpha \le F(d) \\ \frac{c}{1-\alpha}\left\{\mathrm{E}(X\wedge u) - \mathrm{E}(X\wedge F_{\alpha}^{-1}) \right. & \text{if } F(d) \le \alpha < F(u) \\ \quad \left. +(1-\alpha)(F_{\alpha}^{-1} - d)\right\} & \\ c(u-d) & \text{if } F(u) \le \alpha \end{cases} . \tag{2.11}$$

ⓘ *Under the Hood.* See the *Online Supplement* for the Section 2.2.3 ES development.

From equations (2.10) and (2.11), we see that if $F(u) \le \alpha$, then the value at risk and expected shortfall are the same. That is, they are equivalent for small values of upper limits, at least smaller than the α quantile. This observation will be useful in Chapter 7 when interpreting risk retention frontiers.

Extending this for the range value at risk, $RVaR$, for retained risks, one can show that

$$RVaR_{(\alpha,\beta)}[g(X;\boldsymbol{\theta})]$$
$$= \begin{cases} 0 & \alpha+\beta < F(d) \\ \frac{c}{\beta}\left\{(1-\alpha-\beta)(d - F_{\alpha+\beta}^{-1}) \right. & \alpha < F(d) < \alpha+\beta < F(u) \\ \quad \left. +[\mathrm{E}(X\wedge F_{\alpha+\beta}^{-1}) - \mathrm{E}(X\wedge d)]\right\} & \\ \frac{c}{\beta}\left\{(1-\alpha-\beta)(d-u) + [\mathrm{E}(X\wedge u) - \mathrm{E}(X\wedge d)]\right\} & \alpha < F(d) < F(u) < \alpha+\beta \\ \frac{c}{\beta}\left\{(1-\alpha)(F_{\alpha}^{-1} - u) + \beta(u-d) \right. & F(d) < \alpha < F(u) < \alpha+\beta \\ \quad \left. +[\mathrm{E}(X\wedge u) - \mathrm{E}(X\wedge F_{\alpha}^{-1})]\right\} & \\ \frac{c}{\beta}\left\{(1-\alpha)(F_{\alpha}^{-1} - F_{\alpha+\beta}^{-1}) + \beta(F_{\alpha+\beta}^{-1} - d) \right. & F(d) < \alpha < \alpha+\beta < F(u) \\ \quad \left. +[\mathrm{E}(X\wedge F_{\alpha+\beta}^{-1}) - \mathrm{E}(X\wedge F_{\alpha}^{-1})]\right\} & \\ c(u-d) & F(d) < F(u) < \alpha \end{cases} . \tag{2.12}$$

ⓘ *Under the Hood.* See the *Online Supplement* for the $RVaR$ development.

Although complex, equations (2.11) and (2.12) show how to compute a retained risk version of the ES and $RVaR$ using expressions involving standard limited expected value quantities.

Example 2.3. Property Fund Claims Distribution - Continued. Let us continue Example 2.1 where we examined empirical estimates of the distribution using 1,377 property damage claims. We now impose an upper limit u but, for simplicity, assume no deductible ($d = 0$) and no coinsurance ($c = 1$). A confidence level of $\alpha = 0.99$ is used for this illustration. For these data, it turns out that $\widehat{VaR}_{0.99} = 236{,}427$ which is marked with dashed vertical lines in Figure 2.4. The left-hand panel of this figure shows a change in the slope for the VaR whereas the right-hand panel shows that the ES is smoother at this point.

2.3 Risk Retention Changes

Risk managers would like to minimize both the uncertainty of retained risks and the cost of transferring risks. However, there is a natural trade-off between the risk retained, $g(X;\boldsymbol{\theta})$, and the amount transferred, $X - g(X;\boldsymbol{\theta})$. To get more insights into this trade-off, I consider a risk retention function $g(\cdot)$ that changes only through its parameters. With these potential changes, risk managers have substantial flexibility to adjust both the level of uncertainty

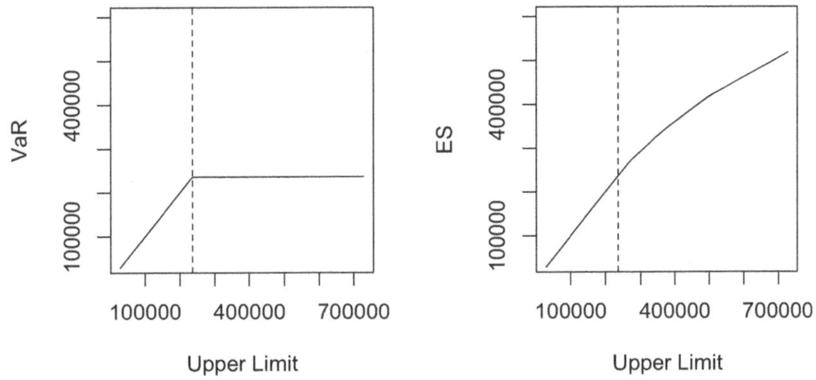

FIGURE 2.4
Property Fund VaR and ES Plots for Various Upper Limits. The left-hand panel shows the retained risk VaR over different upper limits and the right-hand panel gives similar information for the expected shortfall (ES). The dashed vertical line marks $\widehat{VaR}_\alpha$.

and costs to suit their preferences. For example, for an upper limit policy, one can let u approach zero so that no risk is retained, resulting in full transfer. Conversely, one can let u approach infinity, resulting in full retention.

Changes in risk retention parameters lead to changes in both the uncertainty of the retained risk and the associated cost of risk transfer. This section summarizes these two changes as a *ratio*, specifically by examining the change in the uncertainty of risk retained, the risk measure, per unit change in the risk transfer cost.

The rationale for investigating this ratio originates from the observation that risk managers possess an intuitive understanding of risk retention parameters, such as comparing two upper limits (e.g., 250,000 and 500,000). They can also readily interpret changes in risk transfer costs, as these represent firm expenditures (e.g., the cost savings associated with shifting from an upper limit of 250,000 to 500,000). Conversely, interpreting a risk measure, and changes therein, proves to be more complex. The aim is to facilitate interpretation by expressing these changes in risk measures as a "standardized unit," per unit change in the risk transfer cost. Chapter 3 outlines alternative approaches for summarizing the trade-off between the risk transfer cost and the uncertainty of the retained risk.

2.3.1 Discrete Changes

To help with the interpretation, I begin by examining discrete changes of risk retention summary measures for a Pareto distribution which is commonly used in insurance.

Example 2.4. Changes of Risk Retention Summary Measures Using a Pareto Distribution. Assume that X follows a Pareto distribution with scale parameter γ and shape parameter η, as in equation (2.5). The α quantile is

$$F_\alpha^{-1} = \gamma\left((1-\alpha)^{-1/\eta} - 1\right).$$

One can use this, with equation (2.10), to get the value at risk for the retained risk.

TABLE 2.1
Risk Transfer Costs and *VaR*s for Selected Risk Retention Parameters.

Deduct d	Coins c	Upper Limit u	RTC	VaR	$\frac{\Delta VaR}{\Delta RTC}$	ES	$\frac{\Delta ES}{\Delta RTC}$
0	1.0	Inf	0.00	2684.03	NaN	4526.05	NaN
100	1.0	Inf	86.78	2584.03	-1.15	4430.27	-1.10
0	0.9	Inf	50.00	2415.63	-5.37	4102.66	-8.47
0	1.0	2000	55.56	2000.00	-12.31	2000.00	-45.47

Note:
As with the statistical package 'R' conventions, 'Inf' means 'infinity' and 'NaN' means 'not a number.'
The symbol Δ means 'change in'.

The Pareto mean turns out to be $\mathrm{E}(X) = \frac{\gamma}{\eta-1}$ (if $\eta > 1$). Further, from equation (2.7), the expected value of retained risks for the Pareto distribution is

$$\mathrm{E}[g(X;\boldsymbol{\theta})] = \mathrm{E}[g(X;d,c,u)] = \frac{c\gamma}{\eta-1}\left[\left(\frac{\gamma}{d+\gamma}\right)^{\eta-1} - \left(\frac{\gamma}{u+\gamma}\right)^{\eta-1}\right].$$

This is based on standard limited expected value calculations documented in, for example, Actuarial Community (2020), Chapter 18. Limited expected values, together with equation (2.11), also allow one to get the expected shortfall for the retained risk.

For this example, I use the risk transfer cost as the expected risk minus this expected value of retained risks, $RTC(\boldsymbol{\theta}) = \mathrm{E}(X) - \mathrm{E}[g(X;\boldsymbol{\theta})]$. To be even more specific, I further assume that $\eta = 3$, $\gamma = 1000$, and $\alpha = 0.98$. Table 2.1 illustrates the effect on risk transfer costs and risk measures for selected risk retention parameters.

As a baseline, assume deductible $d = 0$, coinsurance $c = 1$, and upper limit $u = \infty$. Then, the cost of ceding the risk is $RTC(d = 0, c = 1, u = \infty) = 0$ and the level of uncertainty is $VaR_{\alpha=0.98}[g(X;\boldsymbol{\theta})] = 2{,}684.03$.

Suppose that one decides to impose a deductible $d = 100$. Then, the risk transfer cost increases to $RTC(d = 100, c = 1, u = \infty) = 86.78$. By paying 86.78, the risk owner is only responsible for losses in excess of 100 and not responsible at all for losses less than 100. With this deductible, the uncertainty level goes down to 2,584.03. This is a reduction of uncertainty level per unit change of transfer cost which turns out to be -1.15.

Table 2.1 summarizes changes for all three risk retention parameters. For this example, the upper limit option yields the greatest reduction in uncertainty level (2,684.03 -2,000.00 = 684.03) per change in transfer cost (0.00 -55.56 = -55.56) that is summarized as a ratio, -12.31.

Happily, these results are consistent with the ES measure. That is, if the manager opted to instead use the ES instead of the VaR, again the upper limit option would result in the greatest reduction per risk transfer cost change. This is comforting in the sense that the manager may not have a deep intuitive appreciation for either measure yet both recommend the same action meaning that any decision about which measure to rely upon is of less importance.

Code. The *Online Supplement* provides Example 2.4 illustrative statistical code.

2.3.2 Differential Changes

Naturally, the functions that we work with are highly nonlinear and so proportional changes vary considerably according to the separation of two comparison points and their location.

Moreover, discrete changes become awkward to interpret as one moves to a world of multivariate risks, which we do beginning in Chapter 4. So, I now introduce differential changes to ease the interpretation.

As motivated by Section 2.3.1 on discrete changes, we look to the ratio of differential change in the quantile per unit differential change in risk transfer costs. For simplicity, this chapter focuses on risk transfer costs using expectations so that $RTC(\boldsymbol{\theta}) = \mathrm{E}(X) - \mathrm{E}[g(X;\boldsymbol{\theta})]$. As overall mean losses do not change with risk retention parameters, this means that any differential of risk transfer costs is simply minus one times the differential of mean retained claims. To illustrate for the coinsurance parameter c, from equation (2.10) and the discussion following equation (2.7), we have

$$\frac{\partial_c \; VaR_\alpha[g(X;\boldsymbol{\theta})]}{\partial_c \; RTC(\boldsymbol{\theta})} = \begin{cases} 0 & \text{if } \alpha < F(d) \\ -\frac{F_\alpha^{-1}-d}{\mathrm{E}(X\wedge u)-\mathrm{E}(X\wedge d)} & \text{if } F(d) \le \alpha < F(u) \\ -\frac{u-d}{\mathrm{E}(X\wedge u)-\mathrm{E}(X\wedge d)} & \text{if } \alpha \ge F(u) \end{cases}.$$

This uses the short-hand notation $\partial_c = \frac{\partial}{\partial c}$. Results for other parameters appear in Table 2.2.

Example 2.5. Changes of Risk Retention Summary Measures Using a Pareto Distribution - Continued. This is a continuation of Example 2.4. Table 2.3 summarizes differential changes for selected risk retention parameters. From this table, one can see that:

- In the baseline case where $d = 0$, $c = 1$, and $u = \infty$, a finite reduction in the required risk capital is produced by the coinsurance parameter c.
- In the case where $d = 100$, $c = 1$, and $u = 10000$, the differential for changing the u parameter is zero. This is because at this value of u, we have $\alpha = 0.98 < F(10000)$, meaning that the upper limit is in a region where the retained risk is constant in u.
- From the third and fourth rows, the largest (negative) relative change is for the parameter u. That is, at these parameter values, if one were to move any of the risk retention parameters, the greatest decline in the uncertainty per unit of risk transfer cost is produced by changes in the upper limit parameter.

Code. The *Online Supplement* provides Example 2.5 illustrative statistical code.

2.4 Convexity and Smoothness of Uncertainty Measures

Numerical optimization algorithms often rely heavily on concepts such as convexity and smoothness (e.g., differentiability) of functions. Figure 2.4 illustrates how different risk measures can exhibit varying degrees of smoothness. Hence, in the simple univariate case, I delineate these characteristics for the measures that summarize the distribution, specifically, the distribution function, the value at risk, and the expected shortfall.

According to the definition of convexity, a linear (convex) combination of a function evaluated at points should be greater than or equal to the function evaluated at the same linear combination of points. Specifically, a function h is convex if $h[cu_1 + (1-c)u_2] \le ch(u_1) + (1-c)h(u_2)$ for $0 \le c \le 1$. In other words, the line segment between any two points on the graph lies at or above the graph. In the same way, a function h is concave if $-h$ is convex. For a concave function, the line segment between any two points on the graph lies at or below the graph. This is an abstract definition that can be applied in

TABLE 2.2
Differential Changes of Risk Transfer Cost, Value at Risk, and Expected Shortfall.

Summary	Parameter (θ)			
Measure	**Deduct d**	**Coins c**	**Upper Limit u**	**Range of α**
$\partial_\theta\ RTC(\boldsymbol{\theta})$	$c\{1-F(d)\}$	$-\{\mathrm{E}(X\wedge u)-\mathrm{E}(X\wedge d)\}$	$-c\{1-F(u)\}$	
	0	0	0	$\alpha < F(d)$
$\partial_\theta\ VaR_\alpha[g(X;\boldsymbol{\theta})]$	$-c$	$F_\alpha^{-1}-d$	0	$F(d)\le\alpha<F(u)$
	$-c$	$u-d$	c	$F(u)\le\alpha$
	$\frac{-c}{1-\alpha}\{1-F(d)\}$	$\frac{1}{1-\alpha}\{\mathrm{E}(X\wedge u)-\mathrm{E}(X\wedge d)\}$	$\frac{c}{1-\alpha}\{1-F(u)\}$	$\alpha < F(d)$
$\partial_\theta\ ES_\alpha[g(X;\boldsymbol{\theta})]$	$-c$	$\frac{1}{1-\alpha}\{\mathrm{E}(X\wedge u)-\mathrm{E}(X\wedge F_\alpha^{-1}) +(1-\alpha)(F_\alpha^{-1}-d)\}$	$\frac{c}{1-\alpha}\{1-F(u)\}$	$F(d)\le\alpha<F(u)$
	$-c$	$u-d$	c	$F(u)\le\alpha$

TABLE 2.3
Differential Relative Changes of VaR.

d	c	u	RTC	VaR	$\frac{\partial_d VaR}{\partial_d RTC}$	$\frac{\partial_c VaR}{\partial_c RTC}$	$\frac{\partial_u VaR}{\partial_u RTC}$
0	1.0	Inf	0.00	2684.03	–1.00	–Inf	NaN
100	1.0	10000	86.78	2584.03	–1.33	–28.42	0.00
500	0.9	2000	300.00	1965.63	–3.37	–3.86	–27.00
1000	0.9	1500	459.50	450.00	–8.00	–0.98	–15.62

many contexts. As will be discussed in Section 4.4, convexity is a very useful property for numerical optimization. Because of this, our interest will be mainly in thinking about the "points" as retention parameters such as an upper limit u, not the potential losses X.

To visualize the convexity and smoothness, Figure 2.5 shows the distribution function, value at risk (quantile), and the expected shortfall as a function of the upper limit u. In this figure, the deductible is set as $d = 0$ and the coinsurance as $c = 1$.

The left-hand panel shows the discontinuity, or jump, of the distribution function in equation (2.8). As a function of the upper limit u, the distribution function is neither convex nor concave. (This is also evident in the more complex Figure 2.3.) The middle panel shows the value at risk based on equation (2.10). This function is concave in u and is also linear except at the point F_α^{-1}. At this point, there is a sharp discontinuity in the slopes, indicating a type of "non-smoothness." The right-hand panel shows the expected shortfall based on equation (2.11). The figure suggests that the ES is also concave and is smooth in u.

Some readers will find it instructive to take differentials in u just less than and greater than $u = F_\alpha^{-1}$ using equation (2.11). This gives another way to get insights into the smoothness of ES for retained risks. To explain the smoothness of the ES in u, return to equation (2.11) and take a partial derivative with respect to u to get

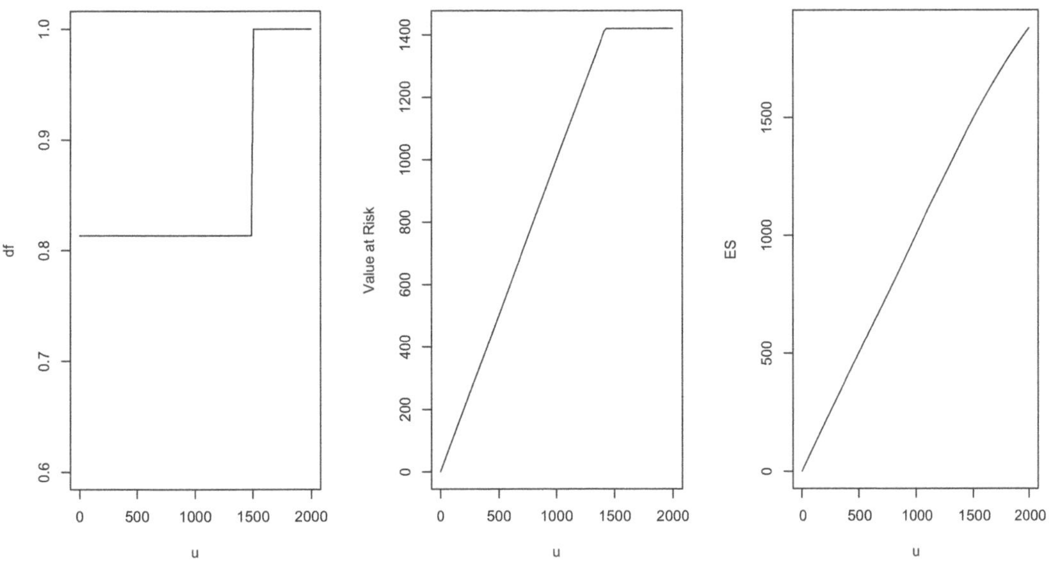

FIGURE 2.5
Uncertainty Measures versus Upper Limit u. The measures of uncertainty are the distribution function, value at risk, and ES.

$$\partial_u \; ES_\alpha[g(X;\boldsymbol{\theta})] = \begin{cases} \frac{c}{1-\alpha}\{1-F(u)\} & \text{if } \alpha < F(d) \\ \frac{c}{1-\alpha}\{1-F(u)\} & \text{if } F(d) \le \alpha < F(u) \\ c & \text{if } F(u) \le \alpha \end{cases} .$$

using the relation $\partial_u \mathrm{E}(X \wedge u) = 1 - F(u)$. This expression shows that the derivative is smooth in u; even at the point $\alpha = F(u)$, it has the same derivative from the left and from the right.

Although the risk measures are typically smooth and convex (or concave) in the presence of a single retention parameter, this is not true in general with multiple parameters. This is best illustrated with a simple example.

Example 2.6. Pareto Distribution. I use the same assumptions as in Example 2.4 but with $\gamma = 2000$ and $\alpha = 0.85$ (to better display graphically the effects we want to show). Now select two sets of retention parameters, $(d_1 = 0, u_1 = 1000)$ and $(d_1 = 1000, u_1 = 2500)$. Using equation (2.11), I evaluate the expected shortfall at each point and at linear combinations of the form $(d = cd_1 + (1-c)d_2, u = cu_1 + (1-c)u_2)$. Figure 2.6 shows the results, establishing that the expected shortfall is neither convex nor concave.

R *Code.* The *Online Supplement* provides Example 2.6 illustrative statistical code.

This lack of convexity will be revisited in Section 5.1 in the context of multiple risks.

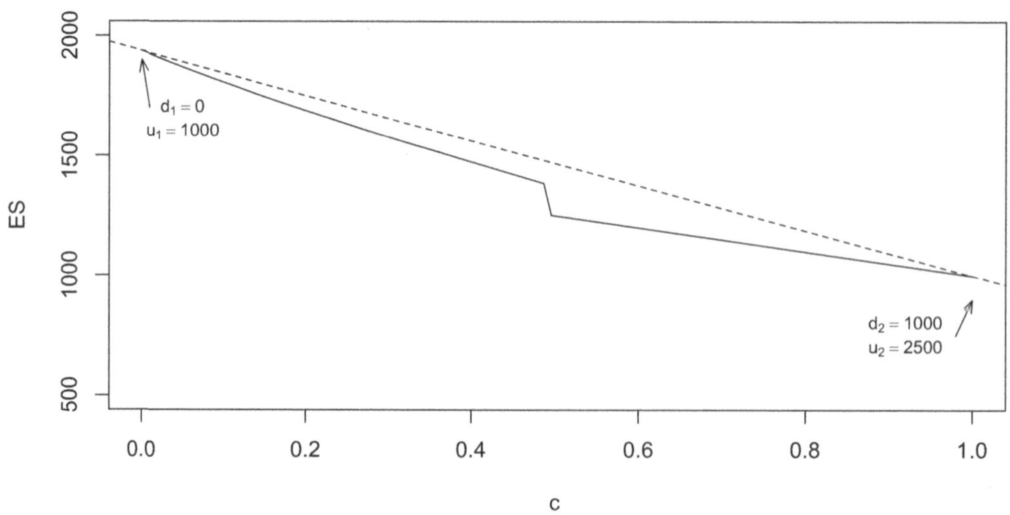

FIGURE 2.6

Expected Shortfalls are Neither Convex nor Concave. The solid line shows the *ES* evaluated at linear combinations of two points, $(d_1 = 0, u_1 = 1000)$ and $(d_1 = 1000, u_1 = 2500)$. The straight dashed line shows the linear combination of the *ES*s evaluated at the two points.

2.5 Supplemental Materials

2.5.1 Further Resources and Readings

Readers wanting a quick refresher of some basic statistical and actuarial concepts may refer to the online book, Actuarial Community (2020). In particular, Chapter 3 describes calculation of risk transfer costs for policies with coverage modifications, Chapter 10 provides an introduction to risk measures, and Chapter 18 describes limited expected value calculations.

For example, Chapter 10 of Actuarial Community (2020) describes alternative versions of the expected shortfall ES. That is, closely related to the ES are the *average value at risk*, also known as the *tail value at risk* and the *conditional value at risk*, among others. In fact, under mild continuity assumptions, they are all equivalent. For the purposes of this text, I assume such continuity assumptions hold and use their different alternative expressions as needed. Naturally, there are some applications where this is not the case. To understand their differences, you can begin with the distinctions drawn in Chapter 10 of Actuarial Community (2020).

As summarized in Wang and Zitikis (2021), both VaR and ES have solid axiomatic foundations and "appear in the banking regulation frameworks of Basel III/IV, as well as in the insurance regulation frameworks of Solvency II and the Swiss Solvency Test." Thus, although their application to insurable risk portfolio problems considered in this book is novel, their usefulness in determining adequate solvency for banks and insurers suggest that they are appropriate measures upon which to base this work.

For additional discussion of the ratio of differential changes, see Frees (2017). Instead of risk transfer costs, that paper describes the differential change in a risk measure (RM) per unit change in expected retained premiums, that is, $\{\partial_\theta \; RM[g(X;\boldsymbol{\theta})]\} \,/\, \{\partial_\theta \; \mathrm{E} \; g(X;\boldsymbol{\theta})\} = -\,\{\partial_\theta \; RM[g(X;\boldsymbol{\theta})]\} \,/\, \{\partial_\theta \; RTC(\theta)\}$, and calls this a *risk measure relative marginal* change, or RM^2 change, for short.

2.5.2 Exercises

See the *Online Supplement* for selected exercise solutions.

Section 2.1 Exercises

Exercise 2.1. Comparing Parametric Value at Risk Measures. In Example 2.1, we compared nonparametric versions of the value at risk VaR and expected shortfall ES. Subsequently, in Example 2.2 we determined parametric versions of an extension, the range value at risk $RVaR$, using the Pareto distribution. In this exercise, we compare the VaR and ES using different parametric assumptions. As a baseline, we assume the Pareto distribution with scale parameter $\gamma = 2300$ and shape parameter $\eta = 0.999$ (rounded from Example 2.2 for simplicity).

a. Using the `R` function `mpareto()`, verify that the mean for this distribution is infinite. This means that we cannot use the method of moments to compare distributions. (*Hint*: It may be convenient to use the conventions in the online book Actuarial Community (2020).)

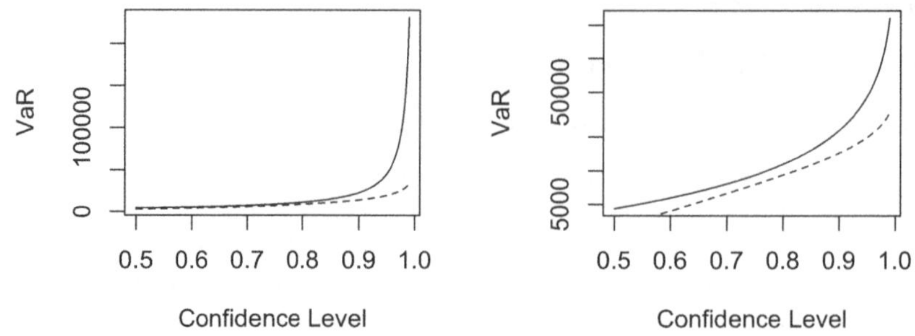

FIGURE 2.7
Value at Risk (VaR) for Pareto and Gamma Distributions. The black solid line is for the Pareto distribution, the dashed vertical line is for the gamma distribution. The left-hand panel is in the original units that emphasizes the differences in VaR for large confidence levels. The vertical axis in the right-hand panel is on the log scale that permits visualization of the differences over the range of the confidence levels.

b. One can use quantiles from the Pareto distribution to determine parameters of the gamma distribution. Specifically, the illustrative code shows how to match the 60th and 70th percentiles. Use this code to match the 20th and 80th percentiles (that results in more stable parameter estimates). This illustrative code:

- Uses the `R` function `multiroot()` to find the root of several (nonlinear) equations, a function that we will use going forward.
- Re-parameterizes parameters to constrain them to be positive. For each parameter, the search is over the entire real line but then the solution is transformed to the positive portion of the real line through exponentiation. This re-parameterization technique saves us from the trouble of constraining parameters.

c. Figure 2.7 is a plot that summarize a comparison of VaR measures for the Pareto and gamma distribution assumptions. Note that the gamma VaR measures are much lower than the corresponding Pareto values for large values of α (confidence levels). Write code to reproduce this plot. (*Hint:* See the *Online Supplement* for illustrative code.)

Exercise 2.2. Comparing Parametric $RVaR$ Measures. In Exercise 2.1, we compared value at risk VaR measures based on different parametric distributions that were equated by the method of matching percentiles. However, we could not compare the ES risk measures because the mean of the distribution was infinite. The purpose of this exercise is to show how to make the comparison using the range value at risk $RVaR$ measure.

Specifically, compare the Pareto and gamma distributions with the same parameters determined in Exercise 2.1. Use the `R` function `integrate()` and equation (2.4) to compute $RVaR$ values. To check your work, it turns out that $RVaR(\alpha = 0.80, \beta = 0.10)$ is 11,518.43 for the gamma distribution and is 13,673.63 for the Pareto distribution. Figure 2.8 provides a rough idea of other values of $RVaR$ to calibrate your work.

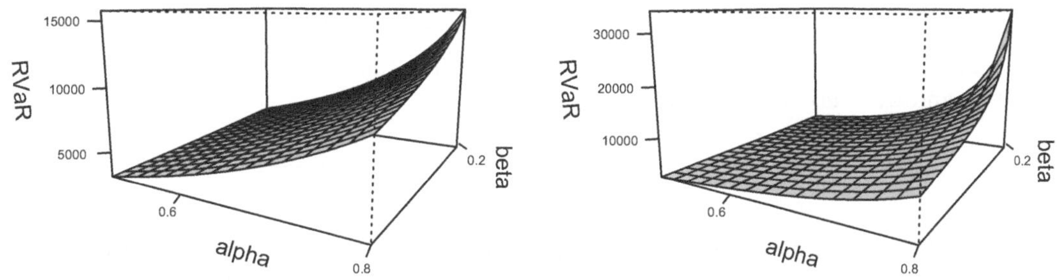

FIGURE 2.8
Range Value at Risk ($RVaR$) for Gamma and Pareto Distributions. The left-hand figure is for the gamma distribution and the right-hand is for the Pareto distribution. Values of $RVaR$ are given by level of confidence alpha and range parameter beta.

Exercise 2.3. Risk Measures for the Sum of Two Risks. Consider the following two non-identical random variables:

- X_1 has a gamma distribution with shape parameter 2 and scale parameter 5,000. Thus, the mean is 10,000 and the 95th percentile is 23,719.
- X_2 has a Pareto distribution with shape parameter 3 and scale parameter 2,000. Thus, the mean is 1,000 and the 95th percentile is 1,714.

Assume that the risks are independent and consider their sum $S = X_1 + X_2$ (extensions to dependence and more than two risks begin in Chapter 4).

a. Determine $\Pr(S \leq 18000)$. *Hint:* Use the convolution formula $\Pr(S \leq y) = \int_0^y F_1(y-z) f_2(z) dz$ and the `R` function `integrate()`.

b. Determine the 95th percentile of S. *Hint:* Use the `R` function `uniroot()` and solve the equation $\Pr(S \leq y) = 0.95$.

c. Determine the expected shortfall of S for confidence level $\alpha = 0.95$. *Hint:* Use the limited expected value expression in Display (2.3).

Section 2.2 Exercises

Exercise 2.4. Risk Retention for the Sum of Two Risks. Consider the sum of two risks S introduced in Exercise 2.3 and let $g(S) = g(S; d = 100, c = 0.8, u = 30000)$ be the retained risk. For confidence level $\alpha = 0.95$, determine:

a. the value at risk $VaR_\alpha[g(S)]$,

b. the expected shortfall $ES_\alpha[g(S)]$,

c. the range value at risk $RVaR_{\alpha,\beta=0.04}[g(S)]$, and

d. find the value of a newer risk measure, the $GlueVaR$.

The $GlueVaR$ was introduced by Belles-Sampera et al. (2014); it is a flexible risk measure and so exhibits many of the same qualities as the $RVaR$. Based on this work, we consider the expression

$$GlueVaR = \omega_1 VaR_\alpha + \omega_2 ES_\alpha + \omega_3 ES_{\alpha+\beta},$$

where ω_1, ω_2, and ω_3 are weights lying with [0,1] that sum to one. For this exercise, use $\beta = 0.04$ and $\omega_1 = \omega_2 = \omega_3 = 1/3$.

Exercise 2.5. Portfolio Management and Simulation. You are the Chief Risk Officer of a telecommunications firm and responsible for its captive insurer. The firm has several property and liability risks. We will consider:

- X_1 - buildings, modeled using a gamma distribution with mean 200 and scale parameter 100.
- X_2 - motor vehicles, modeled using a gamma distribution with mean 400 and scale parameter 200.
- X_3 - directors and executive officers risk, modeled using a Pareto distribution with mean 1000 and scale parameter 1000.
- X_4 - cyber risks, modeled using a Pareto distribution with mean 1000 and scale parameter 2000.

Denote the total risk as $S = X_1 + X_2 + X_3 + X_4$. You have negotiated to transfer risks in excess of $d = 100$, to an upper limit of $u = 5000$, yet covering only $c = 0.80$ of those claims. That is, the risk transferred is $g(S) = g(S; d = 100, c = 0.80, u = 5000)$ where $g()$ is defined in equation (2.6). Using simulation techniques, you should:

a. Plot the simulated distribution of $g(S)$.

b. Determine the expected value and standard deviation of $g(S)$.

c. Determine the value at risk VaR at confidence levels $\alpha = 0.80, 0.90, 0.95$.

d. Determine the expected shortfall ES at confidence levels $\alpha = 0.80, 0.90, 0.95$.

For readers wanting a refresher in simulation techniques, see the online book Actuarial Community (2020), Chapter 6.

Section 2.3 Exercise

Exercise 2.6. Risk Retention Changes for the Sum of Two Risks. This is a continuation of Exercises 2.3 and 2.4. We now seek to observe how the various risk measures change as retention parameters d, c, u change. For simplicity, we restrict attention to the basic measures, value at risk and expected shortfall. Further, from the discussion in Section 2.2.3, it is clear that the effects of the coinsurance parameter are linear. So, although the changes to this parameter can have important financial implications, because of the linearity they are easy to anticipate and so we do not need to investigate them further here.

a. Quantify the impact of an increase of 500 to the deductible d for both the VaR and ES measures using the baseline scenario in Exercise 2.5. Compare this to an increase of 500 in the upper limit u. For each risk measure, describe the size and direction of the changes.

b. Quantify the impact of increases 10, 50, 100, and 500 to the deductible d for both the VaR and ES measures where the increase is measured from the baseline scenario in Exercise 2.5. Note the change in each risk measure per unit change in the deductible.

c. Use the results summarized in Table 2.2 to calculate the differential change at the baseline scenario in each risk measure for a change in deductible d. Compare this change to the discrete changes in part (b). You should establish that the differential approximations are much easier to compute and are excellent proxies for the discrete changes in the cases considered.

2.5.3 Appendix. Distortion Risk Measures

Risk measures are well known to actuaries and other insurance analysts as they are employed in solvency regimes that base capital adequacy criteria on standard risk measures such as VaR and ES. In this book, I use risk measures to inform readers about insurance retention and portfolio applications. The choice of the risk measure matters.

The focus of this book is on selected commonly used risk measures introduced in this chapter, the VaR, ES, and $RVaR$. However, given their importance in practice, it is not surprising that there is a robust literature on alternative risk measures. The following describes one class that encompasses many (but not all) of these alternatives. For actuaries and other analysts that provide advice on insurable risk portfolios, it is helpful to understand how these alternatives are defined and where one can look if further investigation into alternatives seem warranted

Using notation from Serfling (1980) (page 265), a general risk measure can be expressed as:

$$RM(F) = \int_0^1 F^{-1}(t)\ dK(t) = \int_0^1 J(t)F^{-1}(t)dt + \sum_{j=1}^{m} a_j F^{-1}(\alpha_j). \tag{2.13}$$

This is known as a *distortion* risk measure and encompasses many important measures as special cases, cf. Assa (2015).

- We have already considered the VaR (quantile) risk measure where one chooses $J(t) \equiv 0$, $m = 1$, and $a_1 = 1$.
- The expected value is also a special case of $RM(F)$ by using $J(t) \equiv 1$ and $m = 0$.
- The expected shortfall ES is another special case, where one chooses $m = 0$ and $J(t) = 0$ for $0 \leq t < \alpha$ and $J(t) = 1/(1-\alpha)$ for $\alpha \leq t \leq 1$. Here, α $(0 \leq \alpha < 1)$ is the risk preference level.
- The risk measure from the proportional hazards transform arises when $m = 0$ and $J(t) = \alpha(1-t)^{\alpha-1}$. Here, $0 < \alpha \leq 1$.

Further, from Embrechts et al. (2018) (Appendix, EC.1), we have that $RVaR$ is a special case of the distortion measure with the choice of

$$J(t) = \begin{cases} \min[(I(t > \alpha)\frac{t-\alpha}{\beta}, 1] & \text{if } \beta > 0 \\ I(t > \alpha) & \text{if } \beta = 0 \end{cases}.$$

3

Balancing Retained Risk and Risk Transfer Cost

Chapter Preview. The trade-off between the amount of risk to retain and the cost of offloading risk, documented in Chapter 2, has long been recognized by risk managers. One simple and intuitive method of assessing this trade-off, examining the ratio of changes, was put forth in the earlier Section 2.3. Moreover, diverse methods have emerged to navigate this trade-off over time. Section 3.1 summarizes some important ideas drawing on what we now know from the insurance economics and actuarial literatures.

More recently, industry participants have looked to the concept of a *total cost of risk* as described in Section 3.2. In this approach, both the uncertainty of retained risks and risk transfer costs are combined additively. Combining these two offsetting quantities into a single term, the "total cost," can simplify the development of optimal strategies. This approach is useful for arguing that a manager's strategy is effective in certain well-defined settings.

Nevertheless, as one considers more complex settings of multiple risks with many players, optimizing a single criterion such as a total cost becomes limiting. A more robust strategy is to reduce the uncertainty of retained risks as much as possible limited by a budget for risk transfer. This approach intuitively appeals to risk managers, resembling analyses conducted for asset allocations. Section 3.3 provides an introduction to this approach, which will dominate the remainder of the book.

As with the preceding Chapter 2, this chapter confines attention to a single risk X which can be interpreted as the risk of an entire portfolio. The intent is to lay the groundwork for more complex situations in subsequent chapters that consider multiple risks.

3.1 Risk Retention Fundamentals

This section presents selected findings from the academic literature, offering guidance on the optimal policy for risk transfer. The first two propositions originate from the insurance economics literature, while the second two stem from the actuarial literature. Although their methodologies vary, the conclusions drawn from these findings are largely consistent. We frame these results in the context of transfers from an entity that is a "risk owner" to an "insurer" to underscore their wide applicability.

Risk portfolio managers will find these results beneficial as they serve to reinforce and inform the development of intuition when selecting an appropriate insurance or reinsurance contract. However, in practice, the choice of a contract is influenced by various other factors, including the participants' market experience and the availability of desired contract forms at a reasonable price. These additional factors make it difficult to implement these results directly.

 DOI: 10.1201/9781003469568-3

Purchasing Insurance Coverage. The insurance economics literature has a long history of determining risk retention agreements. Through this framework, an optimal choice of risk retention parameters can be determined through expected utility. For instance, a classic result from insurance economics involves determining the value of a risk retention parameter to maximize expected utility.

Proposition 3.1. Mossin (1968) Type Result. Suppose that:

- The risk owner has a concave utility function U with initial wealth W. Before any risk transfer, the beginning of the period utility of wealth is $U(W)$.
- The risk owner and the insurer enter into a coinsurance contract. At the end of the period, the insurer pays an indemnity $I(X) = cX$ to the risk owner for loss X. At the beginning of the period, it receives premium cP_0 where $P_0 = (1 + \lambda_{SL})\mathrm{E}[X]$. Here, λ_{SL} is a loading for safety and expenses.
- The risk owner's end of period expected utility is given by $\mathrm{E}\{U[W - cP_0 - (1-c)X]\}$, that is, the expected utility after deducting the premium paid and the portion of risk retained. The risk owner seeks to maximize this quantity over different choices of c.

If $\lambda_{SL} \leq 0$ then the choice $c = 1$ maximizes end of the period expected utility.

ⓘ *Under the Hood.* See the *Online Supplement* for the verification of Proposition 3.1.

The result of Mossin in 1968 indicates that full coverage corresponding to $c = 1$ is optimal when insurance is fair, $\lambda_{SL} = 0$. When a loading is in place so that $\lambda_{SL} > 0$, one can also show that a partial insurance coverage corresponding to a value of $c < 1$ is optimal, cf. Schlesinger (2013).

Optimal Insurance Contract. A retention function changes the shape of the distribution of retained risks. What is the best choice? The optimality of upper limit contracts was established by Arrow in 1963.

Proposition 3.2. Arrow (1963) Type Result with Limited Liability. Suppose that:

- The risk owner has a concave utility function U with initial wealth W.
- The insurer pays an indemnity $I(X) = X - g(X)$ to the risk owner for loss X. The indemnity is limited by a fixed amount, u_{lim}.
- The insurer receives premium $P_0 = (1+\lambda_{SL})\mathrm{E}[I(X)]$. Here, $\lambda_{SL} > 0$ is a positive safety loading.
- The risk owner seeks to maximize end of the period expected utility, $\mathrm{E}\{U[W - P_0 - X + I(X)]\}$, over different choices of $I(X)$.

For a fixed P_0, the best choice is $I^*_{opt}(X) = u_{lim} \wedge (X - u^*)_+$, where u^* is such that $P_0 = (1 + \lambda_{SL})E[I^*_{u^*}(X)]$.

ⓘ *Under the Hood.* See the *Online Supplement* for the verification of Proposition 3.2.

To interpret this result, consider the unlimited liability case so that $u_{lim} = \infty$. In this case, the optimal retention function is $g_{opt}(x) = X - I^*_{u^*}(X) = X - (X - u^*)_+ = X \wedge u^*$, a contract with upper limit u^*. This is intuitively appealing - with the upper limit provision, the insurer takes the responsibility for very large losses in the tail of the distribution, not the risk owner.

Coinsurance is Desirable for Insurers. The coinsurance contract is particularly desirable for the party taking on the transferred risk that we interpret to be the insurer. To see this, suppose that a risk owner and an insurer wish to enter into a contract to share total losses X such that the retained loss is $g(X) = cX$. Thus, the risk owner retains the fraction c of losses. The insurer pays an indemnity $I(X) = (1-c)X$ to the risk owner for loss X.

Suppose further that the risk owner is willing to enter into other contracts yet only cares about the variability of retained losses. That is, the risk owner is indifferent to the choice of g as long as $\mathrm{Var}[g(X)]$ stays the same and equals, say, Q. Then, the following result shows that the coinsurance contract minimizes the insurer's uncertainty as measured by $\mathrm{Var}[X - g(X)]$.

Proposition 3.3. Consider the collection of all $g(\cdot)$ such that $\mathrm{Var}[g(X)] = Q$. Then, by letting $c = \sqrt{Q/\mathrm{Var}(X)}$, we have that $\mathrm{Var}[(1-c)X] \leq \mathrm{Var}[X - g(X)]$ for all $g(.)$ in this collection.

ⓘ *Under the Hood.* See the *Online Supplement* for the verification of Proposition 3.3.

The proposition is intuitively appealing - with coinsurance, the risk owner and insurer share the responsibility for very large losses in the tail of the distribution. This arrangement is desirable for insurers. In addition, Section 4.1.2 will introduce the concept of *moral hazard*, the situation where a contract alters the behavior of parties to the agreement. Coinsurance is also desirable because incentives for the risk owner and insurer are aligned, thus mitigating potential moral hazard.

Upper Limit is Desirable for Risk Owners. The upper limit contract can be particularly desirable for the risk owner. To see this, suppose that the risk owner has a budget constraint and is willing to spend up to a constant denoted here as RTC_{max}, the maximal risk transfer cost. With risk transfer costs determined by expectations, the risk owner is willing to consider any risk transfer function $g(\cdot)$ as long as $\mathrm{E}[X - g(X)] = RTC_{max}$. Subject to this budget constraint, assume that the risk owner wishes to minimize uncertainty of the retained risks as measured by the variance. Then, the following result shows that an upper limit insurance contract minimizes the risk owner's uncertainty.

Proposition 3.4. Consider the collection of all $g(\cdot)$ such that $\mathrm{E}[X - g(X)] = RTC_{max}$. Then, by letting u to be such that $\mathrm{E}(X \wedge u) = \mathrm{E}[X] - RTC_{max}$, we have $\mathrm{Var}(X \wedge u) \leq \mathrm{Var}[g(X)]$ for all $g(.)$ in this collection.

ⓘ *Under the Hood.* See the *Online Supplement* for the verification of Proposition 3.4.

As with the Arrow type result in Proposition 3.2, this proposition is intuitively appealing in that the insurer takes the responsibility for very large losses in the tail of the distribution, not the risk owner.

This section summarizes four results from the academic literature on the choice of insurance coverages. In economics, the "risk owner" is typically an individual policyholder and so principles of utility maximization are intuitively appealing. In actuarial science, often the "risk owner" is an insurance company (so that the insurer of an insurance company is a reinsurer). Here, measures that quantify the uncertainty of risk such as variance have historically been appealing due to the ease of application.

Familiarity with these results helps guide the intuition of the risk manager. Nonetheless, as noted above, the actual choice of the contract form will finally often be triggered by intuition and experience as well as simplicity and transparency. For example, one may look for concrete protection against many losses in the case of protection from hail in homeowners insurance. Or, one may look to the protection against large losses from terrorist attacks on sporting venues. Additional factors may include consequences of a reinsurance treaty on taxation, dividends, and so forth.

3.2 Risk Owner's Total Cost

As a way to balance the trade-off between how much risk to retain and the cost of offloading risk, we can look to

$$T_g(X) = g(X) + RTC(g),$$

the *total risk*. Here, $g(X)$ represents the contingent losses borne by the risk owner that are retained after an agreement is in place and $RTC(g)$ represents the (non-random) **risk transfer cost** associated with offloading the risks. The additive structure is intuitive; $T_g(X)$ represents the total monetary losses for the risk owner. Because $T_g(X)$ is a random variable, we summarize its uncertainty with a risk measure RM. Assuming that the risk measure preserves location translations, then we may write

$$TC = RM[T_g(X)] = RM[g(X)] + RTC(g),$$

that we can interpret to be the *total cost* to the risk owner, as introduced by Cai and Tan (2007). As noted in Section 2.2.3, the measures VaR, ES, and $RVaR$ all preserve location translations and so can be used as special cases of RM here.

3.2.1 Special Cases

To begin, consider the risk retention function defined in Section 2.2.1. Moreover, as in Cai and Tan (2007), assume that risk transfer costs are expressed as $RTC(X;\theta) = (1 + \lambda_{SL})\{\mathrm{E}(X) - \mathrm{E}[g(X;\theta)]\}$, where λ_{SL} is a loading for expenses and overhead. Assuming the value at risk is used to quantify the uncertainty, the total cost is

$$TC = VaR_\alpha[g(X;\theta)] + (1 + \lambda_{SL})\{\mathrm{E}(X) - \mathrm{E}[g(X;\theta)]\}. \tag{3.1}$$

To minimize this, one can take a derivative with respect to a retention parameter θ of interest and set this equal to zero to get

$$0 = \partial_\theta\, TC = \partial_\theta\, VaR_\alpha[g(X;\theta)] - (1 + \lambda_{SL})\partial_\theta\, \mathrm{E}[g(X;\theta)]. \tag{3.2}$$

We developed explicit expressions for these derivatives; they are summarized in Table 2.2. For example, for the case $\theta = u$, we have on the set $\{u : F(u) < \alpha\}$,

$$0 = c - (1 + \lambda_{SL})c[1 - F(u)] \Rightarrow \frac{\lambda_{SL}}{1 + \lambda_{SL}} = F(u) \Rightarrow u = F^{-1}\left(\frac{\lambda_{SL}}{1 + \lambda_{SL}}\right).$$

By requiring $\alpha > \lambda_{SL}/(1 + \lambda_{SL})$, we have that the optimal value of u lies in the set $\{u : F(u) < \alpha\}$, thus satisfying our choice of looking at this set.

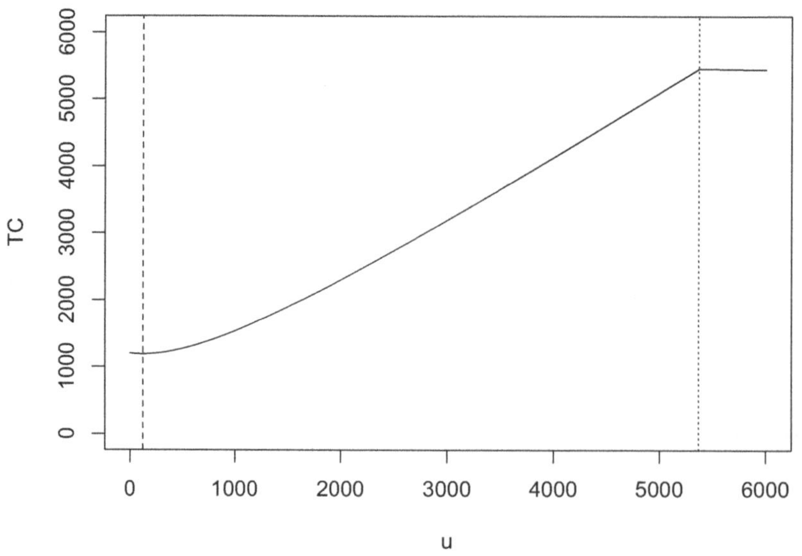

FIGURE 3.1
Total Cost Using Value at Risk for an Upper Limit Policy. The left vertical dashed line marks the point of optimal (lowest) cost. The right vertical dashed line marks that point where the upper limit equals the quantile of the confidence level, $F^{-1}_{0.98}$.

Example 3.1. Total Cost for a Pareto Distribution. Continuing with Example 2.4, we use a Pareto distribution but with parameters $\eta = 3$ and $\gamma = 2000$. We also assume a confidence level $\alpha = 0.98$ and a relative security loading $\lambda_{SL} = 0.20$. For this illustration, we take the deductible $d = 0$ and coinsurance $c = 1$. To promote consistency in the literature, we have used the assumptions of Example 2.2 in Cai and Tan (2007). With these assumptions, the total cost is

$$TC = VaR_\alpha(X \wedge u) + (1 + \lambda_{SL})[\mathrm{E}(X) - \mathrm{E}(X \wedge u)].$$

Figure 3.1 shows the total cost as a function of u and marks the point of lowest cost, $u^* = F^{-1}\left(\frac{0.2}{1.2}\right) = 125.3171$.

There are several insights that we can get from this example.

- Suppose that in equation (3.1), one uses the ES measure instead of the VaR. Then, by examining Table 2.2, one can see that the differential association with u remains unchanged and so the optimal upper limit remains unchanged. This is comforting to managers of risk. The optimal amount of risk retention does not matter when deciding between the value at risk and the expected shortfall.
- Suppose that a risk manager wishes to use the total cost procedure optimizing over the coinsurance parameter (keeping the deductible and upper limit fixed). Then referring to equation (3.2) and Table 2.2, one can see that this is problematic. That is, both the risk measure and the cost of ceding the risk are linear in this parameter so that a unique optimum does not exist.
- In the preceding example, the optimal upper limit (125.3171) seems low especially when compared to the mean risk (1,000 for this example). The result is more intuitively appealing if one considers optimizing over deductibles that, from

equation (3.2) and Table 2.2, gives exactly the same result. That is, if the risk owner wants to outsource the small nuisance claims in a cost-efficient way, then a deductible makes sense and this procedure provides an intuitively plausible result.

Despite the limitations listed above, the total cost approach is attractive because it is simple and intuitively appealing. Now, suppose that we have available a menu of risk transfer options that depend on risk retention parameter θ. Each option has an associated cost $RTC(\theta)$ that may be market driven from an external insurer or reinsurer. Then, the total cost is $TC = RM[\theta] + RTC(\theta)$ that one could minimize by taking derivatives and solving for zero to get

$$\partial_\theta\ RM[\theta] = -\partial_\theta RTC(\theta) \Rightarrow \frac{\partial_\theta\ RM[\theta]}{-\partial_\theta\ RTC(\theta)} = 1.$$

This provides a necessary condition for achieving an optimal total cost. As described in Section 2.3.2, ratios of differentials are intuitively appealing. Moreover, as seen in this section, they also provide information as to when we are close to an optimal total cost.

3.2.2 Total Cost Optimal Designs

This section examines the theoretical literature on optimal design, building on the work of Cai and Chi (2020) in the realm of risk measures. To formalize the preceding discussion, the optimal design aims to minimize $RM[T_g(X)]$ over a set of retention *functions*, with RM representing a given risk measure. This optimal design encompasses three key dimensions:

1. A set of feasible retention functions.
2. A premium principle for calculating $RTC(g)$.
3. A risk owner's objective function RM.

Initially proposed by Cai and Tan (2007), this approach determined the optimal choice of the retention parameter for an upper limit policy for a VaR risk measure (also done for the ES measure). Expanding on this work, the same team of authors in Cai et al. (2008) identified the upper limit policy as optimal when (1) the choice of feasible ceded loss, $X - g(X)$, function is convex increasing, (2) the expected value principle is used for calculating risk transfers, and (3) the objective function is either the VaR or the ES.

Cai and Chi (2020) further trace various contributions that propose alternative optimal functions as the set of retention/ceded loss functions varies. For different sets, with the VaR objective function, the optimal policy may be either upper limit, a one-layer policy of the form $g(x) = x - \min[(x-a)_+, b-a)]$, or a truncated upper limit of the form $g(x) = x - (x-d)_+ I(x \leq m)$, where a, b, d, m are constants. However, when using the ES risk measure as an objective function, it turned out that the choice of feasible functions was robust in the sense that the upper limit policy was optimal for these sets.

Instead of separately discussing the optimality under VaR and ES measures, the literature has subsequently cast both problems as special cases of a general distortion risk measure defined in Section 2.5.3. Not surprisingly, the optimal design is contingent upon the choice of the risk measure.

Constrained Optimization Strategy. As another extension, this literature has separated the total risk into two components, the uncertainty of retained risks and the risk transfer cost. Specifically, the objective is to identify contract types that minimize the uncertainty of the risk $RM[g(X)]$ subject to the constraint $RTC(g) \leq RTC_{max}$ for a fixed *maximal risk transfer cost*, denoted here as RTC_{max}. The solution to this optimization problem is that a piece-wise functional strategy is optimal. Examples of this approach can be found in the budget constrained work of Lo (2017) and Cheung et al. (2019).

This separation is intuitively plausible. One rationale is that the marketplace, and hence prices, for insurance/reinsurance costs differs from the capital requirements faced by the risk owner. The risk owner might achieve the desired protection below a specific "reservation" price, an aspect that is explicitly recognized in a budget constrained set-up. Another justification is the discrepancy between retained risks and risk transfer costs in the timing and accounting. Payments for risk transfer costs are typically known early on, sometimes being made at the beginning of a contract period. In contrast, many risks take years to develop before their full impact is discernible. This difference in timing may also warrant different accounting treatments. Hence, it might be limiting to aggregate these components into a total cost; a separate treatment is more intuitively appealing.

Starting from the following section, we delve into the constrained optimization approach. The work cited in this section primarily concentrates on establishing conditions to optimize the structure of the risk transfer agreement. In contrast, we assume the contract form as known and aim to select the most suitable parameters. An advantage of working with this "reduced" problem is that we are able to make contributions to the study of multivariate risks, a central theme of this book.

3.3 Constrained Optimization

Risk managers seek to reduce the uncertainty of retained risks and minimize the costs of offloading risks. Our approach, consistent with modern asset portfolio analysis, is to treat this as an optimization problem. A general strategy that one might adopt consists of optimizing both quantities simultaneously, known as *vector optimization*, cf. Boyd and Vandenberghe (2004), Section 4.7. Due to the complexity of this strategy, we use the far more widely adopted approach and look to constrained optimization. In the context of this problem treated in this chapter, we seek to find those risk retention parameters in the vector $\boldsymbol{\theta}$ that

$$\boxed{\begin{array}{ll} \text{minimize}_{\boldsymbol{\theta}} & RM[g(X;\boldsymbol{\theta})] \\ \text{subject to} & RTC(\boldsymbol{\theta}) \leq RTC_{max} \quad . \end{array}} \tag{3.3}$$

We seek to minimize the uncertainty in retained risks summarized by $RM[g(X;\boldsymbol{\theta})]$ but are limited to the aptly termed "budget" constraint $RTC(\boldsymbol{\theta}) \leq RTC_{max}$. The collection of those values of $\boldsymbol{\theta}$ that satisfy this constraint is known as the *feasible set*. Additional terminology is collected in Appendix Section 3.6.3.

Before further discussing this problem in the risk retention situation, we first provide further context by describing applications in closely related areas.

3.3.1 Applications in Other Fields

It is likely that many readers are familiar with constrained optimization from their studies of economics, finance, and data science. As a refresher, several well-known problems can be addressed using constrained optimization.

Economics. Numerous fundamental problems in microeconomics can be solved using constrained optimization. For instance, consider a scenario where a consumer selects the most preferred bundle from a budget set. The consumer has a utility function (U) defined over two goods, such as the quantity of milk (C_1) and butter (C_2). The objective is to maximize the utility of these goods, $U(C_1, C_2)$, but is limited by a budget constraint of the form $P_1C_1 + P_2C_2 \leq Budget_0$. Here, P_1 is the price of milk, P_2 is the price of butter, and $Budget_0$ is the maximum amount that the consumer has to spend on these goods.

Another typical microeconomics problem likely familiar to many readers is a cost minimization problem. For example, a firm produces output using amounts of capital Cap and labor $Labor$ according to a Cobb-Douglas production function of the form $f(Cap, Labor) = A_3Cap^{a_1}Labor^{a_2}$. The firm must produce at least $Product_0$ of output. Subject to this constraint, the firm seeks to choose inputs Cap and $Labor$ to minimize costs given by $Cost(r, w) = rCap + wLabor$, where r is a rental rate for capital and w is a wage rate.

Finance. Perhaps the example most well-known to risk managers is the asset allocation problem. Markowitz (1952) proposed a theory to construct a portfolio of assets in order to maximize expected returns with a given level of uncertainty. Because of the relevance of asset allocations to our interests, Section 3.4 provides a detailed introduction to this problem.

Data Science. Optimization is heavily utilized in data science, statistics, and machine learning. Many, if not almost all, methods developed for fitting and prediction can be viewed as some form of optimization. Constrained optimization problems appear frequently, particularly in those methods that seek to develop a balance between the ability to fit a model to data and a desire to reduce model complexity. To illustrate, both lasso and ridge regressions can be viewed as constrained problems that balance fit and complexity (see, for example, Exercise 3.1). Another data science example of constrained optimization is support vector machines, a classification method (see, for example, Hastie et al. (2009)).

3.3.2 An Upper Limit Contract

To demonstrate how the constrained optimization approach works within the realm of risk retention, we initially consider an upper limit contract. In this scenario, only a single parameter varies, making optimization relatively straightforward. This section provides specific details to establish a foundation for tackling more complex settings. We consider only the value at risk measure here but describe other risk measures in the following Section 3.5.

One aspect that simplifies the problem is that the constraint function $RTC(u)$ can be assumed to be monotonic decreasing in the choice variable u. Consequently, as we increase retention by increasing u, the risk transfer cost stays the same or becomes smaller. For instance, in the baseline scenario with no additional surcharges, $RTC(u)$ represents only the expected transferred risk. With the risk retention function from equation (2.6), considering only an upper limit means no coinsuring of risks ($c = 1$) and no deductible ($d = 0$). Thus, with equation (2.7), we have

$$RTC(u) = \mathrm{E}(X) - \mathrm{E}(X \wedge u) = \int_u^\infty [1 - F(z)]\, dz, \tag{3.4}$$

that can readily be seen to be monotonic decreasing in u. Define u_{max} to be the smallest value of u that satisfies the constraint in Display (3.3). Thus, the feasible interval over which we seek to minimize the uncertainty of retained risks is $[u_{max}, \infty)$. Note that we could also use the notation $u_{max}(RTC_{max})$ to remind ourselves that the value of u_{max} depends on the choice of RTC_{max}.

For the uncertainty measure, consider the value at risk. From equation (2.10), we see that the value at risk is $VaR_\alpha(X \wedge u) = \min(F_\alpha^{-1}, u)$ – see the right panel of Figure 3.2. Thus, the problem is to

$$\boxed{\begin{array}{ll} \text{minimize}_u & \min(F_\alpha^{-1}, u) \\ \text{subject to} & u \geq u_{max}. \end{array}} \tag{3.5}$$

As evident from Figure 3.2, like the constraint function, the uncertainty measure is also monotonic in u, but is monotonic increasing, not monotonic decreasing. Thus, we minimize the objective function by making the decision variable u as small as possible. Formally, by changing the value of RTC_{max}, we can think about two cases.

- In the first case, suppose that $F_\alpha^{-1} \geq u_{max}$. In this case, the smallest that the upper limit u can be and still lie within the feasible region is $u^* = u_{max}$ with this as the value at risk.
- In the second case, suppose that $F_\alpha^{-1} < u_{max}$. In this case, for any upper limit u in the feasible region, $u \geq u_{max}$ $(> F_\alpha^{-1})$, has the minimal value at risk F_α^{-1}.

Summarizing, the optimal value at risk is

$$VaR_\alpha(X \wedge u)^* = VaR_\alpha(X \wedge u_{max}) = \begin{cases} u_{max} & \text{if } u_{max} \leq F_\alpha^{-1} \\ F_\alpha^{-1} & \text{if } u_{max} > F_\alpha^{-1} \end{cases}. \tag{3.6}$$

Example 3.2. Optimal Upper Limits Using a Pareto Distribution. Continuing with the example from Section 3.2.1, we use a Pareto distribution but with parameters $\eta = 3$ and $\gamma = 1000$. We also assume a confidence level $\alpha = 0.90$. For this distribution and confidence level, the quantile is $F_{0.90}^{-1} = 1154.435$. The risk transfer cost without loading is given in equation (3.4).

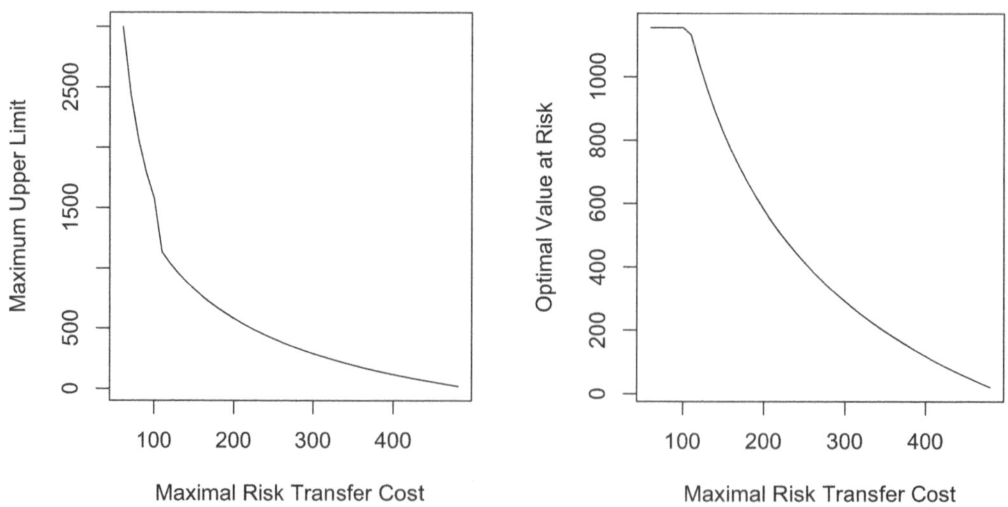

FIGURE 3.2
Maximum Upper Limit and Optimal Value at Risk versus Required Maximal Risk Transfer Cost.

Results are in Figure 3.2. The left-panel shows the maximum upper limit u_{max} as a function of the maximal risk transfer RTC_{max}. The right-hand panel shows the optimal value at risk, from equation (3.6), as a function of the maximal risk transfer. Note that the value at risk is bounded from above by $F^{-1}_{0.90}$; all maximal risk transfers below this point produce the same value at risk.

Code. The *Online Supplement* provides Example 3.2 illustrative statistical code.

3.3.3 The Lagrangian

With constrained optimization problems, one can often employ the *Lagrange method* for optimization. This is a classical approach for minimizing (or maximizing) a function subject to equality constraints. To illustrate, for the risk retention problem in Display (3.3), suppose that we are only interested in risk retention parameters that achieve the maximal risk transfer cost, that is, those values of $\boldsymbol{\theta}$ so that $RTC(\boldsymbol{\theta}) = RTC_{max}$. Then, the corresponding Lagrangian is

$$LA(\boldsymbol{\theta}, LM) = RM[g(X;\boldsymbol{\theta})] + LM\left\{RTC(\boldsymbol{\theta}) - RTC_{max}\right\}. \tag{3.7}$$

In this classical approach, one determines the parameters in $\boldsymbol{\theta}$ and the new parameter LM that optimize the function LA. Thus, the Lagrange method essentially turns a constrained problem into an unconstrained problem.

The constraint is said to be *active* or *binding* when, at the optimum $\boldsymbol{\theta}^*$, we have $RTC(\boldsymbol{\theta}^*) = RTC_{max}$. In many of our problems (but not all), this is a natural consequence of the constraints on the risk transfer cost. If the constraint is binding then we can treat the problem as an equality constrained problem.

To interpret the new parameter LM, we can take derivatives. Specifically, we can take a derivative of the Lagrangian with respect to the ith variable in $\boldsymbol{\theta}$. If we evaluate it at the optimum, then this derivative equals zero. Thus, we see that the Lagrange multiplier can be expressed as

$$LM = -\frac{\partial_{\theta_i}\ RM[g(X;\boldsymbol{\theta})]}{\partial_{\theta_i}\ RTC(\boldsymbol{\theta})} = RM_i^2 \tag{3.8}$$

at the optimal value of $\boldsymbol{\theta}$. This ratio of the differential change is the uncertainty of retained risks per unit differential change in the risk transfer cost, similar to that described in Section 2.3.2. In earlier work, Frees (2017) called this the **risk measure relative marginal** change, or RM^2 change, an intuitively appealing measure. We saw how to compute these quantities in Section 2.3.2.

In constrained optimization problems, it can be helpful to check that both the objective function ($RM[g(X;\boldsymbol{\theta})]$) and the constraint function ($RTC(\boldsymbol{\theta})$) are convex (in $\boldsymbol{\theta}$). If so, then any local optimal point will be a global optimal point. However, as we saw in Section 2.4, the risk retention problem in Display (3.3) is **not** a convex optimization problem in general. This means that the local optimum that we calculate is not guaranteed to be a global optimum point, implying that the analyst must take care in interpreting results.

3.3.4 A General Constrained Optimization Problem

Most of the risk retention problems considered in this text can be written as a special case of the following optimization problem:

$$\boxed{\begin{array}{lll} \text{minimize}_{\mathbf{z}} & f_0(\mathbf{z}) & \\ \text{subject to} & f_{con,j}(\mathbf{z}) \leq 0 & j \in CON_{in} \\ & f_{con,j}(\mathbf{z}) = 0 & j \in CON_{eq}. \end{array}} \tag{3.9}$$

In this display, the objective $f_0(\cdot)$ and the constraint functions $f_{con,j}(\cdot)$ all depend on the decision variables $\mathbf{z} = (z_1, \ldots, z_{p_z})'$. Some of the constraints are of the inequality type and those are members of the set CON_{in}. In the same way, equality constraints are members of the set CON_{eq}. Often, the decision variables are simply the risk retention parameters so $\mathbf{z} = \boldsymbol{\theta}$. However, in some cases we let one or more retention variables be fixed in advance and so they are not decision variables. In addition, for some risk measures we will include additional auxiliary variables as elements of $\mathbf{z}$ in the optimization that are not risk retention parameters. So, we allow decision variables and risk retention parameters to differ with differing notation; we employ a p dimensional vector of risk retention variables and a p_z dimensional vector of decision variables.

Generally, f_0 is a risk measure that we wish to minimize subject to constraints on the risk transfer costs. For some problems, f_0 can be a risk transfer cost that we wish to minimize subject to constraints on a maximum acceptable level of risk determined by the risk measure.

An equivalent problem formulation that is mathematically more elegant can be written as

$$\boxed{\begin{array}{lll} \text{minimize}_{\mathbf{z}} & f_0(\mathbf{z}) & \\ \text{subject to} & f_{con,j}(\mathbf{z}) \leq 0 & \text{for all } j. \end{array}}$$

Display (3.9) reduces to the simpler formulation simply by dropping the equality constraints. However, one can also go the other direction. Starting with the simpler form, if we use two inequality constraints of the form $f_{con,j}(\mathbf{z}) \leq 0$ and $-f_{con,j}(\mathbf{z}) \leq 0$, this is equivalent to an equality constraint.

We work with the seemingly more cumbersome formulation in Display (3.9) because this form is used in many software presentations - it is common for algorithms to separate considerations for equality constraints from those of inequality constraints. Another form, used in many economic applications, involves separating parameter constraints and assuming parameters are non-negative; this aids in interpreting results. For example, in later chapters we will work with multivariate risks and, in some of those applications, $\theta_i = 0$ may mean the full risk transfer of the ith risk, something that is of special importance for risk retention problems. This convention could also easily be adopted, see Section 11.3.1.

In many instances, we are able to solve a constrained optimization problem numerically by relying on software. Nonetheless, to gain insights into how the software works and sometimes reformulate problems into ones that are numerically more stable, it is helpful to think about the underlying theory.

The classical approach using Lagrangians for optimization subject to equality constraints can be extended to include inequality constraints. For Display (3.9), the associated Lagrangian is

$$LA(\mathbf{z}, \mathbf{LMI}, \mathbf{LME}) = f_0(\mathbf{z}) + \sum_{j \in CON_{in}} LMI_j \; f_{con,j}(\mathbf{z}) + \sum_{j \in CON_{eq}} LME_j \; f_{con,j}(\mathbf{z}), \tag{3.10}$$

where $\mathbf{LMI} = (LMI_1, \ldots, LMI_{m_{in}})'$ and $\mathbf{LME} = (LME_1, \ldots, LME_{m_{eq}})'$ are Lagrange multiplier vectors for the inequality and equality constraints, respectively.

3.4 Asset Allocation Problem

Asset allocations can be readily seen as a type of constrained optimization problem.

3.4.1 Markowitz Portfolio Optimization

To describe how portfolios are formed, we first introduce a bit of notation. An investor considers:

- p asset risks $\mathbf{R} = (R_1, \ldots, R_p)'$, each with mean $\mathrm{E}(R_j)$ and variance $\mathrm{Var}(R_j) = \Sigma_{jj}$, $j = 1, \ldots, p$.
- The dependence of a pair of risks is measured by a covariance $\mathrm{Cov}(R_i, R_j) = \Sigma_{ij}$. The $p \times p$ matrix $\boldsymbol{\Sigma}$ collects variances and covariances (with Σ_{ij} as the element in the ith row and jth column).
- Let c_j be the fraction of an investor's wealth in the jth asset and $\mathbf{c} = (c_1, \ldots, c_p)'$ is the vector of investment allocations. These are the choices that the investor makes and are the decision variables for this problem.
- The weighted sum $\sum_{j=1}^{p} c_j R_j = \mathbf{c}'\mathbf{R}$ gives the investor's portfolio return.
- For a given allocation vector $\mathbf{c}$, the portfolio mean is $\mathrm{E}[\mathbf{c}'\mathbf{R}] = \mathbf{c}'\ \mathrm{E}(\mathbf{R})$ and the variance is $\mathrm{Var}[\mathbf{c}'\mathbf{R}] = \mathbf{c}'\ \boldsymbol{\Sigma}\ \mathbf{c}$.

Asset Allocation Problem. The asset allocation problem can be expressed as one of three following forms:

$$\boxed{\begin{array}{ll} \text{maximize}_{\mathbf{c}} & \mathbf{c}'\ \mathrm{E}(\mathbf{R}) \\ \text{subject to} & \mathbf{c}'\ \boldsymbol{\Sigma}\ \mathbf{c} \le Var_0 \end{array}} \tag{3.11}$$

$$\boxed{\text{maximize}_{\mathbf{c}} \quad \mathbf{c}'\ \mathrm{E}(\mathbf{R}) - \lambda_{Port}\ \mathbf{c}'\ \boldsymbol{\Sigma}\ \mathbf{c}} \tag{3.12}$$

$$\boxed{\begin{array}{ll} \text{minimize}_{\mathbf{c}} & \mathbf{c}'\ \boldsymbol{\Sigma}\ \mathbf{c} \\ \text{subject to} & \mathbf{c}'\ \mathrm{E}(\mathbf{R}) \ge Req_0 \end{array}} \tag{3.13}$$

Here, Var_0 is the maximal portfolio variance, λ_{Port} is a penalty parameter that controls the trade-off between the portfolio mean and variance, and Req_0 is the required expected portfolio return. Displays (3.11)-(3.13) are equivalent in the sense that they trace the same frontier by varying the values of Var_0, λ_{Port}, and Req_0.

Example 3.3. Portfolio of Insurance Stock Returns. This is a continuation of Example 1.1. The portfolio frontier is displayed in Figure 3.3 for these data. In addition to the frontier, stocks with the highest mean and lowest standard deviation are displayed for comparison purposes. Also shown is the equally weighted portfolio, where $c_j = \frac{1}{8}$ for each j.

Code. The *Online Supplement* provides Example 3.3 illustrative statistical code.

Three portfolios were created, all subject to the constraint that the allocation choices sum to one, $\sum_j c_j = \mathbf{c}'\mathbf{1} = 1$ (the bold-faced $\mathbf{1}$ is a vector of ones). One is the equally weighted portfolio, the second a "Markowitz" portfolio (taking $\lambda_{Port} = 5$), and the third is a "minimum variance" portfolio. You can think of the third as a special case of the Markowitz portfolio in Display (3.13) where the minimal expected return Req_0 is so small

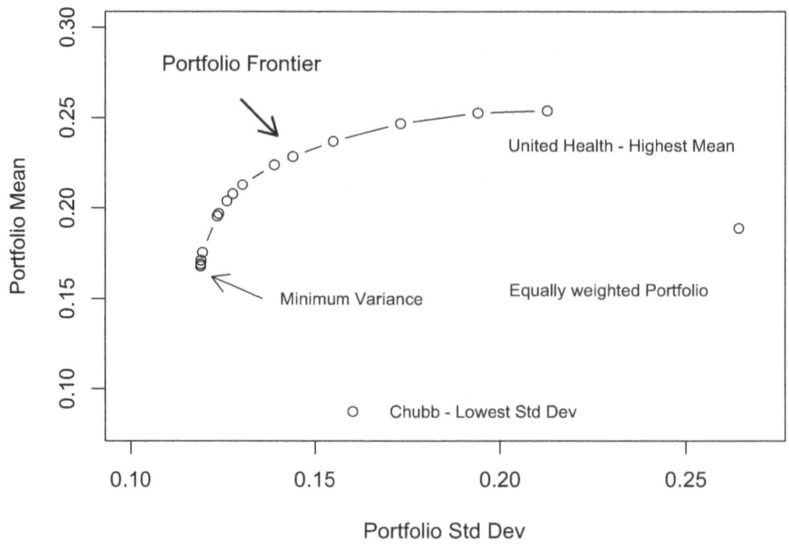

FIGURE 3.3
Frontier of Portfolio Standard Deviations and Means.

that it can be neglected. The three sets of portfolio weights are shown in Figure 3.4. The Markowitz portfolios were calculated using the `R` package `CVXR` - you can learn more about this package at the website https://cvxr.rbind.io.

The first three columns of Table 3.1 summarize the performance of the portfolios on the training sample as measured by the annual return, standard deviation, and the Sharpe ratio. Many financial analysts use the Sharpe ratio, the mean minus a risk-free rate divided by the standard deviation, to summarize portfolio performance. As anticipated, the `MinVar` minimum variance portfolio had the lowest standard deviation. With a zero risk-free rate, the Markowitz portfolio is the strongest performer for the training sample based on the Sharpe ratio.

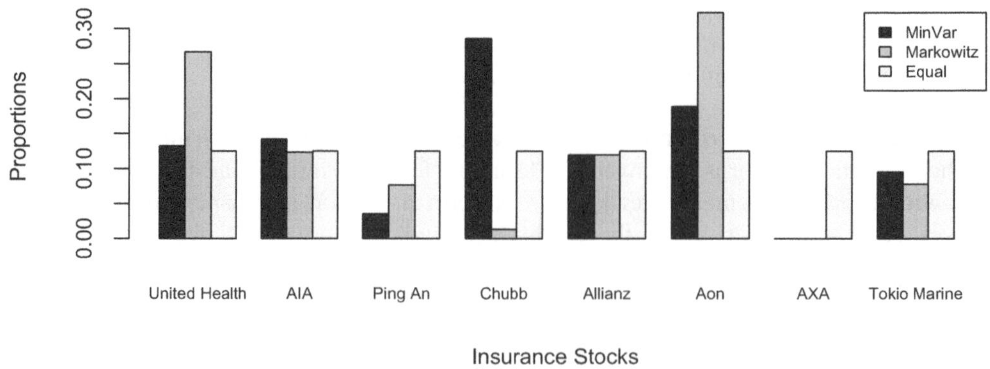

FIGURE 3.4
Three Sets of Portfolio Allocations, by Insurer.

Columns four through six of Table 3.1 summarize the performance of the portfolios on the test sample. Interestingly, the `MinVar` portfolio had a high test standard deviation (nearly the same as the two other portfolios). Based on the Sharpe ratio criterion, it was also the strongest performer.

To get a better sense of the portfolios' performance on the test sample, Figure 3.5 shows cumulative returns. All three are close to one another.

One cannot conclude too much from these results. Practicing analysts will use this approach while conducting more extensive testing by varying the training period, the test period, and the sample of stocks (risks) selected. Readers are invited to experiment with the code to determine their own optimal portfolios, see, for example Exercise 3.3.

3.4.2 Constraining Markowitz Portfolio Optimization

The popularity of the Markowitz portfolio approach is due in part to the fact that there is a simple closed form solution to the problem (see Exercises 4.2 and 4.3).

Moreover, given knowledge of the means, variances, and covariances, the optimization problem summarized in Displays (3.11)-(3.13) is typically convex (assuming a semi-positive definite covariance matrix). As will be discussed in Chapter 4, this means that fast solvers

TABLE 3.1
Portfolio Performance Measures.

	Train Ann Return	Train Std Dev	Train Sharpe (Rf=0)	Test Ann Return	Test Std Dev	Test Sharpe (Rf=0)
MinVar	0.164	0.119	1.378	0.168	0.143	1.178
Markowitz	0.203	0.126	1.614	0.166	0.143	1.160
Equal	0.173	0.130	1.329	0.155	0.142	1.093

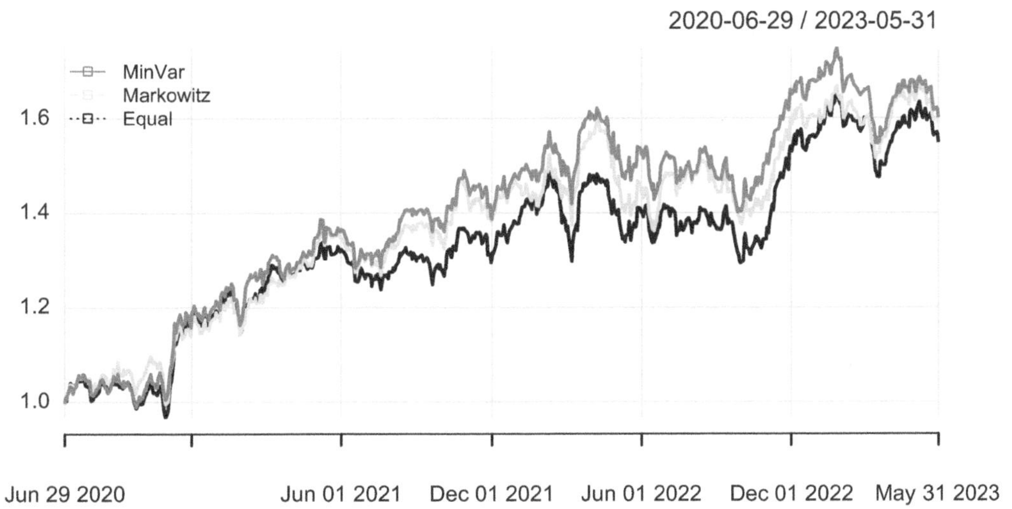

FIGURE 3.5
Cumulative Portfolio Returns (Test Sample).

are readily available for the problem. Convexity will also mean that one can impose additional linear (or convex) constraints on the problem and still achieve quick, reliable solutions. Display (3.14) exhibits several additional constraints regularly employed by practicing analysts:

$$\begin{array}{|ll|}
\hline
\mathbf{c}'\mathbf{1} = 1 & \text{full investment constraint} \\
\hline
c_j \geq 0, c_j \leq 1 & \text{short/long constraints} \\
\mathbf{c}'\mathbf{1} = 0 & \text{dollar-neutral constraint} \\
lower_j \leq c_j \leq upper_j & \text{holding constraint} \\
\hline
\sum_j |c_j| \leq Bound_{leverage} & \text{leverage constraint} \\
\sum_j |c_j - c_{0,j}| \leq Bound_{turnover} & \text{turnover constraint} \\
 & c_{0,j} \text{ represents the} \\
 & \text{current position} \\
\hline
\end{array} . \tag{3.14}$$

The first condition, "full investment constraint," ensures that one can interpret the allocations c_j as fractions of a given wealth amount invested in a portfolio. The second condition excludes "short selling" as is common in financial investment transactions; an investor may not take a negative position on a risk. The other conditions place limits on the portfolio allocations to meet differing investment strategies.

3.5 Constrained Optimization Approach to Risk Retention

To cultivate an understanding of how the constrained optimization approach applies to risk retention problems, let us examine several special cases of the risk retention function (2.6). Working with a single risk and a limited number of risk retention parameters enables us to obtain precise analytical results in many instances, aiding in the development of intuition for the process.

3.5.1 Upper Limits and ES

I now extend the discussion of upper limit contracts begun in Section 3.3.2 to include another uncertainty measure, the expected shortfall. From equation (2.11), the expected shortfall is

$$ES_\alpha(X \wedge u) = \begin{cases} u & \text{if } u \leq F_\alpha^{-1} \\ F_\alpha^{-1} + \frac{1}{1-\alpha}\left\{E[X \wedge u] - E[X \wedge F_\alpha^{-1}]\right\} & \text{if } u > F_\alpha^{-1} \end{cases} .$$

Thus, the problem is

$$\boxed{\begin{array}{ll} \text{minimize}_u & ES_\alpha(X \wedge u) \\ \text{subject to} & u \geq u_{max}, \end{array}} \tag{3.15}$$

where we can use the same maximum upper limit u_{max} for a given maximal risk transfer RTC_{max} as in the value at risk case discussed in Section 3.3.2. Unlike the value at risk case, the objective function is strictly increasing in u (assuming continuity of the distribution function). The smallest that the upper limit u can be and still be within the feasible region is $u^* = u_{max}$. Thus, this represents the optimal choice of u with the expected shortfall evaluated at u^*. We summarize the optimal objective function as

$$ES_\alpha(X \wedge u)^* = ES_\alpha(X \wedge u_{max})$$
$$= \begin{cases} u_{max} & \text{if } u_{max} \leq F_\alpha^{-1} \\ F_\alpha^{-1} + \frac{1}{1-\alpha}\{E[X \wedge u_{max}] - E[X \wedge F_\alpha^{-1}]\} & \text{if } u_{max} > F_\alpha^{-1} \end{cases}. \tag{3.16}$$

Example 3.4. Optimal Upper Limits with ES Using a Pareto Distribution. This is a continuation of Example 3.2. Because the risk transfer cost is the same, we can refer to the left-hand panel of Figure 3.2 to see the maximum upper limit u_{max} as a function of the maximal risk transfer RTC_{max}. The relationship between the maximal risk transfer RTC_{max} and the optimal expected shortfall $ES_\alpha(X \wedge u)^*$ is summarized in Figure 3.6. In this figure, a horizontal dashed line is given at $F_{0.90}^{-1} = 1,154.43$ where this is a break in the expression for $ES_\alpha(X \wedge u)^*$. One can see that this is smooth, at least compared to the similar graph for the value at risk in the right-hand panel of Figure 3.2.

Comparing Value at Risk to Expected Shortfall

From the discussions in Sections 3.3.2 and 3.5.1 and a comparison of equations (3.6) and (3.16), one sees that the optimal upper limit in both cases is u_{max} when $u_{max} \leq F_\alpha^{-1}$. This is emphasized in the Figure 3.7 comparison.

In Exercise 3.2, readers are asked to show that the range value at risk measure $RVaR_{(\alpha,\beta)}(X \wedge u)$ is monotone increasing in u. This result is not surprising given that this risk measure captures the value at risk and the expected shortfall as special cases. One can use this fact to readily use this general risk measure to determine an optimal parameter value for the upper limit case considered in this section.

3.5.2 Other Coverages

This section summarizes some interesting consequences of constrained optimization for other coverages using first principles. Later, in Chapter 11, I show how to obtain similar results using the Karush-Kuhn-Tucker, or *KKT*, conditions that hold for general constrained optimization problems.

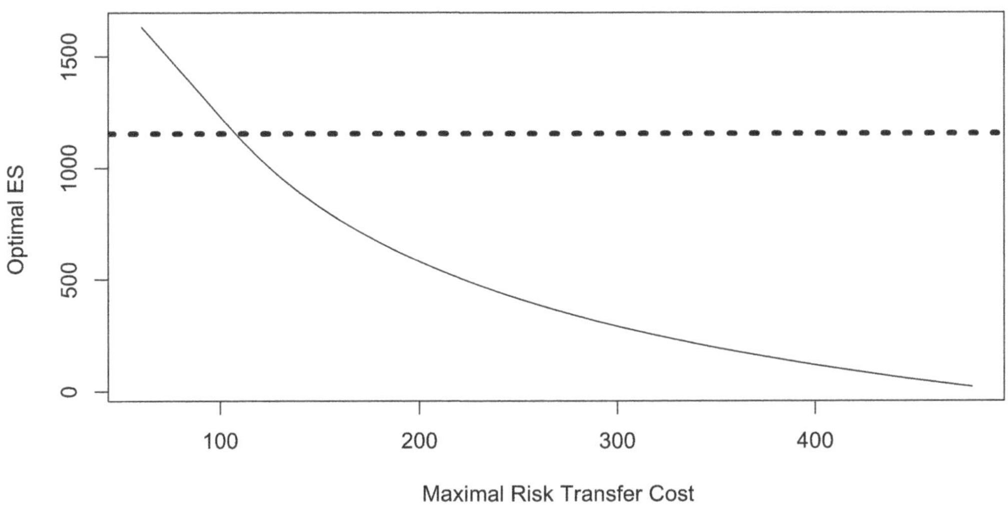

FIGURE 3.6
Optimal Upper Limit and ES versus Maximal Risk Transfer Cost.

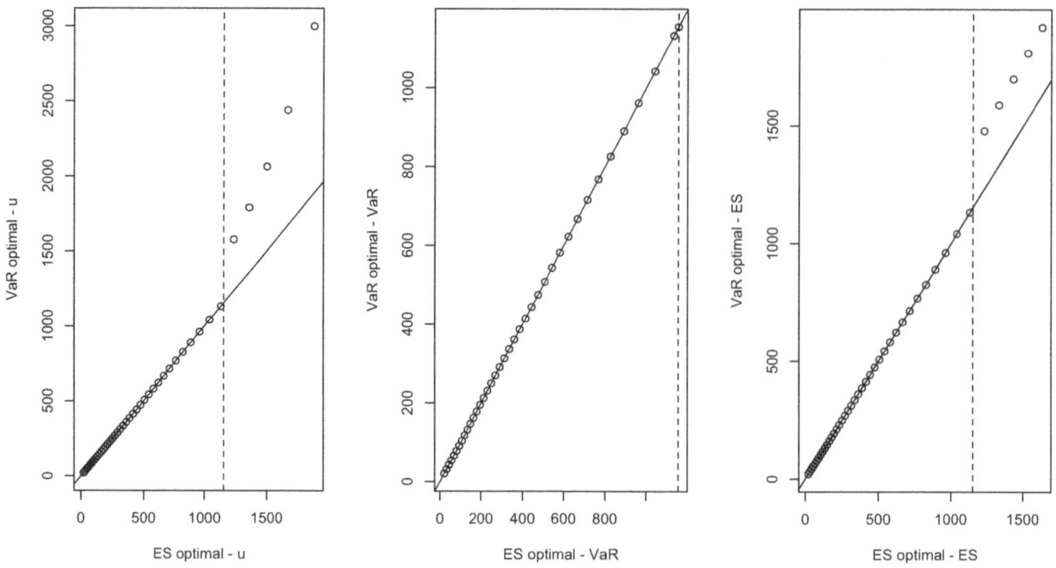

FIGURE 3.7
Comparing Retentions and Summary Measures from VaR and ES Optimization. The vertical dashed line marks the alpha percentile of the distribution function.

Upper Limits with Coinsurance

Now consider a contract with an upper limit u and coinsurance c but assume there is no deductible so $d = 0$. Starting with the value at risk, from equation (2.10) with $d = 0$, one has

$$VaR_{\alpha}\{g[X;\boldsymbol{\theta} = (d = 0, c, u)]\} = c\left[F_{\alpha}^{-1} \wedge u\right].$$

For the risk transfer cost, let us write

$$\begin{aligned} RTC(c, u) &= \mathrm{E}[X - c(X \wedge u)] = \mathrm{E}(X) - c\int_0^u \left[1 - F(z)\right] dz \\ &= \mathrm{E}(X) - cH(u). \end{aligned}$$

Here, $H(\cdot)$ is the integral of the survival function, $1 - F(\cdot)$.

To determine values of c and u that minimize $VaR_{\alpha}[g(X)]$ subject to the constraint $RTC(c, u) \leq RTC_{max}$, one strategy is to fix the value of c, solve the constraint problem, and then use that value of c that produces the smallest value at risk. To simplify the presentation, assume that RTC_{max} is sufficiently small so that the constraint is binding. Now, by equating the risk transfer cost to its maximum value, one can write the maximum upper limit in terms of the maximum cost and the coinsurance parameter as

$$\begin{aligned} RTC(c, u_{max}) = RTC_{max} \Rightarrow &\quad \mathrm{E}(X) - cH(u_{max}) = RTC_{max} \\ \Rightarrow &\quad u_{max}(c) = u_{max} = H^{-1}\left(\frac{\mathrm{E}(X) - RTC_{max}}{c}\right). \end{aligned}$$

Similar to equation (3.6), the solution to this subproblem is

$$VaR_{\alpha}(X \wedge u)_{opt,c} = F_{\alpha}^{-1} \wedge u_{max}(c).$$

This solution is interesting. To begin, for very small values of c, one sees that the value at risk is simply the quantile F_{α}^{-1}. To make $u_{max}(c)$ as small as possible, make c as large as possible which is 1. Thus, the solution to the two parameter problem is the same as the one parameter problem described in Section 3.3.2 with $c = 1$.

Deductibles Only

From the definition of a deductible contract in Section 2.2.1, the risk owner is responsible for

$$g(X; d, c = 1, u = \infty) = \begin{cases} 0 & \text{if } X < d \\ X - d & \text{if } X \geq d \end{cases} = (X - d)_+.$$

That is, the risk owner has responsibility only for losses in excess of d. As written above, this can be denoted using the subscript with a plus symbol '$(\cdot)_+$' meaning to take the positive part.

This can also be expressed as an upper limit as seen by the last equality. From this, we see that the deductible for the risk owner can be viewed as upper limit contract for the "risk taker." The reverse is easier to interpret; an upper limit policy for the risk owner can be interpreted as a deductible policy for the entity that takes on the risk.

Deductible policies can easily be seen to be the mirror image of upper limit policies. That is, as the risk transfer cost is monotone decreasing in d and the risk measures VaR, ES, and $RVaR$ are also monotone decreasing in d. Hence, the constrained optimization techniques developed for upper limit policies work equally well with deductible policies. Our treatment focuses on upper limit policies as these are more useful when the risk owner is a firm.

Deductibles and Upper Limits

Introducing a deductible and an upper limit at the same time would be unusual if the risk owner is a firm but is common for risk owners that are insurers or, as will be discussed in Chapter 6, reinsurers. So, we now consider the case where the risk owner has no responsibility for small losses (less than d) and transfers excess large losses (losses in excess of u). For the objective function, let us now focus on the ES. Although more complex, this function is smoother than the VaR measure and so may be easier to analyze.

The risk transfer cost from an expected value principle (without safety loadings) is

$$RTC(d, u) \quad = \mathrm{E}(X) - \{\mathrm{E}(X \wedge u) - \mathrm{E}(X \wedge d)\}.$$

The feasible region consists of those values of (d, u) that satisfy $RTC(d, u) \leq RTC_{max}$. From the expression for $RTC(d, u)$, we see that as RTC_{max} becomes large and approaches $\mathrm{E}(X)$, the upper limit u becomes close to the lower limit d, although there is no restriction on the value of d. Conversely, as RTC_{max} becomes small and approaches 0, then d approaches 0 and u approaches ∞.

Further, for the transfer to be feasible, we must have $RTC(d, \infty) \leq RTC_{max}$. That is, even if the risk owner retains all large losses, small losses must also be limited and so d must not be too large. For the largest possible value of the lower limit, define d_{max} to be the solution of the equation $RTC(d, \infty) = RTC_{max}$. So, deductibles $d^* \leq d_{max}$ are in this feasible region. Now, we know that $ES(d, u)$ is non-decreasing in u (see, for example, Table 2.2), so we want to choose u to be as small as possible. So, define $u^* = u^*(d^*)$ to be the solution of the equation $RTC(d^*, u) = RTC_{max}$.

How does the value of u^* depend on d^*? Differentiating this equation yields

$$\begin{aligned} 0 = \partial_{d^*} RTC(d^*, u^*) \quad &= \partial_d RTC(d, u)|_{d=d^*, u=u^*} \\ &\quad + \partial_u RTC(d, u)|_{d=d^*, u=u^*} \times \partial_{d^*} u^* \\ &= \partial_d \mathrm{E}(X \wedge d)|_{d=d^*} - \partial_u \mathrm{E}(X \wedge u)|_{u=u^*} \times \partial_{d^*} u^* \\ &= [1 - F(d^*)] - [1 - F(u^*)] \times \partial_{d^*} u^* \\ \Rightarrow \qquad & \partial_{d^*} u^* = \tfrac{1 - F(d^*)}{1 - F(u^*)}. \end{aligned}$$

From this expression, we see that $\partial_{d^*} u^* \geq 0$, so u^* decreases as d^* decreases.

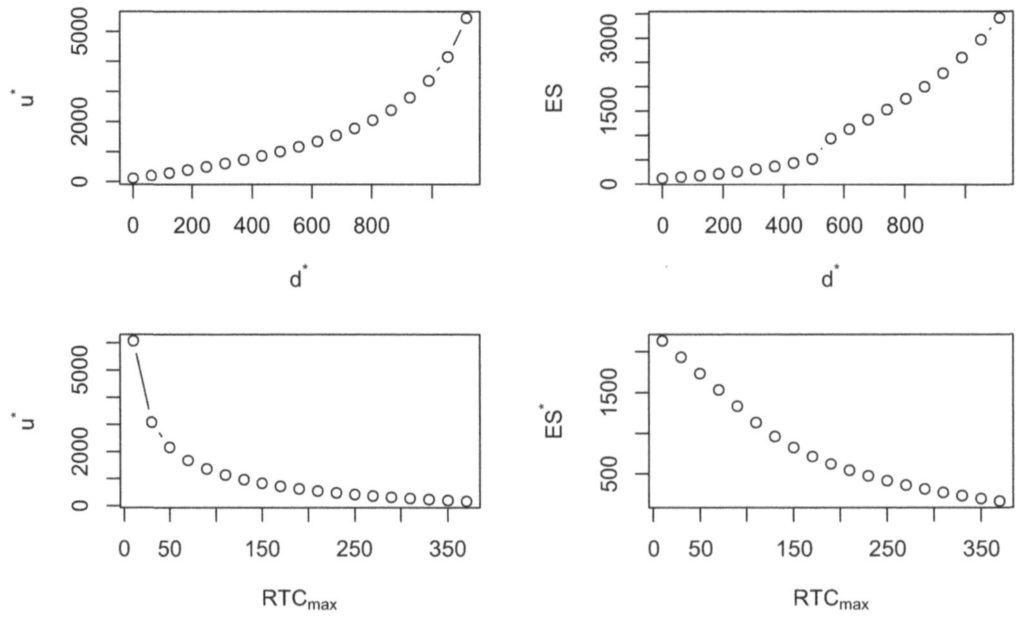

FIGURE 3.8
Deductible and Upper Limit Policy.

Figure 3.8 visualizes these relationships. The upper left-hand panel shows the value of upper limits that achieve the max cost RTC_{max} for a given deductible. As determined theoretically, the slope is non-negative. The upper right-hand panel shows the corresponding ES where we see that the ES is increasing in d.

It can be shown that $RTC(d^*, u^*)$ is nondecreasing in d^* (see the online demonstration noted below). This means that the minimizing ES can be achieved by choosing $d^* = 0$. For this value of d^*, the optimal upper limit $u.opt$ is the solution of the equation $RTC(0, u) = RTC_{max}$. The lower left-hand panel of Figure 3.8 gives this upper limit as a function of the maximal risk transfer cost and the lower right-hand panel gives the corresponding value of the ES. From this panel, we see that as the maximal risk transfer cost increases the uncertainty of retained risks as measured by ES decreases, as anticipated.

Under the Hood. See the *Online Supplement* for the derivative of the optimal ES.

3.6 Supplemental Materials

3.6.1 Further Resources and Readings

Readers wanting a quick refresher to some basic statistical and actuarial concepts may refer to the online book, Actuarial Community (2020). In particular, Chapter 3 describes calculation of risk transfers for policies with coverage modifications, Chapter 10 provides an introduction to risk measures, and Chapter 18 summarizes limited expected value calculations.

The constrained optimization approach has long been known in the actuarial academic literature. See early contributions of Bühlmann (1970) and Daykin et al. (1994). The work of

Bühlmann (1970) (page 114) traces its origins in the actuarial literature to contributions by Vajda (1962) and De Finetti (1940). See Frees (2017) and Lee (2017) for earlier investigations of the constrained optimization approach to risk retention most related to the approach in this book.

The supplemental Sections 3.6.4 and 3.6.5 provide additional descriptions of the best insurance designs from the economics and optimal reinsurance literatures.

3.6.2 Exercises

See the *Online Supplement* for selected exercise solutions.

Exercise 3.1. Ridge and Lasso Regressions. As is customary in regression problems, assume that we have $p+1$ variables from a sample of size n. Denote observation i as $(x_{i1}, \ldots, x_{ip}, y_i) = (\mathbf{x}_i, y_i)$ and think of y_i as a "response" variable to be explained or predicted by $x_{i1}, \ldots, x_{ip}$. The traditional ridge and lasso regression estimators can be computed as the solution of

$$\boxed{\begin{array}{ll} \text{minimize}_{\boldsymbol{\beta}} & \frac{1}{n}\sum_{i=1}^{n}(y_i - \mathbf{x}_i'\boldsymbol{\beta})^2 \\ \text{subject to} & \sum_{j=1}^{p}|\beta_j|^s \leq c_{regression}, \end{array}}$$

where $s = 2$ is for ridge regression and $s = 1$ is for lasso regression. Here, $c_{regression}$ is a constant that limits the size of the regression coefficients $\beta_1, \ldots, \beta_p$.

a. Use equation (3.10) to show that one can express the Lagrangian as

$$LA(\boldsymbol{\beta}, LMI) = \frac{1}{n}\sum_{i=1}^{n}(y_i - \mathbf{x}_i'\boldsymbol{\beta})^2 + LMI\sum_{j=1}^{p}|\beta_j|^s.$$

We can interpret the first part of the right-hand side as the goodness of fit and the second part as a penalty for size of the regression coefficients. When some of the regression coefficients are zero, this reduces model complexity.

b. Show that the ridge regression coefficient estimator has a closed-form expression

$$\hat{\boldsymbol{\beta}}^{ridge} = [\mathbf{X}'\mathbf{X} + n\ LMI\ \mathbf{I}]^{-1}\mathbf{X}'\mathbf{y},$$

where $\mathbf{y}$ is a vector of responses, $\mathbf{X} = (\mathbf{x}_1', \ldots, \mathbf{x}_n')'$ is a matrix of explanatory variables, and $\mathbf{I}$ is an identity matrix.

c. The idea underpinning ridge regression is not difficult. When the explanatory variables are highly collinear, the ordinary least squares solution is not unique because $\mathbf{X}'\mathbf{X}$ is not invertible. To handle this case, the ridge regression component adds a ridge (diagonal matrix) to $\mathbf{X}'\mathbf{X}$ to permit a unique solution. To further interpret this estimator, show that the ridge regression estimator can be expressed as a (matrix) weighted average between the ordinary least squares estimator and zero for the form

$$\hat{\boldsymbol{\beta}}^{ridge} = \mathbf{W}\hat{\boldsymbol{\beta}}^{ols} + (\mathbf{I} - \mathbf{W})\mathbf{0},$$

and determine the weight matrix $\mathbf{W}$. In this sense, the ridge regression estimator is said to be "shrunk towards zero."

d. Ridge regression estimators impose stability by shrinking estimators towards zero but do not actually remove variables by making regression coefficients equal to zero. Model simplification can be accomplished using the lasso estimator. Figure 3.9 shows a well-known figure for contrasting lasso and ridge estimators.

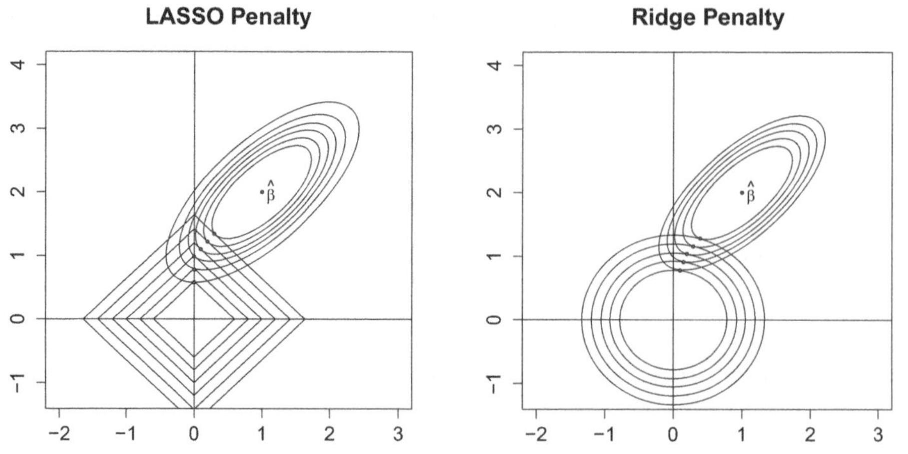

FIGURE 3.9
Lasso and Ridge Regression Comparison from Frees and Lee (2015).

Here, the concentric ellipsoids in the upper right show the goodness of fit for different values of $\boldsymbol{\beta}$. The solid figures about the origin (a diamond on the left for the lasso and a circle on the right for the ridge) show the parameter constraint regions. Explain how this figure demonstrates that the lasso optimization can result in zero regression coefficients, thus reducing model complexity.

Exercise 3.2. Upper Limit and *RVaR*. For the upper limit contract, show that the range value at risk measure $RVaR_{(\alpha,\beta)}(X \wedge u)$ is monotone increasing in u.

Exercise 3.3. Alternative Training and Testing Periods for Portfolio of Insurance Stock Returns. In this exercise, modify the set-up and code in Examples 1.1 and 3.3 to see the variation in results for alternatives training and testing periods. Specifically, the training session is a 6.5-year period, 5 January 2015 to 31 May 2021 (instead of 5 January 2015 to 31 December 2019), and the test session is a one-year period, 1 June 2020 to 31 May 2022 (instead of the three-year period 1 June 2020 to 31 May 2023). In addition to changing the relative lengths of training and test, we now consider the outlying behavior induced by the Covid epidemic.

Reproduce the results in Table 3.2 and comment on differences from those in the benchmark Table 3.1.

3.6.3 Appendix. Constrained Optimization Terminology

Constrained optimization is used in many fields. So, it is natural that different sets of terminology have developed over time. Here, we set the language, or terminology, used in this book, while at the same time trying to make connections with other literatures.

- The *objective function*, $f_0(\cdot)$, is a quantitative measure of the performance of the system under study that we want to minimize or maximize.

TABLE 3.2
Portfolio Performance Measures for Alternative Training and Testing Periods.

	Train Ann Return	Train Std Dev	Train Sharpe (Rf=0)	Test Ann Return	Test Std Dev	Test Sharpe (Rf=0)
MinVar	0.168	0.154	1.098	0.060	0.147	0.407
Markowitz	0.193	0.157	1.229	0.037	0.151	0.247
Equal	0.163	0.162	1.003	0.055	0.141	0.392

  - The *domain* of the objective function is the set of all points $\mathbf{z}$ for which the function is defined.
  - The objective function is a mapping of a set of characteristics to the real line. These characteristics, $\mathbf{z} = (z_1, \ldots, z_{p_z})'$ are sometimes called decision variables or unknowns.

- There are typically restrictions on the decision variables through functional relationships known as *constraints.*
  - If the constraint is of the form $f_{con}(\mathbf{z}) \leq 0$, then it is said to be an *inequality constraint.*
  - If the constraint is of the form $f_{con}(\mathbf{z}) = 0$, then it is said to be an *equality constraint.*
  - A point is said to be *feasible* if it is in the domain and the constraints are satisfied.
- *Optimization* is the minimization or maximization of an objective function subject to constraints on its parameters.
  - Our convention is to minimize an objective function $f_0(\cdot)$. However, one can alternatively maximize $f_0(\cdot)$ by minimizing $-f_0(\cdot)$, so we lose no generality by focusing on minimizing functions. In economic applications one typically uses the convention of maximizing functions such as those based on the utility of resources. However, here we seek to minimize risks and costs and so use the minimization convention.
  - *Unconstrained optimization* refers to an optimization problem where no constraints are present.
  - *Constrained optimization* refers to an optimization problem where constraints are present.
- An *optimal point*, or *global solution*, is the point with lowest function value among all feasible points. That is, $\mathbf{z}^*$ is a global solution if $f_0(\mathbf{z}^*) \leq f_0(\mathbf{z})$ for all feasible $\mathbf{z}$. Note that a global solution need not be unique.
  - The *optimal value* is the value of the objective function at a global solution.
  - A *local solution* is a point at which the objective function is smaller than at all other feasible nearby points.
  - An inequality constraint $f_{con}(\mathbf{z}) \leq 0$ is said to be *active*, or *binding*, at $\mathbf{z}^*$ if $f_{con}(\mathbf{z}^*) = 0$.

- The term *convex* can be applied both to sets and to functions.
  - A set $S \in R^n$ is a *convex set* if the straight line segment connecting any two points in S lies entirely inside S. Formally, for any two points $\mathbf{z}_1 \in S$ and $\mathbf{z}_2 \in S$, we have $c\mathbf{z}_1 + (1-c)\mathbf{z}_2 \in S$ for all $c \in [0,1]$.
  - The function f is a *convex function* if its domain S is a convex set and if for any two points $\mathbf{z}_1$ and $\mathbf{z}_2$ in S, the following property is satisfied:

$$f(c\mathbf{z}_1 + (1-c)\mathbf{z}_2) \leq cf(\mathbf{z}_1) + (1-c)f(\mathbf{z}_1), \text{ for all } c \in [0,1].$$

  - A function f is said to be *concave* if $-f$ is *convex*.
- For a function of n variables, $f(x_1, \ldots, x_n)$, the *gradient* ∇f is its vector of first partial derivatives and the *Hessian* ∇f^2 is a matrix of second partial derivatives of the form:

$$\nabla f = \begin{pmatrix} \partial_{x_1} f \\ \vdots \\ \partial_{x_n} f \end{pmatrix} \quad \text{and} \quad \nabla^2 f = \begin{pmatrix} \partial^2_{x_1} f & \partial_{x_1,x_2} f & \cdots & \partial_{x_1,x_n} f \\ \vdots & \vdots & \ddots & \vdots \\ \partial_{x_n,x_1} f & \partial_{x_n,x_2} f & \cdots & \partial^2_{x_n} f \end{pmatrix}.$$

  - Here, we use the short-hand notation $\partial_{x_1} f = \frac{\partial}{\partial x_1} f$, $\partial^2_{x_1} f = \frac{\partial^2}{\partial x_1^2} f$, and $\partial_{x_1,x_2} f = \frac{\partial^2}{\partial x_1, x_2} f$.
- A point $\mathbf{z}$ is *stationary* if its gradient is zero, that is, $\nabla f_0(\mathbf{z}) = 0$.
- For a multivariate function of n variables having m outcomes $(f_1, \ldots, f_m)$, the *Jacobian* is a matrix of first partial derivatives of the form:

$$Jf = \begin{pmatrix} \partial_{x_1} f_1 & \partial_{x_2} f_1 & \cdots & \partial_{x_n} f_1 \\ \vdots & \vdots & \ddots & \vdots \\ \partial_{x_1} f_m & \partial_{x_2} f_m & \cdots & \partial_{x_n} f_m \end{pmatrix}.$$

3.6.4 Appendix. Designs Based on Insurance Economics

As discussed in Section 3.1, when a security loading is in place, $\lambda_{SL} > 0$, then a partial insurance coverage corresponding to a value of $c < 1$ is optimal. Schlesinger (2013) also notes that this insurance result has a portfolio interpretation. One can define $A = w_0 - (1 + \lambda_{SL})\mathrm{E}\ (y)$ to be a non-risky asset and the weighted average, $Y(c) = (1-c)A + c(A+y)$, to be a combination of a non-risky and risky asset y. In this sense, the insurance choice problem is equivalent to the portfolio problem in financial economics.

A set of related results involve the notion of a "background risk," where the initial wealth is no longer treated as a known constant but is itself a random variable. In the portfolio context, this has a natural interpretation where $Y(c) = (1-c)A + c(A+y)$ represents a combination of two risky assets. Typically, results on optimal coverage assume independence between two risky assets. See the review article of Schlesinger (2013) for more background and additional references.

The insurance economics approach to selection of risk retention parameters is useful but, by design, relies on individual-specific preferences. As noted by Schlesinger (2013) (page 167), "... whereas most financial assets are readily tradeable and have a risk that relates to the marketplace, insurance is a contract contingent on an individual's personal wealth. This personal nature is what distinguishes it from other financial assets." In contrast, this book considers instances of insurance traded from one company to another, where wealth and risk preferences play a smaller role. There is also the literature on non-optimality of the aggregate loss policy beginning with Gollier and Schlesinger (1995); see also the discussion in Gollier (2013), page 111.

3.6.5 Appendix. Designs Based on Risk Measures

As briefly introduced in Section 3.2.2, there is a vibrant literature that seeks to determine the best risk transfer designs employing risk measures to summarize uncertainty of retained risks. By and large, this literature seeks to expand the boundaries of the theory of risk retention by identifying optimal risk retention functions in conjunction with choices of risk premium principles for the cost of transferring risk and risk measures for assessing the uncertainty of retained risks. As mentioned in Section 3.2.2, these works provide insights about the types of optimal design but, due to their general nature, are difficult to implement in concrete practical settings. As remarked in Lo (2017), "In spite of the vast literature on optimal reinsurance, the majority of prior studies can be justifiably criticized for being in a vacuum detached from practicalities and externalities." This section highlights some of the works in this literature most closely related to the ideas in this book to give readers a starting point for their own explorations into this literature.

Extending earlier work in Cai and Tan (2007), Tan and Weng (2014) utilized a constrained optimization approach, minimizing retained risk evaluated using a risk measure subject a budget constraint. Their work uses an empirical approximation of the risk retention function, thus "reducing" the problem from an infinite dimensional optimization to one involving the number of observations, a number that is large yet finite. By not imposing a specific nonlinear retention function, they were able to use convex optimization techniques to solve problems of interest.

Subsequently, this work was extended in Asimit et al. (2018) in several aspects, including considering multivariate risks. Their motivation was from an enterprise risk management perspective and so they referred to this as "multiple lines of business." As with Tan and Weng (2014), this paper adopts an empirical approach, solving finite yet very large dimensional problems. Thus, it is not surprising that the paper restricts considerations to those problems that could be solved using convex optimization approaches. Interestingly, this paper also investigates *robust optimization* approaches to addressing parameter uncertainty. This topic will be taken up later in Section 9.4.

4

Transferring Multiple Risks Including Reinsurance

Chapter Preview. This chapter begins our study of **multivariate risks**. As emphasized in Chapter 1, any risk owner, be it a person, firm, or insurer, faces a multitude of interrelated risks related to the risk owner itself, its property, as well as its obligations and responsibilities.

In the historical development of risk retention, the focus has largely been on either individual consumers or insurers as the risk owners. When an insurer transfers risk to another insurer, it is referred to as *reinsurance*. Reinsurance not only holds historical significance but also serves as a natural context to introduce significant risk retention agreements. As demonstrated in Section 1.1, reinsurance continues to be an active commercial activity, and so, Section 4.1 delves deeper into its specific agreements.

With several risks, one cannot disregard their interrelationships. Section 4.2 begins our investigations into dependent risks; dependency impacts diversification, a critical feature when constructing portfolios. This section introduces *copulas* as tools to account for interdependencies among risks. The subsequent Section 4.3 illustrates their application in various reinsurance agreements.

As modeling becomes more complex, the special case of independence, while not always realistic, serves as a valuable benchmark for comparison. Similarly, in this chapter, the *variance* operator is used to gauge the uncertainty of risks in a multivariate context, offering interpretability and serving as a benchmark. However, we have already seen the limitations of the variance operator, and in subsequent chapters, we will employ quantile-based risk measures. This means that calculations will become more complex, and so, Section 4.4 introduces numerical tools that can be utilized to solve constrained optimization problems.

Another intriguing aspect of reinsurance is that it naturally suggests the introduction of more than two parties taking responsibility for a risk. Even in a simple setting, a firm might have a contract with an insurer to take on a portion of its risks, and this insurer has a separate contract with another insurer (a reinsurer) that may include ceding a portion of these risks. Thus, there is a sharing of the risks among the three parties. In more complex settings, it is common to have more parties involved. Section 4.5 concludes the chapter with a comprehensive discussion of risk sharing among multiple parties known as a *risk exchange*.

4.1 Reinsurance

As noted above, reinsurance is simply insurance purchased by an insurer. Insurance purchased by non-insurers is sometimes known as *primary insurance* to distinguish it from reinsurance. Reinsurance differs from personal insurance purchased by individuals, such as auto and homeowners insurance, in contract flexibility. Like insurance purchased by major corporations, reinsurance programs are generally tailored closely to the needs of the buyer. In contrast, in personal insurance buyers typically cannot negotiate on the contract terms although they may have a variety of different options (coverages) from which to choose.

DOI: 10.1201/9781003469568-4

The two broad types of agreements are proportional and non-proportional reinsurance. A *proportional reinsurance* contract is an agreement between a reinsurer and a *ceding company* (also known as the *reinsured*) in which the reinsurer assumes a given percent of losses and premiums. Non-proportional agreements are simply everything else. As examples of non-proportional agreements, this chapter looks first to stop-loss and excess of loss contracts and then to more complex arrangements.

A reinsurance contract can also be categorized as *facultative* or as a *treaty*. Facultative reinsurance contracts typically cover a single, large complex risk that cannot be readily grouped with other risks. In contrast, a treaty is a reinsurance contract that covers more than a specific risk, typically a specified share of all the insurance policies issued by the ceding company which come within the scope of the contract.

4.1.1 Proportional Reinsurance

The simplest example of a proportional treaty is called *quota share*. In a quota share reinsurance treaty, the reinsurer receives a flat percent, say 50%, of the premium for the book of business reinsured. In exchange, the reinsurer pays the same flat percent of losses, including allocated loss adjustment expenses.

For loss sharing, this is exactly the same as the risk retention agreement with a coinsurance parameter c introduced in Section 2.2 where the insurer cedes a proportion $1 - c$ of losses and retains the fraction c. However, in a quota share treaty, the premium is also specified as part of the treaty. Moreover, the insurer, or ceding company, has already incurred expenses for acquiring the business. Therefore, it is customary for the reinsurer to also pay the ceding company a ceding commission which is designed to reflect the differences in underwriting expenses incurred.

Quota share reinsurance treaties are particularly suitable for growing insurers or established companies which are new to a certain class of business. Quota share also makes sense for primary insurers who are seeking capital relief in light of solvency considerations or protection against random fluctuations across an entire portfolio. Because of their automatic nature, quota share treaties work well with high frequency lines of business. From the point of view of the reinsurer, quota share is desirable because the interests of the insurer and reinsurer are aligned and so there is little opportunity for moral hazard issues (such as sloppy claims handling).

From a theoretical viewpoint, we have already seen in Section 3.1 (Proposition 3.3) the desirability of the quota share contract for the reinsurer. There, we learned that the quota share reinsurance treaty minimizes the reinsurer's uncertainty, measured by the variance, when subject to some mild constraints on the insurer's variance of retained losses. Intuitively, quota share insurance is appealing because the insurer shares the responsibility for large losses in the tail of the distribution. As will be seen, this is in contrast to non-proportional agreements where reinsurers take responsibility for large losses.

Optimizing Quota Share Agreements for Insurers

As a variation, now assume p risks in the portfolio, $X_1, \ldots, X_p$, so that the portfolio sum is $S = X_1 + \cdots + X_p$. For simplicity, we now focus on the case of independent risks. Extensions to the dependent case are in Section 4.3. In a variation of the basic quota share agreement, the amount retained by the insurer may vary with each risk, say c_i. Thus, the insurer's portion of the portfolio risk is $Y_{retained} = \sum_{i=1}^{p} c_i X_i$. What is the best choice of the proportions c_i? This is a question raised by De Finetti (1940).

To formalize this question, one seeks to find those values of c_i that minimize $\mathrm{Var}(Y_{retained})$ subject to the constraint that $\mathrm{E}(Y_{reinsurer}) = RTC_{max}$. The requirement that $E(Y_{reinsurer}) = RTC_{max}$ suggests that the insurer wishes to spend a fixed quantity, say RTC_{max}, on the cost of reinsurance. Subject to this budget constraint, the insurer wishes to minimize the uncertainty of the retained risks as measured by the variance. Framed this way, this question is a constrained optimization problem. As only equality constraints are invoked, a solution can be derived using the Lagrange method.

The optimal proportions can be expressed as

$$c_i^* = \frac{LME^*}{2} \frac{E(X_i)}{\mathrm{Var}(X_i)}, \tag{4.1}$$

where

$$LME^* = 2\left[\sum_{i=1}^{p} E(X_i) - RTC_{max}\right] / \left[\sum_{i=1}^{p} \frac{[E(X_i)]^2}{\mathrm{Var}(X_i)}\right]. \tag{4.2}$$

Under the Hood. See the *Online Supplement* for the verification of these optimal retention proportions.

Thus, the optimal proportion for the ith risk, c_i^* is proportional to $E(X_i)/\mathrm{Var}(X_i)$. This is intuitively appealing. Other things being equal, a higher revenue as measured by $E(X_i)$ means a higher value of c_i^*. In the same way, a higher value of uncertainty as measured by $\mathrm{Var}(X_i)$ means a lower value of c_i^*. The proportional scaling factor is determined by the revenue requirement $E(Y_{retained}) = RTC_{max}$. The following example helps to develop a feel for this relationship.

Example 4.1. Quota Sharing of Three Pareto Risks. Consider three risks that have a Pareto distribution, each having a different set of parameters (so they are independent but non-identical). Specifically, use the parameters:

- $\alpha_1 = 3$, $\theta_1 = 1,000$ for the first risk X_1,
- $\alpha_2 = 3$, $\theta_2 = 2,000$ for the second risk X_2, and
- $\alpha_3 = 4$, $\theta_3 = 3,000$ for the third risk X_3.

Figure 4.1 gives optimal values of c_1, c_2, and c_3 as the required risk transfer cost RTC_{max} varies. Note that each coefficient decreases linearly with RTC_{max} in accordance with equations (4.1) and (4.2).

Code. The *Online Supplement* provides Example 4.1 illustrative statistical code.

A limitation of this procedure is evident from Figure 4.1 where we see that some of coefficients exceed one. For example, for $RTC_{max} = 500$, we have $c_1^* = 1.143$, $c_2^* = 0.571$, and $c_3^* = 0.857$. De Finetti (1940) was aware of this limitation and derived a complex expression for the optimal coefficients under the constraints $0 \le c_j \le 1$. See Exercise 11.1 for more details.

However, with tools available today, we now recognize the problem:

$$\begin{array}{ll} \text{minimize}_{c_1,\ldots,c_p} & \sum_i c_i^2 \ \mathrm{Var}(X_i) \\ \text{subject to} & \sum_i (1-c_i)\ E(X_i) = RTC_{max} \\ & 0 \le c_i \le 1, \qquad i = 1,\ldots,p, \end{array}$$

as a *quadratic programming* problem where the objective function is a quadratic function and the constraints are linear in the coefficients. This is a type of convex problem, and so

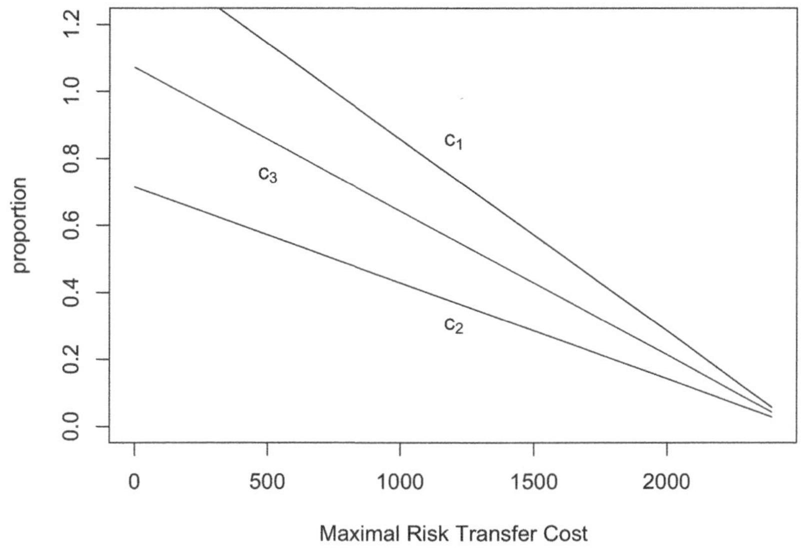

FIGURE 4.1
Quota Share Proportions versus Required Maximal Risk Transfer Cost.

fast solvers are available. I describe this class of problems in Section 4.4.1. For now, the associated R code demonstrates a ready solution for an otherwise complicated problem. For Example 4.1, the solution turns out to be $c_1^* = 1.000$, $c_2^* = 0.600$, and $c_3^* = 0.900$. Under the additional constraints, the objective function is 3,450,000. This is greater than the best value of the objective function without the additional constraints which turns out to be 3,428,571. This is as one would expect, when one introduces additional constraints, the region over which the function is to be minimized is smaller and so the optimal value must stay the same or increase.

Surplus Share Proportional Treaty

Another proportional treaty is known as *surplus share*; this type of contract is common in commercial property insurance. This is essentially a type of variable quota share treaty where the sharing proportion c varies by risk type. As applied, typically the ceding insurer's proportion is larger for smaller policies. In this way, the insurer can retain virtually all of the risk for small policies while seeking relief from the reinsurer for large risk types. Here is a common implementation:

- The ceding company retains all risks with an insured amount below a given amount (the *retained line*).
- The reinsurer is involved in risks where the insured amount exceeds the retained line. Up to a given limit (expressed as a multiple of the retained line, or number of lines), the ceding company's proportion of the risk is the retained line as a fraction of the insured amount.
- For risks in excess of the upper limit, the reinsurer's proportion is the limit as a fraction of the insured amount.

For example, let the retained line be 100,000 and the given limit be six lines (600,000). Then, the proportion retained by the ceding company $c(IA)$ is a function of the insured amount (IA), as follows:

Insurance Amount *IA*	**Reinsurer proportion** $1 - c(IA)$	**Insurer proportion** $c(IA)$
$IA \leq 100000$	0	1
$100000 < IA < 600000$	$\frac{IA - 100000}{IA}$	$\frac{100000}{IA}$
$IA > 600000$	$\frac{500000}{IA}$	$1 - \frac{500000}{IA}$

As with quota share reinsurance, it is customary for the reinsurer to also pay the ceding company a ceding commission which is designed to reflect the differences in underwriting expenses incurred. Given the competitiveness of the marketplace for primary insurance, this premium is also insufficient causing primary insurers to operate at a loss. Thus, many reinsurers are adopting a procedure of returning to the ceding insurer only that part of the original premium that was not paid out in losses, cf. Swiss.Re (2013) (page 27).

Example 4.2. Surplus Share Retention for the Property Fund. Let us continue using the Property Fund data introduced in Example 2.1. In 2010, there were 1,095 policyholders that had positive coverage amounts. To illustrate, assume a retained line equal to 100 (in thousands) and a limit of six lines. For these data, it turns out that there are 981 policyholders with coverage below the retained line, 3 above the limit, and 111 in between. Figure 4.2 shows the surplus share retention as a function of insurance amount. Note the relatively low retention fractions for the three large policyholders.

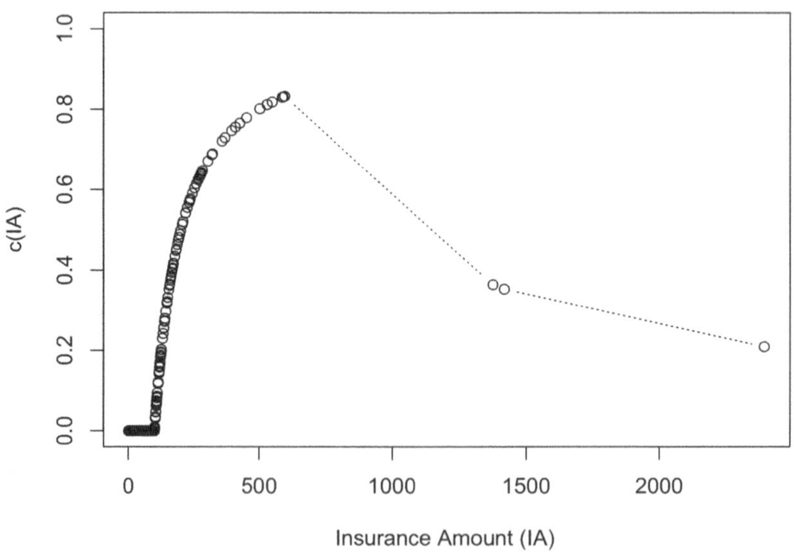

FIGURE 4.2
Insurer's Surplus Share Retention as a Function of Insurance Amount.

4.1.2 Non-Proportional Reinsurance

The simplest example of a non-proportional treaty is called a *stop-loss* treaty. This acts as the same as the upper limit risk retention agreement introduced in Section 2.2. That is, the insurer sets a retention level u and pays in full total losses for which $X \leq u$. Further, for total losses for which $X > u$, the direct insurer pays u and the reinsurer has responsibility for the remaining amount $X-u$. Thus, the insurer retains an amount u of the risk. Summarizing, the amounts paid by the direct insurer and the reinsurer are $Y_{retained} = X \wedge u$ and $Y_{reinsurer} = \max(0, X-u) = (X-u)_+$. As before, note that $Y_{retained} + Y_{reinsurer} = X$. From a reinsurer's perspective, note that u could be considered a deductible. However, this book adopts the risk owner's perspective and so u represents the upper limit of its obligation.

Stop-loss, in its pure form, is not employed extensively in practice. This is likely due to its potential moral hazard weakness. With a pure stop-loss treaty, insurers have few incentives to carefully underwrite and monitor policies, as well as settle associated losses that contribute to experience above the stop-loss limit. Recall from basic economics that the phrase *moral hazard* is used to describe the situation where a contract alters the behavior of parties to the agreement. This would certainly be a concern to reinsurers that offer a stop-loss coverage. An exception to this is coverage between connected entities, such as a firm and its captive insurer, where incentives are aligned (see Section 6.2.2).

Nonetheless, at least conceptually, the stop-loss type of contract is particularly desirable for the insurer. As we learned in Section 3.1 (Proposition 3.4), the stop-loss reinsurance treaty minimizes the insurer's uncertainty as measured by the variance subject to a mild budget constraint. Intuitively, with stop-loss insurance, the reinsurer takes responsibility for very large losses in the tail of the distribution, not the insurer.

Excess of Loss

A closely related form of non-proportional reinsurance is the *excess of loss* coverage. Under this contract, the total risk can be thought of as composed as p separate risks $X_1, \ldots, X_p$ where each risk is subject to an upper limit, say, u_i. So the insurer retains

$$Y_{retained} = \sum_{i=1}^{p} X_i \wedge u_i = S(u_1, \ldots, u_p),$$

and the reinsurer is responsible for the excess, $Y_{reinsurer} = S - Y_{retained}$. (Recall that the portfolio sum is $S = X_1 + \cdots + X_p$.) The retention limits may vary by risk or may be the same for all risks, that is, $u_i = u$ for all i. The development in this book focuses on this contract and so I introduce the notation $S(u_1, \ldots, u_p)$ to represent the sum of losses that are limited by the retention parameters $u_1, \ldots, u_p$.

In addition to arguments from theory, there are important practical considerations that recommend the use of non-proportional insurance to an insurer. Non-proportional insurance can limit an insurer's liability with upper limits that reflect its willingness and capacity to bear risk. Smaller losses are the sole responsibility of the insurer and are not shared with the reinsurer, suggesting an insurer may focus its expertise on this portion of the portfolio and thus enhancing its earnings potential. In particular, administration costs can be lower for both parties as the insurer no longer needs to calculate the proportion of a loss for each risk as under surplus share reinsurance.

Optimal Choice for Excess of Loss Retention Limits

What are the best choices of the excess of loss retention limits u_i? One way to formalize this question is to find those values of u_i that minimize $\mathrm{Var}(Y_{retained})$ subject to the constraint that $\mathrm{E}(Y_{reinsurer}) = RTC_{max}$. Subject to this budget constraint, the insurer wishes to minimize the uncertainty of the retained risks (as measured by the variance).

As before, only equality constraints are invoked and so a solution can be derived using the Lagrange method. Specifically, define the Lagrangian

$$\begin{aligned} LA &= \mathrm{Var}(Y_{retained}) + LME\{\mathrm{E}(Y_{reinsurer}) - RTC_{max}\} \\ &= \sum_{i=1}^{p} \mathrm{Var}(X_i \wedge u_i) + LME\{[\sum_{i=1}^{p} \mathrm{E}(X_i) - \mathrm{E}(X_i \wedge u_i)] - RTC_{max}\}. \end{aligned}$$

Optimizing the function LA over choices of upper limits provides insights into properties of the optimal retention limits u_i^*. In particular, it turns out that the optimal retention limit less the insurer's expected retained losses, $u_i^* - E[X_i \wedge u_i^*]$, is the same for *all* risks. Gray and Pitts (2012) (Section 5.8) attribute the discovery of this result to Bruno de Finetti (1906-1985). Section 11.3.1 will provide extensions to additional contracts.

Under the Hood. See the *Online Supplement* for the verification of optimal excess of loss limits.

Example 4.3. Excess of Loss for Three Pareto Risks. Consider three risks that have a Pareto distribution, each having a different set of parameters (so they are independent but non-identical). Use the same set of parameters as in Example 4.1.

By minimizing the Lagrangian, one can numerically determine the optimal retention limits u_1^*, u_2^*, and u_3^*. This is done in the associated `R` code using the package `alabama`. Results from the numerical optimization show that each optimal retention limit minus expected insurer's losses, $u_i^* - E[X_i \wedge u_i^*]$, is the same for all risks, as derived theoretically. In addition, one can graphically compare the distribution of total risks to that retained by the insurer and by the reinsurer, as displayed in Figure 4.3. This graph summarizes the distributions of possible outcomes that result from the optimal retention policy.

Code. The *Online Supplement* provides Example 4.3 illustrative statistical code.

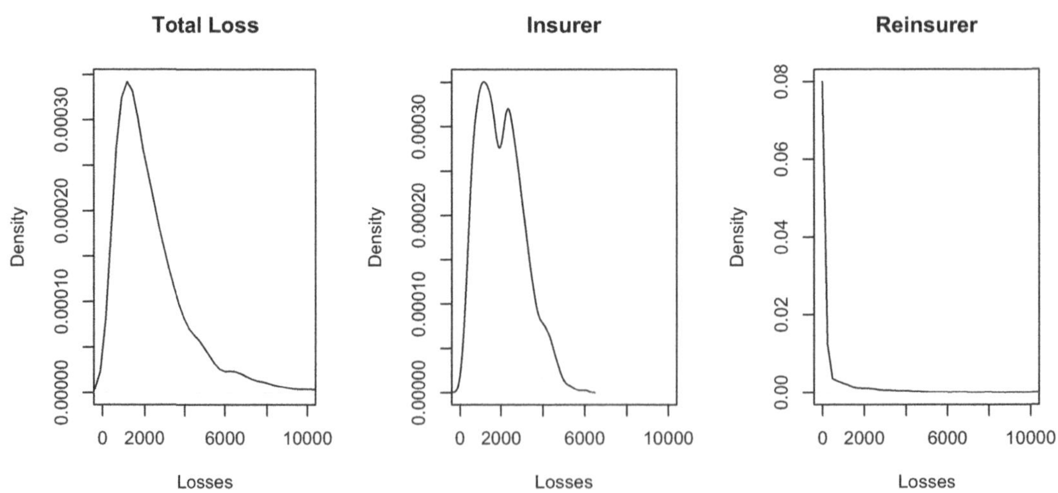

FIGURE 4.3
Density Plots for Total Loss, the Insurer, and the Reinsurer.

4.1.3 Layers of Coverage

Typically excess of loss contracts limit the exposure for a reinsurer, as illustrated in the following example.

Example 4.4. Consider a specific risk of an insurer. For this risk, the reinsurer is responsible for losses in excess of 500 up to a limit of 100,000, that is, $Y_{reinsurer} = \min[(X-500)_+, 100000]$. The insurer retains the remaining risk, $Y_{retained} = X - Y_{reinsurer}$.

From the reinsurer's point of view, in the language of Section 2.2 one would think of 500 as a "deductible" and 100,000 as its "upper limit" of coverage.

In reinsurance, it is common to partition loss amounts into multiple layers. This partitioning of losses is easiest to think about when there are more than two parties responsible for a risk, a concept further developed in Section 4.5. For example, instead of simply an insurer and reinsurer or an insurer and a policyholder, think about the situation with all three parties, a policyholder, insurer, and reinsurer, who agree on how to share a risk. More generally, consider k parties. If $k = 3$, it could be an insurer and two different reinsurers.

Example 4.5. Layers of Coverage for Three Parties.

- Suppose that there are $k = 3$ parties. The first party is responsible for the first 100 of losses, the second responsible for losses from 100 to 3000, and the third responsible for losses above 3000.
- If there are four losses in the amounts 50, 600, 1800 and 4000, then they would be allocated to the parties as follows:

Layer	Claim 1	Claim 2	Claim 3	Claim 4	Total
(0, 100]	50	100	100	100	350
(100, 3000]	0	500	1700	2900	5100
(3000, ∞)	0	0	0	1000	1000
Total	50	600	1800	4000	6450

To handle the general situation with k groups, partition the positive real line into k intervals using the cut-points

$$0 = u_0 < u_1 < \cdots < u_{k-1} < u_k = \infty.$$

Note that the ith interval is $(u_{i-1}, u_i]$. Now let Y_i be the amount of risk shared by the ith party. To illustrate, if a loss X is such that $u_{i-1} < X \le u_i$, then

$$\begin{pmatrix} Y_1 \\ Y_2 \\ \vdots \\ Y_i \\ Y_{i+1} \\ \vdots \\ Y_k \end{pmatrix} = \begin{pmatrix} u_1 - u_0 \\ u_2 - u_1 \\ \vdots \\ X - u_{i-1} \\ 0 \\ \vdots \\ 0 \end{pmatrix}.$$

More succinctly, we have

$$Y_i = \min(X, u_i) - \min(X, u_{i-1}).$$

With this expression, one can see that the ith party is responsible for losses in the interval $(u_{i-1}, u_i]$. With this, it is easy to check that $X = Y_1 + Y_2 + \cdots + Y_k$. To illustrate, applying this structure to Example 4.4, we see that the primary insurer has responsibility for the first and third layers (Y_1 and Y_3) whereas the reinsurer has responsibility for the second layer (Y_2). Thus, the parties taking on different layers need not be different.

For an example used in practice, see the Wisconsin Property Fund site for an illustration of layers of reinsurance.

See Appendix Section 4.6.4 for an overview of a mathematical framework that provides conditions under which a layered coverage scheme is optimal.

4.1.4 Complexities of Reinsurance Coverage

Historically, reinsurance agreements were contracts negotiated based on individual risks. These contracts, done on a case-by-case basis, today are known as *facultative* reinsurance and continue to represent about 15% of the marketplace (in terms of premium volume), cf. Albrecher et al. (2017). The remainder of the market has an *obligatory treaty* form, where both parties are bound by an obligation to cede and assume a contractually agreed share of a risk portfolio.

Because of their experience in risk modeling, reinsurers are key drivers in the adoption of better risk management practices by many organizations. Given their technical expertise and the extensive marketplace for unique contract arrangements, it is not surprising that reinsurance contracts can easily become quite complicated. This section describes a few of the more popular contracts.

To illustrate, some excess of loss contracts have in addition an (annual) aggregate deductible AAD so that the reinsurer's obligation may be written as

$$Y_{reinsurer} = \left[\left(\sum_{i=1}^{p} \min[(X_i - u_{1,i})_+, u_{2,i}]\right) - AAD\right]_+ .$$

Here, $u_{1,i}$ represents a reinsurer's deductible and $u_{2,i}$ represents is upper limit for the ith risk. Deductibles and upper limits may vary by risk or be the same. In the same way, excess of loss contracts may have an (annual) aggregate limit AAL, so that

$$Y_{reinsurer} = \min\left(\sum_{i=1}^{p} \min[(X_i - u_{1,i})_+, u_{2,i}], AAL\right).$$

Naturally, a contract may have both an overall deductible and an upper limit. Albrecher et al. (2017) refer to these as types of *global protections.*

As another variation, a reinsurance cover may be defined on a *per event,* or *per occurrence of a claim,* basis in lieu of the *per risk* basis that has been described so far. This approach is very handy to protect against, for example, natural catastrophes such as earthquakes or hurricanes where there may be many losses due to a common event. For example, a reinsurer's obligation may be of the form

$$Y_{reinsurer} = \sum_{j \in \text{event}} \min[(X_j - u_{1,j})_+, u_{2,j}].$$

When combining several reinsurance features into a single contract, proportional arrangements such as quota share arrangements are easy to include. As an example, it is common for an insurer to have simultaneous quota share contracts on the same portfolio with different reinsurers. Many other variations are routinely used in practice.

4.2 Modeling Dependence with Copulas

4.2.1 Copulas Defined

Copulas are widely used in insurance and many other fields to model the dependence among multivariate outcomes. A *copula* is a multivariate distribution function with uniform marginals. Specifically, let $\{V_1, \ldots, V_p\}$ be p uniform random variables on $(0,1)$. Their joint distribution function

$$C(v_1, \ldots, v_p) = \Pr(V_1 \le v_1, \ldots, V_p \le v_p)$$

is a copula. Of course, applications are based on more than just uniformly distributed data. Thus, consider arbitrary *marginal* distribution functions $F_1(\cdot),\ldots,F_p(\cdot)$. Then, one can define a multivariate distribution function F using the copula such that

$$F(y_1, \ldots, y_p) = C[F_1(y_1), \ldots, F_p(y_p)]. \tag{4.3}$$

Sklar (1959) showed the converse, that is, *any* multivariate distribution function can be written in the form of equation (4.3), that is, using a copula representation. *Note on Notation:* We use V and v for uniform random variables and their realizations in order to distinguish them from other terms, including utility functions U and upper limits u.

As an example with $p = 2$ variables, let us look to the copula due to Frank (1979),

$$C_F(v_1, v_2) = \frac{1}{\gamma} \log \left(1 + \frac{(\exp(\gamma v_1) - 1)(\exp(\gamma v_2) - 1)}{\exp(\gamma) - 1}\right). \tag{4.4}$$

This is a bivariate distribution function with its domain on the unit square $[0,1]^2$. It is easy to see that $C_F(v_1, 1) = v_1$. From this, one can argue that the marginals have a uniform (on $[0,1]$) distribution and so $C_F(\cdot,\cdot)$ is a copula. Further, one can interpret γ to be a dependence parameter in that the extent of dependence is controlled by the parameter γ.

The copula due to Frank (1979) has a simple expression in equation (4.4) (at least simple relative to other copulas) and enjoys several desirable properties. However, it does not readily extend to multiple dimensions (a point of emphasis in this book). So, as an alternative, let us also introduce the normal, or Gaussian, copula. In two dimensions, this is given as

$$C_N(v_1, v_2) = \Phi_2\left[\Phi^{-1}(v_1), \Phi^{-1}(v_2)\right]. \tag{4.5}$$

The expression in equation (4.5) appears more forbidding than Frank's copula but is just as easy to evaluate. The symbol $\Phi(\cdot)$ represents the cumulative standard normal distribution and $\Phi^{-1}(\cdot)$ is the corresponding inverse function, or quantile. In `R`, these can be evaluated using the functions `pnorm` and `qnorm`, respectively. The symbol $\Phi_2(\cdot)$ represents a bivariate normal distribution function that, in `R`, can be evaluated using the function `mvtnorm`.

Because both the Frank and normal copulas are used extensively in applications, it is not surprising that they are both available as `R` functions (e.g., in the package `copula`, see examples that follow). So, users do not really need the mathematical expressions in equations (4.4) and (4.5). Nonetheless, it is helpful to have some familiarity with the mathematical underpinnings of the technique.

One important strength of the copula approach is evident from equation (4.3). We will be able to work with multiple risks, perhaps the first one being gamma distribution, the second having a Pareto distribution, the third having a normal distribution. Even though each marginal distribution may be very different from one another, one can link (or *couple*) them using a copula.

4.2.2 Copulas, Risk Retention, and Simulation

Copulas are well known to work in many arenas of scientific applications but how do they work in risk retention problems? To get some insights, consider the following example.

Example 4.6. Standard Deviation of an Excess of Loss Policy with Two Risks. Consider two risks:

- That X_1 has a gamma distribution with mean 10000 and scale 5,000. The 95th percentile of X_1 is 23,719.
- That X_2 has a Pareto distribution with mean 1000 and scale 2,000. The 95th percentile of X_2 is 3,429.

Their relationship is governed by a normal copula with dependence parameter $\rho_S = 0.30$.

The risk owner retains $S(u_1, u_2) = X_1 \wedge u_1 + X_2 \wedge u_2$ and calibrates uncertainty through

$$\begin{aligned}\mathrm{Var}[S(u_1,u_2)] &= \mathrm{E}[S(u_1,u_2)]^2 - [\mathrm{E}S(u_1,u_2)]^2 \\ &= \mathrm{E}[(X_1 \wedge u_1)^2] + \mathrm{E}[(X_2 \wedge u_2)^2] + 2\mathrm{E}[(X_1 \wedge u_1)(X_2 \wedge u_2)] \\ &\quad - [\mathrm{E}(X_1 \wedge u_1) + \mathrm{E}(X_2 \wedge u_2)]^2.\end{aligned}$$

As in Section 4.1.2, we utilize the relationships

$$\mathrm{E}(X \wedge u) = \int_0^u [1 - F(x)]dx \quad \text{and} \quad \mathrm{E}(X \wedge u)^2 = 2\int_0^u x[1 - F(x)]dx.$$

These functions are easy to compute and, for example, are readily available in `R` through the package `actuar`. One can evaluate the expected cross-product of limited variables using

$$\begin{aligned}\mathrm{E}[(X_1 \wedge u_1)(X_2 \wedge u_2)] &= \int_0^{u_2}\int_0^{u_1} \Pr(X_1 > x_1, X_2 > x_2)\ dx_1 dx_2 \\ &= \int_0^{u_2}\int_0^{u_1} [1 - F_1(x_1) - F_2(x_2) + F(x_1,x_2)]\ dx_1 dx_2.\end{aligned} \tag{4.6}$$

ⓘ *Under the Hood.* See the *Online Supplement* for the verification of equation (4.6).

Although the double integral in equation (4.6) can be readily evaluated for a single point (u_1, u_2), this becomes burdensome as we go through many iterations seeking optimal values through iterative constrained optimization procedures beginning in the next section. (Chapter 5 will demonstrate this complexity.) A simpler approach is the following:

- Using simulation, create a random realization of the bivariate uniform random variates (V_{1r}, V_{2r}). In the `R` package `copula`, one can do this with the function `rCopula`.

- Use the inverse of the distribution function to create random realizations of the random variables, $X_{1r} = F_1^{-1}(V_{1r})$ and $X_{2r} = F_2^{-1}(V_{2r})$.
  - For F_1^{-1}, in R use the function `qgamma`.
  - For F_2^{-1}, in R use the function `qpareto`.
- Create a random realization of the limited sum, $S_r(u_1, u_2) = X_{1r} \wedge u_1 + X_{2r} \wedge u_2$.
- Repeat this for a large number, R, of replications. Summarize using the approximation

$$\widehat{Var}[S_R(u_1, u_2)] = \frac{1}{R}\sum_{r=1}^{R} \left[S_r(u_1, u_2) - \bar{S}_R(u_1, u_2)\right]^2,$$

where $\bar{S}_R$ is the average value of the replicates $S_r(u_1, u_2)$.

Code. The *Online Supplement* provides Example 4.6 illustrative statistical code.

Figure 4.4 summarizes results. As anticipated, the uncertainty of retained risk increases as the upper limits increase.

4.3 Incorporating Dependence

Now return to the basic reinsurance treaties and let us see how they fare in the face of dependence. Let us start with a simple example to show how dependence affects risk measures. After this, one can then address the question as to how these optimal allocations change as the dependence changes. We develop this strategy first for quota share, then for excess of loss contracts.

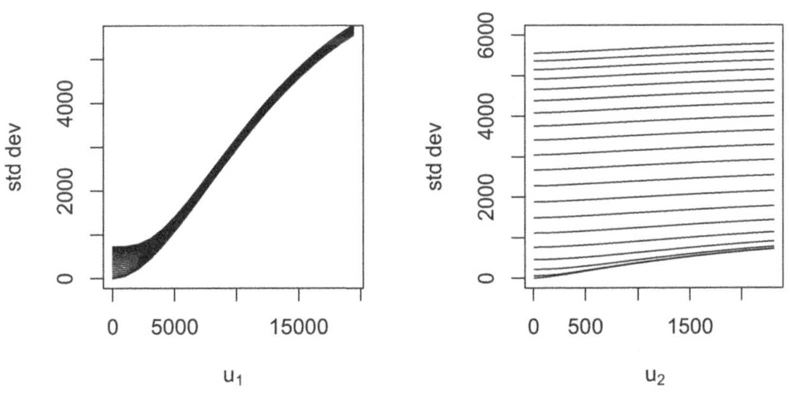

FIGURE 4.4
Standard Deviation of Retained Risk. The left panel show the standard deviation of the retained risk (its square, the variance, is the objective function) as a function of the first upper limit. Here, different lines represent different values of the second upper limit. The right panel compares the standard deviation to the second upper limit.

TABLE 4.1
Portfolio Expected Values, Standard Deviations, and Quantiles (VaR_α).

	Mean	Std Dev	$VaR_{0.90}$	$VaR_{0.95}$	$VaR_{0.99}$
Retained (Ind)	269	48	300	300	300
Reinsurer (Ind)	2333	3907	4487	6269	12644
Total (Ind)	2602	3909	4760	6541	12907
Retained (Dep)	269	48	300	300	300
Reinsurer (Dep)	2336	4067	4498	6283	12772
Total (Dep)	2605	4068	4770	6556	13052

4.3.1 Effects of Dependence on Risk Measures

It is not surprising that dependence affects our measures of uncertainty. In general, positive dependence leads to greater uncertainty, and hence larger values of risk measures. To be concrete, let us see how these general ideas work in a specific example.

Example 4.7. Effects of Dependence on Portfolio Summary Measures. Consider a portfolio management situation outlined in Exercise 4.1 that assumed that the property and liability risks are independent. Now, let us incorporate dependence by allowing the four lines of business to depend on one another through a Gaussian copula. For simplicity, assume a constant parameter equal 0.5 among the four lines.

Table 4.1 summarizes the results for the dependent portfolio in the lower three rows. For comparison, the top three rows provide results for the independence case. As anticipated, the mean losses are virtually identical when comparing the independent to dependent portfolios. The standard deviations (square root of our variance measure) for the dependent portfolio are larger for the reinsurer and in total. This larger uncertainty is also reflected in the values at risk at different levels of confidence.

4.3.2 Quota Share

I now extend the discussion in Section 4.1.1 to include dependence.

For the quota share agreement, recall that the insurer's portion of the portfolio risk is $Y_{retained} = \sum_{i=1}^{p} c_i X_i = \mathbf{c}'\mathbf{X}$. The objective is to find those values of c_i that minimize $\mathrm{Var}(Y_{retained})$ subject to the budget constraint $\mathrm{E}(Y_{reinsurer}) = RTC_{max}$.

This is similar to the Markowitz asset allocation problem introduced in Section 3.4. At least in the asset allocation version, the problem is well known and so is the solution, details are in Exercise 4.3. Nonetheless, we give the solution here to provide a solid foundation so that we can use a similar approach later for more complicated problems. For the quota share version of the problem, define the Lagrangian

$$
\begin{aligned}
LA &= \mathrm{Var}(Y_{retained}) + LME\ [\mathrm{E}(Y_{reinsurer}) - RTC_{max}] \\
&= \mathbf{c}'\boldsymbol{\Sigma}\mathbf{c} + LME\ [\mathrm{E}(\mathbf{X})'(\mathbf{1} - \mathbf{c}) - RTC_{max}].
\end{aligned}
$$

Taking derivatives yields

$$
\begin{aligned}
&\frac{\partial}{\partial \mathbf{c}} LA = 2\boldsymbol{\Sigma}\mathbf{c} - LME\ [\mathrm{E}(\mathbf{X})] = \mathbf{0} \\
&\Longrightarrow \mathbf{c}^* = \frac{LME^*}{2}\ \boldsymbol{\Sigma}^{-1}\ \mathrm{E}(\mathbf{X}).
\end{aligned}
$$

This is consistent with the case of independence in equation (4.1). From this, one can determine LME^* as the solution of

$$RTC_{max} = (\mathrm{E}\ \mathbf{X})'[\mathbf{1} - \mathbf{c}^*] = (\mathrm{E}\ \mathbf{X})'[\mathbf{1} - \frac{LME}{2}\ \mathbf{\Sigma}^{-1}\ \mathrm{E}(\mathbf{X})]$$

$$\Longrightarrow LME^* = 2\frac{(\mathrm{E}\ \mathbf{X})'\mathbf{1} - RTC_{max}}{\mathrm{E}(\mathbf{X})'\ \mathbf{\Sigma}^{-1}\ \mathrm{E}(\mathbf{X})} .$$

For the quota share problem, the coinsurance parameters c_j need not sum to one in contrast to the asset allocation problem in Exercise 4.3. However, it is common to require that these parameters be in the unit interval $[0, 1]$. This can be formulated as a constrained optimization problem although the solution is not as tidy, see Exercise 10.1.

Example 4.8. Quota Share Example with Three Risks. This demonstration assumes that the third risk has the same distribution as the first. For the value of RTC_{max}, we use 4,200, which is 20% of the sum of expected losses. In this example, one can use association parameters and a Gaussian copula as inputs. One can then approximate the variance-covariance matrix $\mathbf{\Sigma}$ using simulation.

Table 4.2 summarizes results for the baseline case.

Now, repeat assuming zero correlations among risks to see the difference. Table 4.3 summarizes these results. There is a large difference in the retention parameters $\mathbf{c}$ when the dependence changes, particularly for the Pareto risk.

Table 4.4 shows that one does not have to look that hard to find a **negative** coefficient associated with a risk. Although common in the investment world (shorting an asset), we have little interest in this possibility in insurance settings.

TABLE 4.2
Quota Share Summary.

	Mean	Std Dev	Corr 1	Corr 2	Corr 3	c
Gamma1	10000	7071	1.0	-0.2	0.8	0.758
Pareto	1000	1732	-0.2	1.0	-0.3	0.378
Gamma2	10000	7071	0.8	-0.3	1.0	0.884

TABLE 4.3
Quota Share Summary—with Zero Correlations.

	Mean	Std Dev	Corr 1	Corr 2	Corr 3	c
Gamma1	10000	7071	1	0	0	0.829
Pareto	1000	1732	0	1	0	0.076
Gamma2	10000	7071	0	0	1	0.843

TABLE 4.4
Quota Share Summary—with Positive Correlations.

	Mean	Std Dev	Corr 1	Corr 2	Corr 3	c
Gamma1	10000	7071	1.0	0.2	0.4	0.537
Pareto	1000	1732	0.2	1.0	0.8	-0.705
Gamma2	10000	7071	0.4	0.8	1.0	1.214

4.3.3 Excess of Loss

As in Section 4.1.2, for the excess of loss setting we consider p risks $X_1, \ldots, X_p$ where each risk is subject to an upper limit, say, u_i. With this, the insurer retains $Y_{retained} = \sum_{i=1}^{p} X_i \wedge u_i$ and the reinsurer is responsible for the excess, $Y_{reinsurer} = S - Y_{retained}$. The retention limits may vary by risk or may be the same for all risks, $u_i = u$, for all i.

What is the best choice of the excess of loss retention limits u_i? To formalize this question, we seek to find those values of u_i that minimize $\mathrm{Var}(Y_{retained})$ subject to the constraint that $\mathrm{E}(Y_{reinsurer}) = RTC_{max}$. That is, subject to this budget constraint, the insurer wishes to minimize the uncertainty of the retained risks (as measured by the variance).

Section 4.4 will outline several approaches for solving this problem numerically. Let us see how one might do this using the root finding methods that will be described in Section 4.4.4. In part, we explore this now simply to demonstrate the difficulties associated with this approach. To implement this approach, assume that the budget constraint is binding. Further, one needs derivatives. So, start with the Lagrangian

$$LA(u_1, \ldots, u_p, LME) = \mathrm{Var}(Y_{retained}) + LME\ [\mathrm{E}(Y_{reinsurer}) - RTC_{max}].$$

It is convenient to introduce an integrated tail function

$$\begin{aligned} IT_{ij}(a, b) &= \int_0^b \{1 - F_i(a) - F_j(z) + F_{ij}(a, z)\}\ dz \\ &= \int_0^b \Pr(X_i > a, X_j > z)\ dz, \end{aligned}$$

where F_{ij} is the joint distribution function of the random variables X_i and X_j. After some pleasant calculus, one can check that the partial derivatives are

$$\begin{aligned} \partial_{u_i}\ LA(u_1, \ldots, u_p, LME) &= 2\left[u_i - \sum_{j=1}^{p} \mathrm{E}(X_j \wedge u_j) - \frac{LME}{2}\right][1 - F_i(u_i)] \\ &\quad + 2\sum_{j \neq i} \{IT_{ij}(u_i, u_j)\}, \end{aligned} \tag{4.7}$$

for $i = 1, \ldots, p$, and

$$\partial_{LME}\ LA(u_1, \ldots, u_p, LME) = \mathrm{E}(S) - \sum_{i=1}^{p} \mathrm{E}(X_i \wedge u_i) - RTC_{max}. \tag{4.8}$$

ⓘ *Under the Hood.* See the *Online Supplement* for the verification of equations (4.7) and (4.8).

Setting these derivatives to zero, there are $p+1$ equations and $p+1$ unknowns which can be solved numerically. Numerical solutions require starting values; solutions can sometimes be sensitive to the choice of starting values. Fortunately, for this problem, one can look to the case of independence developed in Section 4.1.2 for starting values.

Special Case - Independent Risks. It is convenient to first define an integrated distribution function $H_{1i}(x) = \int_0^x F_i(z)dz$. With this notation, it can be shown we may determine LME^* as the solution of the equation

$$\mathrm{E}(S) - RTC_{max} = \sum_{i=1}^{p} \left\{ H_{1i}^{-1}\left(\frac{LME}{2}\right) - \frac{LME}{2} \right\}.$$

TABLE 4.5
Excess of Loss Dependent Risks Summary.

	Mean	Optimal Ind u^*	Ind u^* Prop	Optimal Dep u^*	Dep u^* Prop
Gamma1	10000	11806	0.683	11938	0.689
Pareto	1000	4775	0.974	1214	0.759
Gamma2	10000	11806	0.683	12673	0.72
LME		7725		8942	

With a value of LME^*, one can calculate the optimal retention limits using $u_i^* = H_{1i}^{-1}(\frac{LME^*}{2})$.

Under the Hood. See the *Online Supplement* for the independence case details.

Example 4.9. Optimal Upper Limits with Excess of Loss for Three Dependent Risks. Now return to the same set of marginal distributions as in the quota share Example 4.8. Also as in this example, let $RTC_{max} = 4,200$, which is 20% of the sum of expected losses. For the Gaussian copula parameters, we use

$$\rho = \begin{pmatrix} 1 & \rho_{12} & \rho_{13} \\ \rho_{12} & 1 & \rho_{23} \\ \rho_{13} & \rho_{23} & 1 \end{pmatrix}$$

with $\rho_{12} = 0.95$, $\rho_{13} = 0$, and $\rho_{23} = 0$. Note that the rho parameters are based on the copula, not (Pearson) correlations among the risks.

Table 4.5 shows the optimal upper limits under both models of independence and dependence. Comparing these results, it appears that there can be important differences between the optimal limits based on the dependence.

In addition, another way to view these is to think about their position in the marginal distribution. Specifically, the optimal retention parameters can also be expressed as proportions for the corresponding marginal distribution functions. So, for example, for the best choice of u_1 under the model of independence on the original scale is 11,806; this is the 68.3 percentile of this gamma distribution. Extending this logic to the second row, the upper limits are 97.4 percentile for the independence model versus 75.9 percentile for the dependence model. Thus, when viewed as a percentile or on the original scale, 4,775 versus 1,214, it appears that the dependence assumption can influence the optimal upper limits.

4.4 Numerical Optimization

Recall the general constrained optimization problem in Display (3.9) from Section 3.3.4,

$$\boxed{\begin{array}{lll} \text{minimize}_{\mathbf{z}} & f_0(\mathbf{z}) & \\ \text{subject to} & f_{con,j}(\mathbf{z}) \leq 0 & j \in CON_{in} \\ & f_{con,j}(\mathbf{z}) = 0 & j \in CON_{eq}. \end{array}}$$

Here, $f_0(\cdot)$ is the objective function that we wish to minimize subject to conditions on the constraints $f_{con,j}(\cdot)$ in the inequality (CON_{in}) and equality (CON_{eq}) sets.

4.4.1 Convex Optimization

In today's computing environment, the optimization problem described in Display (3.9) is relatively straightforward if we can identify the problem as **convex**. There are some minor variations in the definition of convex optimization in the literature; I use the conditions put forward in Boyd and Vandenberghe (2004). Specifically, let us require that the objective function $f_0(\mathbf{z})$ and inequality constraints $f_{con,j}(\mathbf{z}), j \in CON_{in}$ are convex and that the equality constraints $f_{con,j}(\mathbf{z}), j \in CON_{eq}$ are affine. By affine (or linear), I mean that one can write $f_{con,j}(\mathbf{z}) = \mathbf{A}\mathbf{z} - \mathbf{b}$ for some matrix $\mathbf{A}$ and vector $\mathbf{b}$. (For a refresher, refer to the Appendix Section 3.6.3 for definitions of convexity and other terms.)

Classical problems based on proportional (e.g., quota sharing) rules with the variance as a measure of risk are typically convex. As an example, consider the quota sharing problem introduced in Section 4.1.1. The quota share coefficients are the parameters of interest and so $\mathbf{z} = \mathbf{c}$. The variance of the portfolio risk, $f_0(\mathbf{z}) = \mathrm{Var}(\mathbf{z}'\mathbf{X}) = \mathbf{z}'\boldsymbol{\Sigma}\mathbf{z}$, is the objective function to be minimized. This is quadratic in the parameters and hence typically convex. In the same way, the risk transfer condition can be seen to be linear in the parameters. Moreover, it will be straightforward to include parameter constraints of the form $c_i = z_i \geq 0$ and $c_i = z_i \leq 1$ as these are linear and hence also convex. Recall that Section 4.3.2 described this problem in more detail.

For convex problems, there are many algorithms that are computationally efficient and that fare remarkably well even when the number of decision variables (the dimension of $\mathbf{z}$) is large. To illustrate this in our demonstrations, we use the `R` package `CVXR`; see Exercise 4.8. Moreover, as emphasized by Boyd and Vandenberghe (2004), local solutions to convex problems are also global solutions, a property that is not generally available for constrained optimization problems. Because of this property, many convex algorithms do not even require starting values.

Special Case: Asset Allocations

Asset allocations, summarized in Section 3.4, can be cast as a convex optimization problem. In addition, mathematically the asset allocation problem is similar to the quota share problem described in Section 4.3.2. Naturally, the practices are very different; the former considers allocation concerns of investors whereas the latter is about diversifying liabilities that is typically only done by insurers and major companies. A common simplifying feature of both problems is that the decision variables, $c_1, \ldots, c_p$, are linear in the risks. This means that the problem is convex in many situations. In this section, I expand the discussion and use the expected shortfall ES as our risk measure.

To see the convexity of the problem, the approach is to define an expanded version of the expected shortfall function,

$$
\begin{aligned}
ES1(z_0, \mathbf{c}) &= z_0 + \tfrac{1}{1-\alpha} \left\{ \mathrm{E}[\mathbf{c}'\mathbf{X}] - \mathrm{E}[\mathbf{c}'\mathbf{X} \wedge z_0] \right\} \\
&= z_0 + \tfrac{1}{1-\alpha} \mathrm{E}[\mathbf{c}'\mathbf{X} - z_0]_+ .
\end{aligned}
$$

For fixed values of $\mathbf{c}$, it is easy to check that the solution of the unconstrained minimization problem is the quantile. Moreover, the minimum value is the expected shortfall. This point will be developed in more detail in Section 7.2.2.

Moreover, we verify below that the function $ES1(z_0, \mathbf{c})$ is convex in the decision variables z_0 and $\mathbf{c}$. Because the minimum of a convex function is also convex (see, for example, Boyd and Vandenberghe (2004), Section 3.2.5), this means that the ES function is convex.

Under the Hood. See the *Online Supplement* for the verification of the convexity.

Earlier in Display (3.13), we wrote the asset allocation problem as minimizing the portfolio variance subject to constraints on the mean return. Now, we extend this by using the expected shortfall as our risk measure. Specifically, we consider

$$\boxed{\begin{array}{ll} \text{minimize}_{z_0,\mathbf{c}} & ES1(z_0,\mathbf{c}) \\ \text{subject to} & \mathbf{c}' \, \mathrm{E}(\mathbf{R}) \geq Req_0. \end{array}} \tag{4.9}$$

Example 4.10. Optimizing a Portfolio of Insurance Stock Returns. This is a continuation of Examples 1.1 and 3.3. We are now ready to compute the optimal portfolio specified in Display (4.9) with the `CVXR` package.

As in Example 3.3, we determine optimal allocations using the training data. With these allocations, various summary measures are computed. Then, the optimal allocations are used on the test data and several summary measures are determined.

Table 4.6 summarizes performance for both the training and test samples. For comparison purposes, the performance measures summarized earlier in Example 3.3 are also included.

Code. The *Online Supplement* provides Example 4.10 illustrative statistical code.

Another way of presenting the performance of the differing portfolios is to track their performance over time. In Figure 4.5, we can see that all portfolios behave similarly over the test period.

As emphasized in Example 3.3, one cannot conclude too much from these results. Practicing analysts will use this approach while conducting more extensive testing by varying the training period, the test period, and the sample of stocks (risks) selected. In addition, many of the constraints on allocations introduced in Section 3.4.2 will come into play (note that these allocation constraints are linear and so can readily be incorporated into the convex optimization problem). Readers are invited to experiment with the code to determine their own optimal portfolios.

4.4.2 Constrained Optimization without Coding Derivatives

Many risk retention problems cannot be classified as convex and so we need a more general approach. For some problems, one only needs to code the objective and constraint functions and use these specifications in a numerical optimization program. Many modern programs now have the ability to calculate approximate derivatives using finite difference methods and use these approximations in lieu of analytical expressions for the derivatives. For many

TABLE 4.6
Portfolio Performance Measures, with *ES*.

	Train Ann Return	Train Std Dev	Train Sharpe (Rf=0)	Test Ann Return	Test Std Dev	Test) Sharpe (Rf=0)
MinVar	0.164	0.119	1.378	0.168	0.143	1.178
Markowitz	0.203	0.126	1.614	0.166	0.143	1.160
Equal	0.173	0.130	1.329	0.155	0.142	1.093
ES-$\alpha = 0.95$	0.173	0.120	1.436	0.158	0.144	1.099
ES-$\alpha = 0.99$	0.181	0.128	1.412	0.155	0.152	1.020

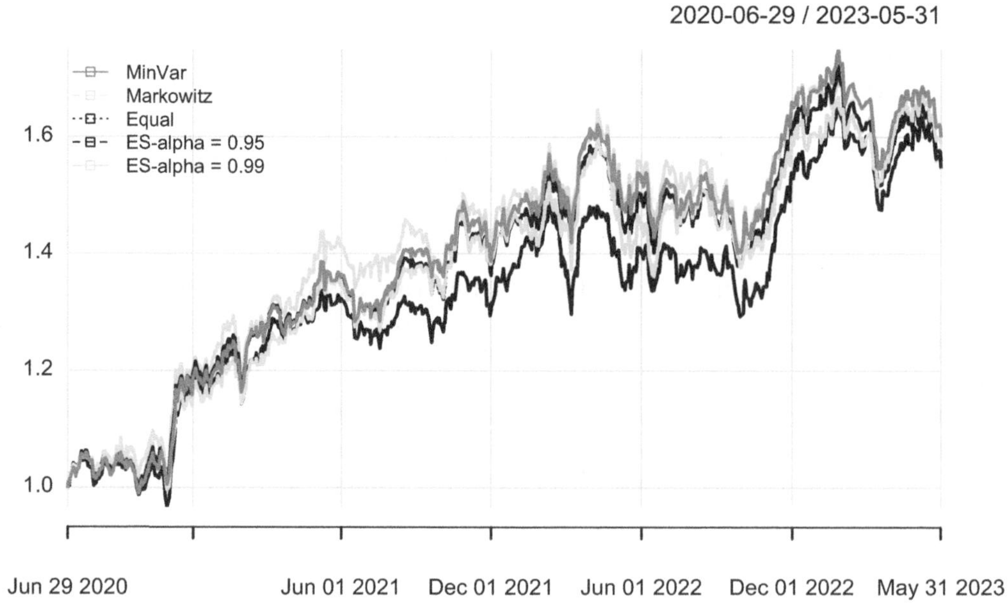

FIGURE 4.5
Cumulative Portfolio Returns (Test Sample), with *ES*.

problems, this approach is much simpler for the user and often works well particularly when there are small numbers of decision variables and constraints. Nonetheless, for some problems, they can take much longer for convergence to optimal solutions when compared to methods using analytical derivatives. Further, they can be sensitive to algorithmic parameters such as starting values. I show how to use this approach in `R` using the function `auglag` from the package `alabama`.

However, in applied risk management problems even the evaluation of objective and constraint functions can be difficult. As seen in Example 4.6, this difficulty can be addressed using **simulation** as a method of approximating these functions. Virtually all numerical methods exhibit deviations from true values when evaluating a solution and simulation is certainly no exception. A strength of the simulation approach is that the approximation error can be reduced simply by increasing the number of simulated replicates. So, if the analyst is satisfied with a result based on the analysis of a simulated trial, then accuracy can be increased by replicating the simulated trial with a larger number of replicates. Let us see how the simulation approach is valuable in the following example.

Example 4.11. Constrained Optimization for Excess of Loss with Two Risks. Let us continue from Example 4.6. The risk owner retains $S(u_1, u_2) = X_1 \wedge u_1 + X_2 \wedge u_2$ and desires to minimize $f_0(u_1, u_2) = \mathrm{Var}[S(u_1, u_2)]$. This is constrained by the budget requirement $f_{con,1}(u_1, u_2) = RTC(u_1, u_2) - RTC_{max} \leq 0$ and constraints on the parameters $f_{con,2}(u_1, u_2) = -u_1 \leq 0$ and $f_{con,3}(u_1, u_2) = -u_2 \leq 0$. For simplicity, we use a fair risk transfer cost $RTC(u_1, u_2) = \mathrm{E}[S(\infty, \infty)] - \mathrm{E}[S(u_1, u_2)]$. We use simulation to evaluate the objective function but not the risk transfer cost which is easy to evaluate.

Code. The *Online Supplement* provides Example 4.11 illustrative statistical code.

Recall from the earlier Example 4.6, Figure 4.4 showed the standard deviation of retained risks as a function of each upper limit. Figure 4.6 supplements this by showing the risk transfer cost RTC as a function of each upper limit. As anticipated, the risk trans-

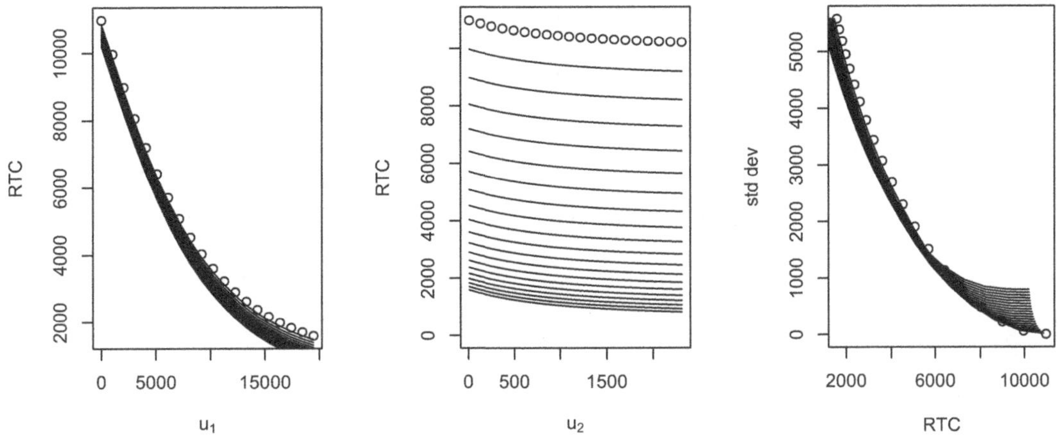

FIGURE 4.6
Excess of Loss Parameters and Objective and Constraint Functions. The left panel shows the *RTC* as a function of the first upper limit, with different lines representing different values of the second upper limit. The middle panel shows it as a function of the second upper limit, with different lines representing different values of the first upper limit. The right panel compares the *RTC* to the standard deviation.

fer cost *RTC* decreases as each upper limit increases. The right-hand panel of Figure 4.6 demonstrates the trade-off between the risk transfer cost and the uncertainty of retained risk.

With these objective and constraint functions, one can then solve the constrained optimization problem that is demonstrated in the associated code. This is done for a specific value of RTC_{max}, which is then varied. This shows how the solution changes as the maximal risk transfer cost changes. Figure 4.7 summarizes results. Here, we see that as RTC_{max} increases, the optimal upper limits decrease, as does the uncertainty of retained risk.

Code. The *Online Supplement* provides Example 4.11, without derivatives, illustrative statistical code.

4.4.3 Constrained Optimization with Coding Derivatives

Another approach is to use a numerical optimization program that allows one to include derivatives of objective and constraint functions. This approach requires more upfront effort for the analyst, since you have to derive and code the derivatives, but often provides more stable and robust results. However, as seen in the following example, even developing the derivatives can be difficult.

Example 4.11. Excess of Loss with Two Risks – Continued. The derivatives of the constraint functions are relatively straightforward. We have

$$\begin{aligned}\partial_{u_j} f_{con,1}(u_1,u_2) &= \partial_{u_j}\left\{\mathrm{E}(S(\infty,\infty)) - \mathrm{E}(S(u_1,u_2)) - RTC_{max}\right\}\\ &= -\partial_{u_j}\left\{\mathrm{E}(X_1 \wedge u_1) + \mathrm{E}(X_2 \wedge u_2)\right\}\\ &= -\left\{1 - F_j(u_j)\right\},\end{aligned}$$

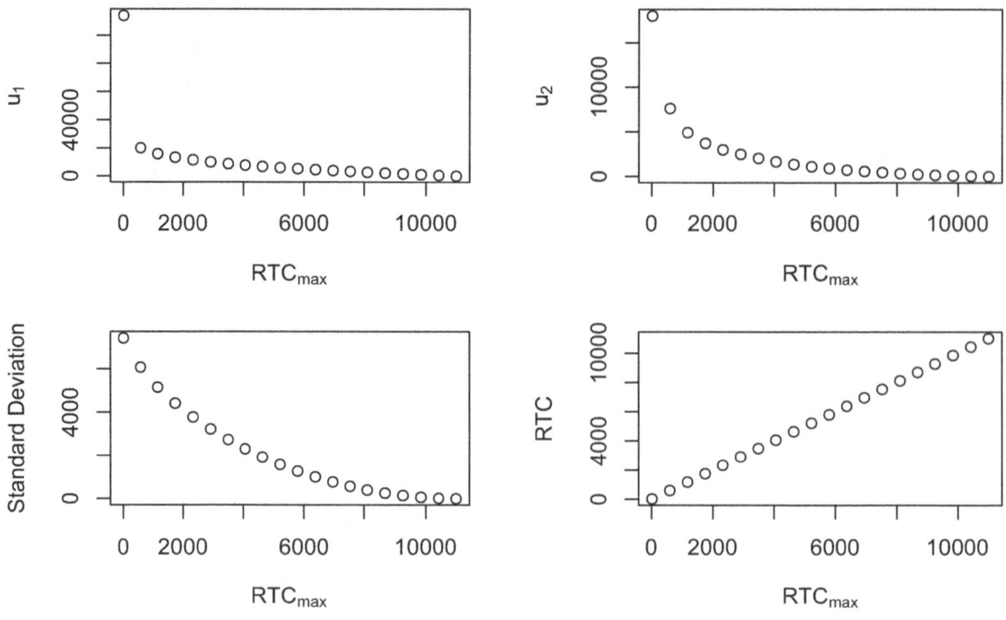

FIGURE 4.7
Optimization Results as a Function of RTC_{max}. The upper left panel shows the optimal first upper limit as a function of the maximal risk transfer cost, RTC_{max}. The upper right panel gives the same information for the second upper limit. The lower left panel shows the best value of the standard deviation of retained risks as a function of the RTC_{max}. The lower right panel shows that the constraint is binding at each optimum.

as well as

$$\begin{aligned}
\partial_{u_1} f_{con,2}(u_1,u_2) &= \partial_{u_1}(-u_1) = -1, & \partial_{u_2} f_{con,2}(u_1,u_2) &= \partial_{u_2}(-u_1) = \quad 0, \\
\partial_{u_1} f_{con,3}(u_1,u_2) &= \partial_{u_1}(-u_2) = \quad 0, & \partial_{u_2} f_{con,3}(u_1,u_2) &= \partial_{u_2}(-u_2) = -1.
\end{aligned}$$

For the objective function, a partial derivative can be shown to be

$$\begin{aligned}
\partial_{u_1} f_0(u_1,u_2) &= \partial_{u_1} \mathrm{Var}[S(u_1,u_2)] \\
&= 2\left\{(u_1+u_2-\mathrm{E}[S(u_1,u_2)])[1-F_1(u_1)] \right. \\
&\quad \left. - \int_0^{u_2} [F_2(x_2) - F(u_1,x_2)]\, dx_2 \right\}.
\end{aligned} \tag{4.10}$$

The result for the other gradient, $\partial_{u_2} f_0(u_1,u_2)$, follows in the same way, simply switch the indices.

ⓘ *Under the Hood.* See the *Online Supplement* for the verification of the derivative of the objective function.

The associated `R` code for constrained maximization with derivatives uses the `auglag` function. This code provides essentially the same results as those without derivatives and so the output is suppressed (the exception is when RTC_{max} is very small and there are numerical difficulties). For this example, it takes longer than the results without derivatives.

This is because the numerical evaluation of the integral above required for each step of the minimization process. The code numerically evaluates the partial derivatives of the constraints and objective function yet uses a simulation approximation for the objective function. This is because the calculation of the variance requires double integration which can be computationally intensive when done as part of an iterative optimization procedure.

Code. The *Online Supplement* provides Example 4.11, with derivatives, illustrative statistical code.

4.4.4 Root-Finding Methods (with Derivatives)

With derivatives developed and coded, we can also employ a more basic approach. Simply take the derivatives of the Lagrangian, set these equal to zero, and solve a system of equations using a root-finding algorithm. That is, the beauty of the Lagrangian approach is that it turns a constrained optimization problem into an unconstrained one. And, we know from calculus, for many smooth problems, you simply take a derivative of the function and solve for zeros to get its critical values. The limitation of this approach, mentioned earlier, is the working assumption that the constraints are active, or binding, at the optimum.

This approach can be implemented numerically in `R` using the function `multiroot` from the `rootSolve` package, see the associated `R` code. Like any root-finding algorithm, this function only finds critical values (of which there may be many) and it is up to the analyst to ensure that they minimize the objective function over the entire feasible region. Further, the function requires starting values which, when varied, can lead to different critical values.

Example 4.11. Excess of Loss with Two Risks – Continued. The associated `R` code shows how to solve the problem using root finding methods. As before, because the results are similar to those from other approaches, the findings are not displayed.

Code. The *Online Supplement* provides Example 4.11, root-finding, illustrative statistical code.

As another example of root-finding methods, you may wish to revisit the code in Example 4.9.

For analysts who wish to take a deeper dive into the intricacies of optimization problems in `R`, visit CRAN Task View: Optimization and Mathematical Programming. Among many other packages and approaches, this page has further links to three packages that focus on variations of the Lagrange method:

- The package `alabama` provides an implementation of the augmented Lagrange multiplier method for solving nonlinear optimization problems with nonlinear equality and inequality constraints. The acronym stands for *Augmented Lagrangian Adaptive Barrier Minimization Algorithm.* This is the primary package used in this book.
- The package `nloptr` provides access to various nonlinear optimization algorithms, including the augmented Lagrangian approach for nonlinear constraints.
- The package `Rsolnp` also provides an implementation of the augmented Lagrange multiplier method for solving problems with equality and inequality constraints.

4.5 Risk Exchanges

We have focused on a risk transfer defined to be the change of possession of an uncertain obligation, a risk, from one party to another in exchange for (certain) monetary compensation. This definition includes commercial insurance as well as many other mechanisms. For example, you might be the firm risk manager seeking to transfer risks due to unforeseen events that could damage the company's property, financial health due to cyber events, and the well-being of the workforce due to accidents at the work facility. Corporate insurance is one way to transfer these potential liabilities but there are other possibilities, including self-insurance, utilizing a captive insurer, participation in a pool or group, insurance from a governmental entity, and other alternative risk transfer mechanisms. Section 6.2 will provide more detail on these alternatives. Although our primary focus is on insurance as a motivation for risk transfer, we aim to develop frameworks that are sufficiently general to handle other mechanisms as well.

Risk owners seek ways to offload some or all of their risk. Instead of simply ceding a portion of the risk, it is helpful to think about a general framework of **exchanging** risks. Specifically, we consider n exchange participants known as *agents*. These agents may be risk owners, such as a firm or an individual, or they may be another party, such as an insurer or a pool. For example, an agent may be a reinsurer that brings no risk to the exchange but agrees to take on risks in exchange for a fee. Associated with each agent, say the ith one, is a set of p risks $\mathbf{X}_i = (X_{i1}, \ldots, X_{ip})$. A risk may be a (degenerate) random variable that has no uncertainty, such as cash.

When terms of a risk exchange have been agreed upon, a set of rules or functions g_i is established, such that

$$\mathbf{Y}_i = g_i(\mathbf{X}_1, \ldots, \mathbf{X}_n), \quad i = 1, \ldots, n, \tag{4.11}$$

defines the collection of risks of the ith agent *after* the exchange has taken place (but before the random variables have been realized). The goal of the risk exchange, also known as the *risk-sharing*, literature is to establish guidance for suitable choices of the rules g_i. For example, one basic principle in selecting the rules is to ensure that $\sum_{i=1}^{n} \mathbf{Y}_i = \sum_{i=1}^{n} \mathbf{X}_i$. This ensures that the sum of risks from all risk owners is equal before and after the exchange. In financial transactions, this principle is known as "clearing the market" and so *clearing the exchange* seems appropriate here.

Another fundamental goal is to achieve an improvement in the risk position for all parties. That is, the measure of uncertainty for the ith party after the exchange should not be larger than prior to the exchange. The reasoning is that if a party's uncertainty increases due to the exchange, that party will not enter into the exchange.

In addition to these basic goals, there is another desirable principle known as *Pareto optimality*. An exchange is said to be Pareto-optimal if there is no other exchange that is strictly better. This principle, as with the risk improvement principle, depends on the type of risk measure used to quantify uncertainty, as well as the application area. Section 4.5.2 provides additional discussion on Pareto optimality. In the following three subsections, we provide some guidance based on separate strands of the actuarial literature. In the appendix Section 4.6.4, we describe an alternative approach for determining risk exchanges based on quantile risk measures such as $RVaR$.

4.5.1 Linear Exchanges

We now know that the risk exchange literature likely began with the work of De Finetti (1940); Seal (1969) provides a detailed description of its early history up to its publication date. Based on this long history, we adopt the structure put forth in Feng et al. (2023).

Similar to most approaches, let us focus on only a single type of risk, so $p = 1$. If cash is required to equate a risk-sharing agreement, one regards it as a "side payment." In this section, we use variance as a measure of risk and, consistent with the early literature, emphasize proportional risk-sharing agreements.

Consider n agents who have risks $X_1, \ldots, X_n$ to be shared among them. In the special case where all risks have the same distribution, it is desirable to share risks equally. That is, the share for agent i after the exchange is $Y_i = (X_1 + \cdots + X_n)/n$, the average. When the risks are independent, the uncertainty after the exchange $\mathrm{Var}(Y_i) = \mathrm{Var}(X_i)/n$ is much lower than prior to the exchange.

In general, risks do not have the same distribution nor are they independent. To handle this more complex situation, this subsection considers *linear* exchanges, or sharing rules, where risks are combined in an additive fashion and the proportion of each risk is set as part of the sharing rule. Specifically, define c_{ij} to be the proportion of the jth risk that agent i takes on after a risk-sharing agreement. In sum, the ith agent has responsibility for $Y_i = c_{i1}X_1 + \cdots + c_{in}X_n$, a random linear combination of losses. This can be summarized for all agents as:

$$\mathbf{Y} = \begin{pmatrix} Y_1 \\ \vdots \\ Y_n \end{pmatrix} = \begin{pmatrix} c_{11} & \cdots & c_{1n} \\ \vdots & \ddots & \vdots \\ c_{n1} & \cdots & c_{nn} \end{pmatrix} \begin{pmatrix} X_1 \\ \vdots \\ X_n \end{pmatrix} = \mathbf{CX}.$$

Several conditions potentially apply:

$$\boxed{\begin{array}{cc} \mathbf{C}'\mathbf{1} = \mathbf{1} & \text{clearing the exchange} \\ \mathbf{C}\boldsymbol{\mu} = \boldsymbol{\mu} & \text{no profits} \\ 0 \le c_{ij} \le 1 & \text{no short/long selling} \\ \mathrm{Var}(Y_i) \le \mathrm{Var}(X_i) & \text{risk improvement} \end{array}}\,. \tag{4.12}$$

The first condition, "clearing the exchange," ensures that $\mathbf{1}'\mathbf{X} = \mathbf{1}'\mathbf{CX} = \mathbf{1}'\mathbf{Y}$, so that the sum of risks before the exchange equals the sum of risks after the exchange, for any realization of the random vector $\mathbf{X}$. The second condition on the means ($\mathrm{E}(\mathbf{X}) = \boldsymbol{\mu}$) implies that this is a non-profit arrangement; that is, no agent can expect to make profit (so that $\mathrm{E}(\mathbf{Y}) = \mathrm{E}(\mathbf{X})$). The third condition excludes "short selling" (and "long" positions) as is common in financial investment transactions. An agent may not take a negative position nor more than 100% of a risk. The fourth condition is about improving a risk position; no agent will be worse off by participating in the risk-sharing agreement.

There are many choices of the matrix $\mathbf{C}$ that satisfy the conditions in Display (4.12). The identity matrix is certainly one, ensuring that the set of feasible parameters is not empty. To obtain a unique set of parameters, we focus on minimizing $\sum_i \mathrm{Var}(Y_i)$ over choices of $\mathbf{C}$ subject to the conditions in Display (4.12). Here, one can interpret $\sum_{i=1}^n \mathrm{Var}(Y_i)$ to be a "system variance."

As is common for optimization problems, with only equality constraints one can obtain closed-form expressions for the solutions. This approach is outlined in Exercises 4.5 and 4.6. More generally (including inequality constraints), this is a convex programming problem that can be readily solved numerically. Specifically, it is a *quadratic program*, and so fast solvers are available (see, for example, Boyd and Vandenberghe (2004), Section 4.4). Fast solvers mean that it can be easy to optimize large dimensional problems. As an illustration, Exercise 4.6 outlines a strategy for $n = 50$ agents (that works in about 4 minutes). Nonetheless, to start, first consider the following example.

Example 4.12. Linear Sharing Among Three Agents. Assume that $n = 3$ and the mean vector and variance-covariance matrix are given as

$$\boldsymbol{\mu} = \begin{pmatrix} 20 \\ 2.5 \\ 10 \end{pmatrix} \quad \text{and} \quad \boldsymbol{\Sigma} = \begin{pmatrix} 10 & -4 & -1 \\ -4 & 8 & 1 \\ -1 & 1 & 1 \end{pmatrix}.$$

Prior to the exchange, the sum of all variances is $\sum_{i=1}^{n} \mathrm{Var}(X_i) = 10+8+1 = 19$. We then determine values of the coefficient matrix $\mathbf{C}$ that minimizes the system variance after the exchange, $\sum_{i=1}^{n} \mathrm{Var}(Y_i)$, subject to a sequence of constraints. The results are summarized in Tables 4.7 and Table 4.8 based on the associated `R` code.

Code. The *Online Supplement* provides Example 4.12 illustrative statistical code.

Table 4.7 shows the impact of the exchange on the uncertainty measure for each agent. To illustrate, for Agent 1, the original (prior to the exchange) variance is $\mathrm{Var}(X_1) = 10$. After imposing on the clearing exchange condition, Agent 1's variance is $\mathrm{Var}(Y_1) = 1.2222$. If instead one does the optimization problem with the clear exchange *and* the no profits condition, Agent 1's variance is $\mathrm{Var}(Y_1) = 2.6281$. Repeating, with the constraints clear exchange, no profits, and short sell, Agent 1's variance is $\mathrm{Var}(Y_1) = 2.8881$. Finally, with all four constraints, Agent 1's variance is $\mathrm{Var}(Y_1) = 3.3164$. The bottom row provides the sum of variances for all three agents. After the exchange, we see that the system variance increases as more constraints are imposed as one would expect. Nonetheless, all four options provide a smaller system variance than the original set-up. This illustrates the benefits of diversification.

It is also interesting to observe the uncertainty measure for Agent 3. From the third row of the table, we see that this agent's variance actually increases under the first three exchanges. It is only when we specifically impose the risk improvement condition is this agent at least no worse off.

Table 4.8 provides more details about the optimal coefficients. Each optimization procedure yields nine coefficients, three for each risk. The table shows only those for Risk 3. First note that each column sums to one, as required under the clearing the exchange condition. Also observe the negative coefficient in the `No Profits` column. This corresponds to an optimization problem before we explicitly introduce the short sell condition, indicating the

TABLE 4.7
Risk-Sharing Model Variances.

	Original	Clear Exchange	No Profits	Short Sell	Risk Improvement
Agent 1	10	1.2222	2.6281	2.8881	3.3164
Agent 2	8	1.2222	0.6695	0.3415	0.4148
Agent 3	1	1.2222	1.1359	1.3959	1.0000
Total	19	3.6667	4.4335	4.6256	4.7312

TABLE 4.8
Risk 3 Coefficients.

	Clear Exchange	No Profits	Short Sell	Risk Improvement
To Agent 1	0.3333	0.9286	0.8247	0.7119
To Agent 2	0.3333	-0.2078	0.0000	0.0000
To Agent 3	0.3333	0.2792	0.1753	0.2881

importance of this condition. In fact, from the `Short Sell` column we see that the lower bound is a binding constraint for some coefficients.

4.5.2 Using Utility Functions to Determine Risk Exchanges

Following the work of De Finetti (1940), intellectual development of risk exchanges was subsequently reignited by contributions due to Karl Borch, summarized in Borch (1974). Borch focused on economic modeling using utility theory. To keep this treatment self-contained, a brief overview of utility theory can be found in Section 4.6.3.

In this context, the ith firm or agent has an initial wealth w_i, faces potential losses X_i, and values money using the utility function $U_i(\cdot)$. Before the risk exchange, its expected utility is $\mathrm{E}[U_i(w_i - X_i)]$. As in equation (4.11), a set of risk exchange rules is denoted by g_i, so that after the exchange, one is responsible for the loss $Y_i = g_i(X_1, \ldots, X_n)$. After the risk exchange, the expected utility is $\mathrm{E}[U_i(w_i - Y_i)]$. Recall that a risk exchange is said to be *Pareto optimal* if it is not possible to improve the situation of one company without worsening the situation of at least one other company.

Theorem (Borch). An exchange $\{g_1, \ldots, g_n\}$ is Pareto-optimal if and only if there exist non-negative constants $k_1, \ldots, k_n$ such that, with probability one, $k_i \; U_i'[w_i - g_i(X_1, \ldots, X_n)]$ is the same for all i.

Before discussing the implications of this result, let us first see how it works in a special case.

Example 4.13. Exponential Utility Functions. Suppose that each agent has an exponential utility function of the form $U_i(x) = \alpha_i[1 - \exp(-x/\alpha_i)]$. In this case, the solution of Borch's Theorem is a familiar quota share treaty with agent i's responsibility being $Y_i = c_i \; S + P_i$, where

$$c_i = \frac{\alpha_i}{\alpha_1 + \cdots + \alpha_p} \quad \text{and} \quad P_i = w_i - c_i \sum_{j=1}^{n} \left[w_j + \alpha_j \ln \frac{k_i}{k_j} \right].$$

Thus, each company takes a *proportional* share c_i of the overall risk pool, $S = X_1 + \cdots + X_n$. To equalize the transactions, cash side-payments in the form of P_i are made.

As summarized by Lemaire (1991):

- This kind of exchange is common in reinsurance practice. Each agent will pay a share c_i of each claim, inversely proportional to risk aversion α_i.
- In order to compensate for the fact that the least risk-averse companies will pay greater claim amounts, zero-sum ($\sum_i P_i = 0$) side-payments or "fees" P_i between the participants occur.
- Note that the quota proportions are determined by the risk-aversion parameters only: they are non-negotiable. Hence the bargaining process will involve only the monetary compensations. Stated another way, agents will negotiate about amounts of money, not about abstract bargaining constants k_i.
- It turns out that quota-share treaties with side-payments also occur when all exchange participants use quadratic or logarithmic utilities, and only in those cases. In these cases, both the quotas and the monetary compensations will be subject to negotiation.

Under the Hood. See the *Online Supplement* for the development of the exchange for exponential utility functions.

Example 4.13 helps to underscore some additional features of Borch's theorem from, for example, Lemaire (1991):

First, note that the condition of the theorem does not depend on the loss amount distributions. There is no mention of the shape of the distribution, the number of moments required, or any other restrictions.

Second, in the exponential utility example, we saw that the exchange rules only depend on the sum of losses, $S = X_1 + \cdots + X_n$ although we allowed for more general functions of the form $g_i(X_1, \ldots, X_n)$. This is true more broadly under mild regularity conditions such as, for example, requiring differentiability of g_i.

From Borch (1962), a Pareto-optimal allocation can be achieved by choosing positive constants $k_1, \ldots, k_n$ and trying to maximize $\sum_{i=1}^{n} k_i \ \mathrm{E}[U_i(X_i)]$. We may interpret the positive constants as weighting factors to equalize contributions of each company.

It is important to remark that the whole model relies on two important implicit assumptions:

1. Shared beliefs: All companies use the same joint distribution of claim amounts.
2. Truthfulness: All companies honestly report the utility function that represents their attitude toward risk.

4.5.3 Conditional Mean Risk-Sharing Rule

Another type of risk exchange assumes that all risks go into a central pool. Now denote the pool risk as $S_n = X_1 + \cdots + X_n$, the total amount to be exchanged with conditional mean $g_{i,n}(s) = \mathrm{E}[X_i|S_n = s]$. With the *conditional mean risk-sharing rule*, the ith participant's share of the total risk is $Y_i = g_{i,n}(S_n)$. Denuit and Robert (2021a) refers to this setting as a *self-governing*, or *no insurance company*, model.

Using conditional expectations, it is straightforward to check that this rule satisfies the market clearing and fairness conditions, that is, that $\sum_i X_i = S_n = \sum_i Y_i$ and that $\mathrm{E}(X_i) = \mathrm{E}(Y_i)$. Moreover, from Jensen's inequality for conditional expectations, we have that

$$\mathrm{E}[U_i(w_i - X_i)] \leq \mathrm{E}[U_i(w_i - g_{i,n}(S_n))], \quad i = 1, \ldots, n,$$

for concave utility functions U_i and initial wealth levels w_i. Thus, the conditional mean risk allocation benefits all risk-averse economic agents, thereby improving their risk position. Moreover, under some mild regularity conditions (when $g_{i,n}(s)$ is non-decreasing in s for all i), the conditional mean risk-sharing rule is Pareto-optimal.

As an example of other results, Denuit and Robert (2021a) also shows

$$\mathrm{E}[U_i(w_i - g_{i,n}(S_n))] \leq \mathrm{E}[U_i(w_i - g_{i,n+1}(S_{n+1})], \quad i = 1, \ldots, n.$$

We may interpret this to mean the risk position of the ith participant improves as the exchange pool becomes larger. They also cite asymptotic results about the pool (as it grows larger), such as law of large numbers and central limit theorem type results, that give qualitative information about the behavior of large pools.

Example 4.14. Conditional Mean Risk-Sharing Among Three Agents. This is a continuation of Example 4.12 with the same means and variances for $n = 3$ agents. In Example 4.12, we only needed means and variances but now supplement this by assuming that the first and third agents follow a gamma distribution while the second agent follows a Pareto distribution. Further, the correlations of Example 4.12 are now treated as association parameters for a Gaussian copula to allow for dependencies.

With knowledge of the entire distribution of the losses $\{X_1, X_2, X_3\}$, one can calculate the conditional mean risk-sharing rules via simulation. With simulated observations, I use nonparametric regression techniques as described in, for example, Härdle (1990) (Chapter 3). Specifically:

- For $r = 1, \ldots, R$, generate simulated losses $\{X_1^r, X_2^r, X_3^r\}$ and sum them to get S^r.
- From simulated outcomes $\{S^r\}$, build a nonparametric density estimator $\hat{f}_S(s)$. This appears in the left-hand panel of Figure 4.8.
- For different values of s, calculate the estimator

$$\hat{g}_j(s) = \frac{\sum_{r=1}^{R} X_j^r \times I(s - bw \leq S^r \leq s + bw)}{\sum_{r=1}^{R} I(s - bw \leq S^r \leq s + bw)}, \quad j = 1, 2, 3,$$

  where bw is known as a *bandwidth* for the estimator (we use $bw = 1$ in our example). In this literature, this is sometimes called a *Nadaraya-Watson* estimator. These appear in the middle panel of Figure 4.8. It is also of interest to examine the proportion of the realized risk ($S = s$) taken on by each agent, $\hat{g}_j(s)/s$. These appear in the right-hand panel.

Code. The *Online Supplement* provides Example 4.14 illustrative statistical code.

From Example 4.12, recall that the sum of the means was $20 + 2.5 + 10 = 32.5$; this is consistent with the density plot in the left-hand panel of Figure 4.8. The third risk had a mean of 10 and a variance of 1; so when we examine its risk-sharing behavior as the solid black line, the middle panel shows it is relatively level (about the mean of 10) whereas the right-hand panel shows that its proportion decreases as the sum of (realized) risks, $S = s$, increases. In contrast, the second risk is the more volatile Pareto risk, depicted as the dashed line. This risk's conditional mean proportion increases as s increases. The first

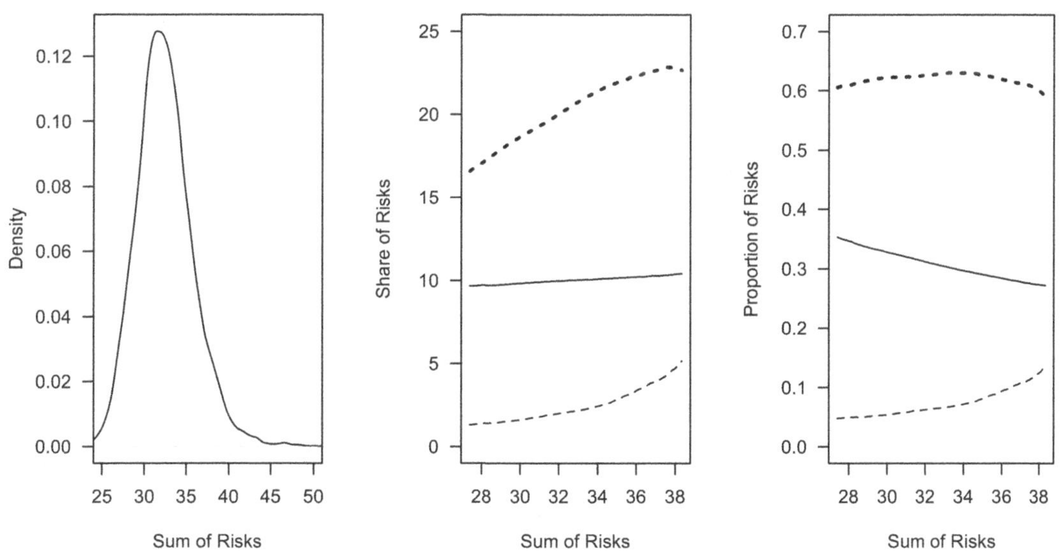

FIGURE 4.8
Sharing Risks Using the Conditional Mean Risk-Sharing Rule. The dotted line corresponds to risk 1, the black solid line is for risk 2, and the dashed line is for risk 3.

TABLE 4.9
Risk Exchange Model Variances.

	Original	Clear Exchange	No Profits	Short Sell	Risk Improve	Conditional Mean
Agent 1	10	1.2222	2.6281	2.8881	3.3164	2.6257
Agent 2	8	1.2222	0.6695	0.3415	0.4148	0.5706
Agent 3	1	1.2222	1.1359	1.3959	1.0000	0.0289
Total	19	3.6667	4.4335	4.6256	4.7312	3.2251

risk, the other gamma distributed random variable, serves as a balancing item for the other two (balancing because the sum of the three risks must add to s). The conditional mean for this risk increases as s increases but decreases when viewed as a proportion of the total risk s.

One can also compare the results from the analysis in Example 4.12 for the Section 4.5.1 proportional risk-sharing rules to those from the conditional mean sharing rules. Table 4.9 provides this comparison, extending the results reported in Table 4.7 to include those from the conditional mean sharing rules. As in Table 4.7, the first column of Table 4.9 gives the original (prior to the exchange) variances for each agent and columns two through five gives optimal variances imposing more conditions on the proportional risk-sharing rules. The right-hand column shows variances of each agent assuming that a conditional mean sharing rule is in place.

Not surprisingly, these compared favorably to other rules considered. Recall that the Section 4.5.1 rules use only information from the first two moments and restrict consideration to proportional sharing rules. In contrast, the conditional mean sharing rule requires knowledge of the entire joint distribution to compute the conditional means.

4.6 Supplemental Materials

4.6.1 Further Resources and Readings

There are many superb treatments of reinsurance in the literature. An outstanding book-long introduction is Albrecher et al. (2017); see also Verlaak and Beirlant (2003).

de Lourdes Centeno and Simões (2009) identify the pioneering work by Bruno de Finetti in 1940, De Finetti (1940), as the first treatment of risk sharing in the academic literature. Especially with the variance as a measure of uncertainty, Beard et al. (1984) summarizes work that influenced generations of actuaries.

The Section 4.4.1 asset allocation problem has already been discussed in Sections 1.3 and 3.4. Additional details on asset allocations can be found in Ingersoll (1987) and Panjer et al. (1998) (Chapter 8).

In addition to the references already cited in Section 4.5.2 on utility based risk exchanges, see Gerber (1979) and Albrecher et al. (2017) for further discussions and references to this fascinating literature. An accessible proof of Borch's Theorem in Section 4.5.2 may be found in Gerber and Pafumi (1998). The Section 4.5 risk exchanges are motivated by pooling of risks that are generally viewed as beneficial for both insurance firms, the suppliers of insurance, as well consumers, who demand insurance. For example, Albrecht and Huggenberger (2017) provide general conditions under which consumers with exchangeable risks will improve their risk positions by pooling (exchangeable risks have the same distribution but may exhibit some positive dependence, a generalization of the idea of "i.i.d." - independent and identically distributed risks). However, as demonstrated in Exercise 4.4, pooling can fail when risks have different distributions, the so-called heterogeneous case.

The conditional mean risk-sharing rule is due to Denuit and Dhaene (2012). The peer-to-peer risk exchange development in Section 4.5.3 follows the work of Denuit and Robert (2021a). Further exploration of this topic appears in Section 6.5.

4.6.2 Exercises

See the *Online Supplement* for selected exercise solutions.

Section 4.1 Exercise

Exercise 4.1. Portfolio Management. Let us consider the same structure as in Exercise 2.5, with risks:

- X_1 - buildings, modeled using a gamma distribution with mean 200 and scale parameter 100.
- X_2 - motor vehicles, modeled using a gamma distribution with mean 400 and scale parameter 200.
- X_3 - directors and executive officers risk, modeled using a Pareto distribution with mean 1000 and scale parameter 1000.
- X_4 - cyber risks, modeled using a Pareto distribution with mean 1000 and scale parameter 2000.

Denote the total risk as $S = X_1 + X_2 + X_3 + X_4$. For simplicity, in this exercise we assume that these risks are independent.

To manage the risk, you seek some reinsurance protection. You wish to manage internally small and intermediate building and motor vehicles amounts, up to u_1 and u_2, respectively. That is, your retained risk is $Y_{retained} = \min(X_1, u_1) + \min(X_2, u_2)$. You seek reinsurance to cover all other risks. Specifically, the reinsurer's portion is

$$Y_{reinsurer} = S - Y_{retained} = (X_1 - u_1)_+ + (X_2 - u_2)_+ + X_3 + X_4.$$

Using upper limits $u_1 = 100$ and $u_2 = 200$, this example will:

a. Determine the expected claim amount of (i) that retained, (ii) that accepted by the reinsurer, and (iii) the total overall amount.

b. Determine the 80th, 90th, 95th, and 99th percentiles for (i) that retained, (ii) that accepted by the reinsurer, and (iii) the total overall amount.

c. Compare the distributions by plotting the densities for (i) that retained, (ii) that accepted by the reinsurer, and (iii) the total overall amount.

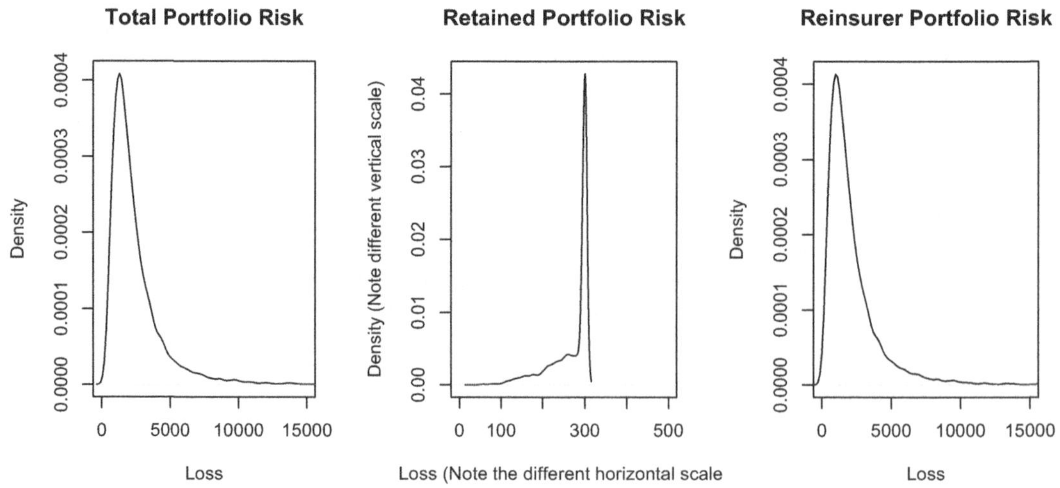

Section 4.3 Exercise

Exercise 4.2. Sensitivity of Optimal Quota Share Proportions. Consider the quota share problem with dependent risks described in Section 4.3.2. How sensitive are these optimal proportions to small changes in $\mathbf{\Sigma}$? One can derive explicit results in the simpler case where we do not impose non-negative constraints on the parameters. The more general problem, with non-negativity constraints, will be taken up in Chapter 10.

Specifically, if we let σ be a generic element of $\mathbf{\Sigma}$, show that

$$\partial_\sigma \mathbf{c} \;= (\partial_\sigma LME)\, \tfrac{1}{2}\, \mathbf{\Sigma}^{-1}\, \mathrm{E}(\mathbf{X}) - \tfrac{LME}{2}\, \mathbf{\Sigma}^{-1}\, (\partial_\sigma \mathbf{\Sigma})\, \mathbf{\Sigma}^{-1}\, \mathrm{E}(\mathbf{X}), \tag{4.13}$$

where

$$\partial_\sigma LME = LME\, \tfrac{\mathrm{E}(\mathbf{X})'\, \mathbf{\Sigma}^{-1} (\partial_\sigma \mathbf{\Sigma}) \mathbf{\Sigma}^{-1}\, \mathrm{E}(\mathbf{X})}{\mathrm{E}(\mathbf{X})'\, \mathbf{\Sigma}^{-1}\, \mathrm{E}(\mathbf{X})}. \tag{4.14}$$

Section 4.4 Exercise

Exercise 4.3. Markowitz Investment Optimal Allocations. Recall the notation from Section 3.4 on the asset allocation problem. In this exercise, we do not require $c_i \geq 0$, so that "short sales" are permitted, yet do require $\sum_i c_i = 1$. With this assumption, show that the optimal allocations can be expressed as

$$\mathbf{c} = \mathbf{c}(LME_1, LME_2) = LME_1\ \mathbf{\Sigma}^{-1}[\mathrm{E}(\mathbf{R})] + LME_2\ \mathbf{\Sigma}^{-1}\mathbf{1}, \tag{4.15}$$

where

$$LME_1 = \frac{A_1 A_3 - A_2^2}{Req_0\ A_3 - A_2} \qquad LME_2 = \frac{A_2^2 - A_1 A_3}{Req_0\ A_2 - A_1} \tag{4.16}$$

and $A_1 = \mathrm{E}(\mathbf{R})'\mathbf{\Sigma}^{-1}\mathrm{E}(\mathbf{R})$, $A_2 = \mathbf{1}'\mathbf{\Sigma}^{-1}\mathrm{E}(\mathbf{R})$, and $A_3 = \mathbf{1}'\mathbf{\Sigma}^{-1}\mathbf{1}$.

Section 4.5 Exercises

Exercise 4.4. Risk Sharing Does not Always work with Heterogeneous Risks. Consider a set of *independent* risks X_j that have the same mean (say μ) with variances σ_j^2 that are possibly different.

a. Suppose that two risks are shared equally so that $Y_1 = (X_1 + X_2)/2 = Y_2$ but that the first risk has a higher variance such that $\sigma_1^2 = 4$ and $\sigma_2^2 = 1$. Show that this risk sharing arrangement improves the risk position for the first risk but not the second. That is, show that

$$\mathrm{Var}(Y_1) \le \mathrm{Var}(X_1) \quad \text{and} \quad \mathrm{Var}(Y_2) \ge \mathrm{Var}(X_2).$$

b. Suppose that n risks are shared equally so that $Y_j = (X_1 + \cdots + X_n)/n$ but that the first risk has a higher variance σ_1^2 and other risks have equal variance, say σ_2^2. Show that the risk position for all other risks improves only if σ_1^2 is not too large relative to the others. Specifically, show that $\mathrm{Var}(Y_j) \le \mathrm{Var}(X_j)$ if $\sigma_1^2 \le (n^2 - n + 1)\sigma_2^2$.

c. For two risks, suppose that the first risk takes $Y_1 = cX_1 + (1-c)X_2$ and the second takes $Y_2 = (X_1 + X_2) - Y_1 = (1-c)X_1 + cX_2$. Show that the variability of each risk must be relatively similar for risk improvement for both risks. Specifically, for $\mathrm{Var}(Y_j) \le \mathrm{Var}(X_j)$, $j = 1, 2$, we require $\sigma_2^2 k_c \le \sigma_1^2 \le \sigma_2^2/k_c$, with the constant $k_c = (1-c)/(1+c)$.

Exercise 4.5. Optimal Linear Exchange Based on a Clearing Exchange Condition. Based on the risk sharing structure described in Section 4.5.1, now use only the exchange clearing condition for this problem. The goal is to determine the set of risk sharing/allocation parameters c_{ij} that

$$\boxed{\begin{array}{ll} \text{minimize}_{\mathbf{C}} & \sum_i \mathrm{Var}(Y_i) \\ \text{subject to} & \mathbf{C}'\mathbf{1} = \mathbf{1}. \end{array}}$$

a. Show that the solution is given as

$$\mathbf{C}_{opt} = \frac{1}{n}\mathbf{1}\mathbf{1}'. \tag{4.17}$$

b. Show that the variance is

$$\mathbf{C}_{opt}\boldsymbol{\Sigma}\mathbf{C}'_{opt} = \frac{k_\Sigma}{n^2}\mathbf{1}\mathbf{1}'.$$

with the constant $k_\Sigma = \mathbf{1}'\boldsymbol{\Sigma}\mathbf{1}$.

Exercise 4.6. Optimal Linear Exchange with Clearing Exchange and No Profits Conditions. Continuing from Exercise 4.5, now add to the exchange clearing the no profits condition. As before, the goal is to find a set of risk sharing/allocation parameters c_{ij} that minimize, for example, the system variance $\sum_{i=1}^n \mathrm{Var}(Y_i)$. In summary,

$$\boxed{\begin{array}{ll} \text{minimize}_{\mathbf{C}} & \sum_i \mathrm{Var}(Y_i) \\ \text{subject to} & \mathbf{C}\boldsymbol{\mu} = \boldsymbol{\mu} \quad \text{and } \mathbf{C}'\mathbf{1} = \mathbf{1}. \end{array}}$$

a. Show that the solution is given as

$$\mathbf{C}_{opt} = \frac{1}{n}\mathbf{1}\mathbf{1}' + \frac{1}{k_\mu}\boldsymbol{\mu}^* \boldsymbol{\mu}'\boldsymbol{\Sigma}^{-1}. \tag{4.18}$$

with the constant $k_\mu = \boldsymbol{\mu}'\boldsymbol{\Sigma}^{-1}\boldsymbol{\mu}$ and the deviations from the means vector $\boldsymbol{\mu}^* = \boldsymbol{\mu} - \bar{\mu}\mathbf{1}$. Note that this vector is zero when the means are constant.

b. Show that the variance is

$$\mathbf{C}_{opt}\mathbf{\Sigma C}'_{opt} = \frac{1}{n^2}k_1\mathbf{11}' + \frac{1}{k_\mu}\bar{\mu}\left[\boldsymbol{\mu}^*\mathbf{1}' + \mathbf{1}\boldsymbol{\mu}^{*\prime}\right] + \frac{1}{k_\mu}\boldsymbol{\mu}^*\boldsymbol{\mu}^{*\prime}. \tag{4.19}$$

where we recall the constant $k_\Sigma = \mathbf{1}'\mathbf{\Sigma}\mathbf{1}$.

Exercise 4.7. Optimal Linear Exchange - Special Case. Based on the risk sharing structure described in Section 4.5.1, now consider a risk exchange where all participants share a proportion of the total risk $\sum_{j=1}^{n} X_j = \mathbf{1}'\mathbf{X}$. That is, the allocation matrix is expressed as $\mathbf{C} = \mathbf{c1}'$ and $\mathbf{c}$ is an $n \times 1$ vector of coefficients. Check that:

a. The implication of clearing the exchange condition is that $\mathbf{c}'\mathbf{1} = 1$, that is, the coefficients sum to one.

b. The implication of the no profits condition is that $\mathbf{c} = \frac{1}{n\bar{\mu}}\boldsymbol{\mu}$, that is, the coefficient vector equals the mean vector, rescaled by the sum of means. Here, the average mean is denoted as $\bar{\mu} = \frac{1}{n}\mathbf{1}'\boldsymbol{\mu}$.

c. The implication of the risk improvement condition is that $k_\Sigma\ c_i^2 \leq \sigma_i^2$ with the constant $k_\Sigma = \mathbf{1}'\mathbf{\Sigma}\mathbf{1}$.

d. Use the parameter values from the Section 4.5.1 example to determine optimal risk exchange variances and coefficients under the

- clear exchange,
- clear exchange and no short sell, as well as the
- clear exchange, no short sell, and risk improvement conditions.

Solutions are provided in Tables 4.10 and 4.11, respectively.

TABLE 4.10
Example 4.5 Risk Sharing Variances.

	Original	Clear Exchange	Short Sell	Risk Improvement
Agent 1	10	1.2222	1.2222	1.3417
Agent 2	8	1.2222	1.2222	1.3417
Agent 3	1	1.2222	1.2222	1.0000
Total	19	3.6667	3.6667	3.6834

TABLE 4.11
Example 4.5 Risk Sharing Coefficients.

	Clear Exchange	Short Sell	Risk Improvement
Agent 1	0.3333	0.3333	0.3492
Agent 2	0.3333	0.3333	0.3492
Agent 3	0.3333	0.3333	0.3015

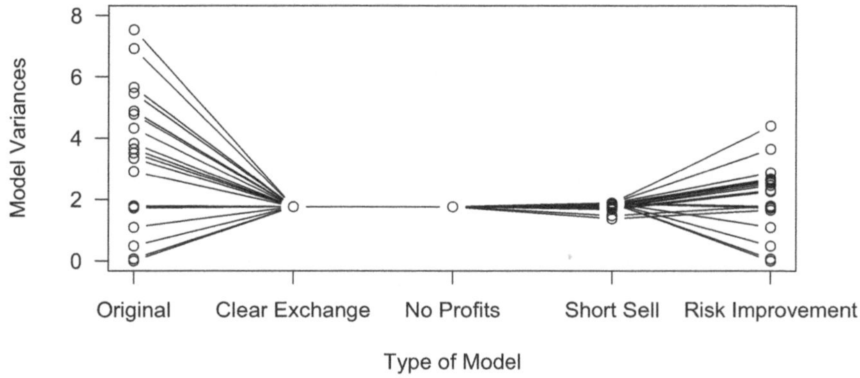

FIGURE 4.9
Risk Sharing Variances under Different Conditions.

Exercise 4.8. Following up on Example 4.14, modify the code to allow for a larger number of risks so that we can demonstrate the quick computational ability of the convex solvers. As the purpose is to demonstrate the utility of numerical convex solvers, use a random number generator to generate generic mean and variance-covariance matrix (although make sure the variance-covariance matrix remains positive definite). Figure 4.9 demonstrates risk sharing variances under different conditions.

Exercise 4.9. Conditional Mean Risk-Sharing Rules. From the notation introduced in Section 4.5.3, we have $w_{i,n} = g_{i,n}(w_n)$ and $B_{i,n} = (w_{i,n} - g_{i,n}(S_n))_+$. Assume that $g_{i,n}(s) = E(X_i|S_n = s)$ is non-decreasing in s. Establish that $(w_n - S_n)_+ = \sum_i B_{i,n}$.

Exercise 4.10. Comparing Proportional to Conditional Mean Risk-Sharing Rules. Following up on Examples 4.12 and 4.14, use the statistical code to compare the proportional sharing rules to the conditional mean sharing rules. Compare these two rules by varying the level of dependence. Also examine the effects of changing the marginal distributions to more short-tail (e.g. exponential) and long-tail (e.g., Pareto). What other assumptions do you think might be important? (e.g., how about the number of risks?)

4.6.3 Appendix. Utility Theory

This is a brief review of utility functions, drawn primarily from Gerber and Pafumi (1998).

To explain certain phenomena, the usefulness of money can be measured on a new scale. Thus, the usefulness of \$x is $U(x)$, the utility (or *moral value*) of \$x. Typically, x is the wealth or a gain of a decision-maker.

Suppose that a utility function $U(x)$ has the following two basic properties:

1. $U(x)$ is an increasing function of x
2. $U(x)$ is a concave function of x.

Usually one assumes that the function $U(x)$ is twice differentiable; then (1) and (2) state that $U'(x) > 0$ and $U''(x) < 0$.

For a given utility function $U(x)$, one can define

$$r(x) = -\frac{U''(x)}{U'(x)} = -\frac{d}{dx} \ln U'(x),$$

known as a *risk aversion function.* Note that properties (1) and (2) imply that $r(x) > 0$. One can check that

$$U(x) = \int_{\xi}^{x} \exp\left(- \int_{\xi}^{z} r(y)dy \right) dz$$

for some ξ, so knowing $U(\cdot)$ and $r(\cdot)$ are equivalent.

Two Applications of Utility Functions

Preference Ordering of Random Gains. Consider a decision-maker with initial wealth w who has the choice between a certain number of random gains. By using a utility function, two random gains can be directly compared: he or she prefers G_1 to G_2, if

$$\mathrm{E}[U(w + G_1)] > \mathrm{E}[U(w + G_2)].$$

Premium Calculation. We consider a company with initial wealth w. The company is to insure a risk and has to pay the total claims S (a random variable) at the end of the period. The expected utility of wealth with the contract should be equal to the utility without the contract:

$$\mathrm{E}[U(w + P - S)] = U(w).$$

This is known as the *principle of equivalent utility.*

Optimality of a Stop-Loss Contract

We consider a company that has to pay the total amount S (a random variable) to its policyholders at the end of the year. We compare two reinsurance agreements:

1. A stop-loss contract with deductible d. Here the reinsurer will pay

$$(S - d)_+ = \begin{cases} 0 & \text{if } S < d \\ S - d & \text{if } S > d. \end{cases}$$

2. A general reinsurance contract, given by a function $h(\cdot)$, where the reinsurer pays $h(S)$ at the end of the year. The only restriction on the function $h(\cdot)$ is that $0 \leq h(x) \leq x$.

We assume that the two contracts are comparable, in the sense that the expected payments of the reinsurer are the same, that is, that $\mathrm{E}[(S - d)_+] = \mathrm{E}[h(S)]$. Furthermore, we make the convenient assumption that the two reinsurance premiums are the same. Then, in terms of utility, the stop-loss contract is preferable:

$$\mathrm{E}[U(w - S + h(S))] \leq \mathrm{E}[U(w - S + (S - d)_+)].$$

In this context, w represents the wealth after receipt of the premiums and payment of the reinsurance premiums. For a proof of this result, see Gerber and Pafumi (1998) or Proposition 3.2 in Section 3.1.

4.6.4 Appendix. Using Quantile-Based Risk Measures to Determine Risk Exchanges

Another approach to sharing risks, taken to be the redistribution of a total risk among multiple participants, utilizes quantile-based risk measures. To describe the problem, we use the approach of Embrechts et al. (2018) who focus on the range value at risk $RVaR$.

Specifically, consider n agents with the ith agent using $RVaR_{\alpha_i,\beta_i}$ to value risks. For notational convenience, suppose that we have aligned the agents so that the last one has the largest β parameter, that is, $\beta_n = \max\{\beta_1, \ldots, \beta_n\}$. Further, assume that $\sum_{i=1}^{n} \alpha_i + \beta_n < 1$ and denote the total portfolio of risks as S.

They prove that an allocation $\{(Y_1, \ldots, Y_n) : \sum_{i=1}^{n} Y_i = S = \sum_{i=1}^{n} X_i\}$ that is chosen to minimize $\sum_{i=1}^{n} RVaR_{\alpha_i,\beta_i}(S)$ is Pareto optimal. Moreover, under mild continuity assumptions , this allocation can be achieved by

$$
\begin{aligned}
Y_i &= S \times I\left(F^{-1}\left[1 - \textstyle\sum_{k=1}^{i} \alpha_k\right] \le S \le F^{-1}\left[1 - \textstyle\sum_{k=1}^{i-1} \alpha_k\right]\right) \quad i = 1, \ldots, n-1 \\
Y_n &= S \times I\left(S \le F^{-1}\left[1 - \textstyle\sum_{k=1}^{n-1} \alpha_k\right]\right),
\end{aligned}
$$

a type of layered stop-loss agreement.

5

Excess of Loss for Two Risks

Chapter Preview. Risk owners want risk transfer agreements that limit the amount of retained risk. This is true for a single risk, where upper limits were introduced in Section 2.2, as well as for several risks, where multivariate excess of loss contracts were introduced in Section 4.1.2. Not only is this feature intuitively appealing, it has also some theoretical basis as seen in Section 3.1. Furthermore, as Chapter 1 noted, upper limit contracts serve as a "basic building block of insurance" (Mildenhall and Major (2022)), making them a fitting starting point for our investigations.

For analysts, upper limits are challenging because of the discrete jumps induced by the minimum operator. For example, even with a single risk we saw in Figure 2.3 how the distribution function changes dramatically above and below an upper limit. These jumps indicate that uncertainty measures might be non-linear and non-smooth (that is, not differentiable). This chapter develops approaches for handling these complications in the context of excess of loss policies.

To focus on fundamental issues, this chapter considers only two risks. Specifically, we now consider two risks X_1, X_2 whose dependence is quantified through a copula C. With u_1, u_2 as the upper limits, the random variable of interest is the retained risk, $S(u_1, u_2) = X_1 \wedge u_1 + X_2 \wedge u_2$. By focusing on only two risks, this chapter leverages deterministic methods to provide solutions for risk retention problems. For problems with more than two risks, beginning in Chapter 7 we will focus on solutions that employ *simulation* techniques.

Even for these more complex problems, the foundations using deterministic methods presented in this chapter can be useful. For example, you might be working with a risk retention problem with ten risks that is far too complex to evaluate using deterministic methods. A common procedure is to combine risks so that only two risk categories remain. These two risks can be evaluated using both deterministic and simulation methods. In this way, the deterministic approach serves as a check on the accuracy of the simulation approach. When both methods agree closely, then the simulation approach can be extended to address important practical problems involving more than two risks.

Because the bivariate excess of loss policy is fundamental to multivariate risk retention problems, this chapter examines it in detail. The beginning Section 5.1 establishes that this is *not* a convex problem meaning that one can only employ numerical approximation methods designed to yield local solutions. Section 5.2 develops the distribution of retained risk and Section 5.3 examines differential changes in the associated risk measures. The visualization in Section 5.4 helps one interpret the optimization problem.

Section 5.5 examines the excess of loss risk retention problem by looking at a special case when the budget constraint is binding. As will be seen, the analysis is very complex even with only two risks. Because of this complexity, it is unlikely that practicing analysts will take an interest in implementing this approach in problems with more than two risks. As another consequence of this complexity, many readers can safely skip this section on their first reading. Nonetheless, this section is important because the dimensionality reduction enjoyed by imposing active constraints may be appealing in some situations. So, by learning

DOI: 10.1201/9781003469568-5

about the potential pitfalls described in this section, analysts will be in a better position to guide consumers on appropriate strategies for minimizing risk uncertainty when subject to budget constraints.

5.1 Lack of Convexity

As we learned in Section 4.4.1, the task of constrained optimization becomes immensely simplified if the problem at hand is convex. With convex optimization problems, solutions are globally optimal, not merely locally. Moreover, there exists a rich suite of algorithms for convex optimization (cf. Boyd and Vandenberghe (2004)) that are reliable and work well on large-dimensional problems.

Our discussion of convexity began in Section 2.4. Recall a function h is convex if a linear (convex) combination of a function evaluated at points is greater than or equal to the function evaluated at the same combination of points. That is,

$$c\, h[\mathbf{u}_a] + (1-c)h[\mathbf{u}_b] \ \geq \ h[c\mathbf{u}_a + (1-c)\mathbf{u}_b] \quad \text{for} \quad 0 \leq c \leq 1.$$

This section demonstrates by means of examples that risk retention problems involving upper limits (and by extension, deductibles) are **not** convex (nor concave). This means, as summarized in Section 4.4, that we will require general numerical methods that do not assume convexity.

Example 5.1. Lack of Convexity of Value at Risk for Excess of Loss. Suppose that X_1 has a gamma distribution with shape parameter 2 and scale parameter 5,000 so that the mean is 10,000. Suppose that X_2 has a Pareto distribution with shape parameter 5 and scale parameter 25,000 so that the mean is 6,250. The two variables are related through a Gaussian copula with parameter $\rho =$ -0.3. We seek to evaluate the value at risk with confidence level $\alpha = 0.90$ for an excess of loss random variable $S(u_1, u_2) = X_1 \wedge u_1 + X_2 \wedge u_2$, denoted as $VaR_{0.9}(u_1, u_2)$. We plot $VaR_{0.9}(u_1, u_2) = VaR_{0.9}(\mathbf{u})$ for various values of $\mathbf{u} = (u_1, u_2)$.

For the left-hand panel of Figure 5.1, define $\mathbf{u}_a = (F_1^{-1}(0.7), F_2^{-1}(0.7))$ and $\mathbf{u}_b = (F_1^{-1}(0.9), F_2^{-1}(0.9))$. Recall the notation that $F_1^{-1}(0.7)$ denotes the 0.7 quantile of the first distribution (that turns out to be 12,196 for this example). This panel shows a plot of $VaR_{0.9}[(1-c)\mathbf{u}_a + c\mathbf{u}_b]$ versus c; the linear interpolation $(1-c)VaR_{0.9}[\mathbf{u}_a] + cVaR_{0.9}[\mathbf{u}_b]$ versus c is superimposed. Because the graph is above the linear interpolation, the VaR function is concave over this region.

For the right-hand panel of Figure 5.1, in the same way define $\mathbf{u}_a = (F_1^{-1}(0.9), F_2^{-1}(0.7))$ and $\mathbf{u}_b = (F_1^{-1}(0.7), F_2^{-1}(0.9))$. Because the graph is below the linear interpolation, the VaR function is convex over this region.

In sum, the value at risk function $VaR_{0.9}(\mathbf{u})$ is neither convex nor concave for all values of $\mathbf{u} = (u_1, u_2)$ meaning that we **cannot** use optimization methods that rely on the convexity of a function.

Code. The *Online Supplement* provides Example 5.1 illustrative statistical code.

Our interest is in the convexity, or lack thereof, of summary measures of a retained risk. It is important to note that we are thinking of these as functions of risk retention parameters, not as functions of potential losses. That is, it is tempting to think of the realized excess of loss random variable, $x_1 \wedge u_1 + x_2 \wedge u_2$, as a function of realized losses x_1 and x_2, as

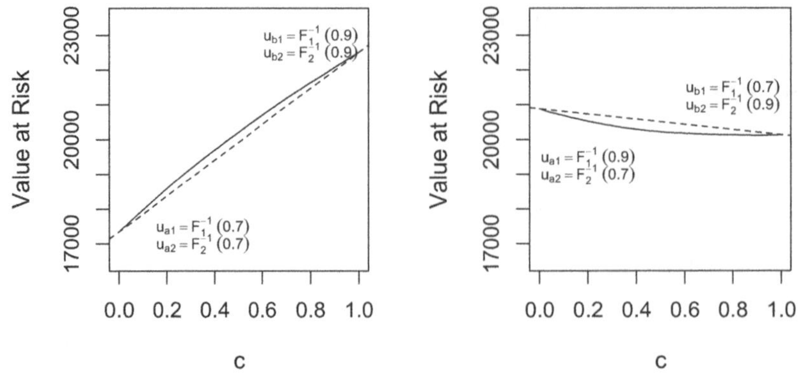

FIGURE 5.1
Lack of Convexity of Value at Risk for Excess of Loss. The left-hand panel shows the value at risk to be concave for selected upper limits. In contrast, for other upper limit values, the right-hand panel shows the function to be convex.

is common in applied statistics. In addition, both Examples 5.1 and 5.2 utilize continuous distributions (gamma and Pareto) for losses, mitigating discreteness issues when confirming convexity in the space of potential losses. However, because the interest is in optimization with respect to the risk retention parameter space, we specifically write the excess of loss random variable as $S(u_1, u_2) = X_1 \wedge u_1 + X_2 \wedge u_2$ to emphasize our interest in retained loss distributions as a function of risk retention parameters.

Example 5.2. Three Measures of Uncertainty. This example shows that each of the three measures of uncertainty (the distribution function, quantile, and ES) is neither convex nor concave for all values of dependence parameters.

To demonstrate this, assume that both X_1 and X_2 have gamma distributions with mean 10,000 and standard deviation 7,071. Their relationship is governed by a normal copula with parameter ρ that is allowed to vary.

We consider two points, $\mathbf{u}_a = (8000, 14000)$ and $\mathbf{u}_b = (16000, 7000)$ and evaluate the distribution function, quantile, and expected shortfall at convex combinations given by $c\mathbf{u}_a + (1 - c)\mathbf{u}_b$. The distribution function is evaluated at 18,000 and the quantile and expected shortfall use $\alpha = 0.75$.

Figure 5.2 shows that the convexity depends on the level of dependence. That is, for each measure, sometimes the function is convex and sometimes concave, depending on the dependence. Further, if you do not know about the dependence, then you do not know whether the function that you are trying to optimize is convex or concave (or neither). As a consequence you may think you are maximizing a function and wind up minimizing it, or vice versa.

Code. The *Online Supplement* provides Example 5.2 illustrative statistical code.

5.2 Excess of Loss Distribution

In the bivariate case, there are two risks X_1, X_2. For $j = 1, 2$, the distribution of X_j is F_j and, when available, the density function is denoted as f_j. The excess of loss retained risk

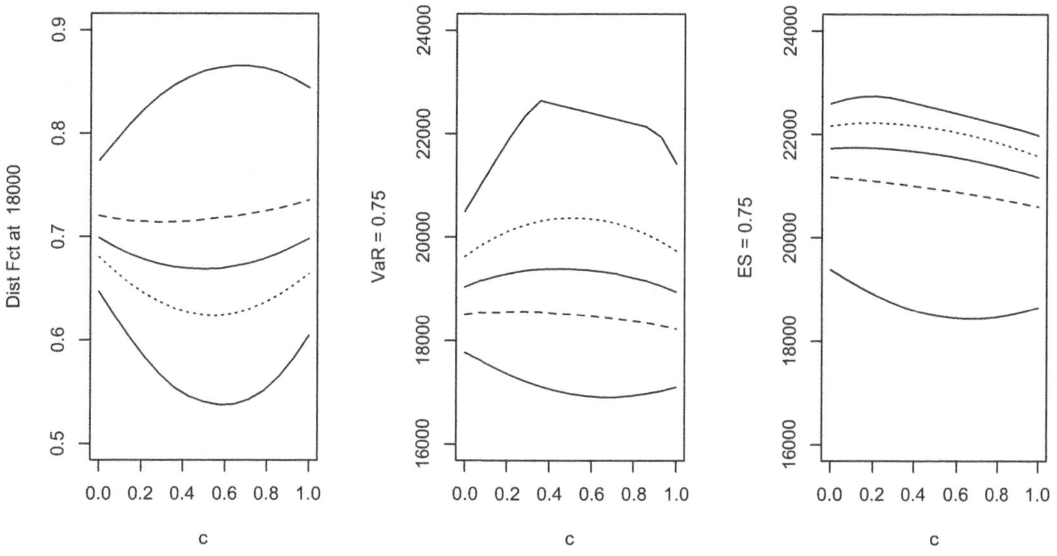

FIGURE 5.2
Measures of Uncertainty Evaluated at Convex Combinations. From the left to right, the panels display the distribution function (at 18,000), the value at risk (at 0.75), and the expected shortfall (at 0.75). The middle solid black line corresponds to the case of independence. For the left-hand panel, the solid top curve is for a positive Spearman correlation 0.9 and the dashed upper curve is for a positive Spearman correlation 0.3. Further, the dotted and solid lower curves are for negative correlations -0.3 and -0.9, respectively. This pattern is reversed for the middle and right-hand panels where the negative associations are on the upper portions and the positive ones are on the lower portions of each figure.

function is $g(X_1, X_2; \boldsymbol{\theta}) = X_1 \wedge u_1 + X_2 \wedge u_2$. For simplicity, this chapter uses the limited sum expression $S(u_1, u_2) = X_1 \wedge u_1 + X_2 \wedge u_2$ for retained risks.

5.2.1 Distribution Function and Quantiles

Distribution Function. In some detail, I develop the distribution function of $S(u_1, u_2)$ that can be evaluated at a generic y. First note that if $y \geq u_1 + u_2$, then $\Pr[S(u_1, u_2) \leq y] = 1$. So, now consider the case where $y < u_1 + u_2$. On this set, one has

$$\begin{aligned}
\Pr[S(u_1, u_2) \leq y] &= \int_0^{F_1(u_1)} \int_0^{F_2(u_2)} I[F_1^{-1}(z_1) + F_2^{-1}(z_2) \leq y] \; c(z_1, z_2) \; dz_2 dz_1 \\
&\quad + F_2(y - u_1) - C[F_1(u_1), F_2(y - u_1)] \\
&\quad + F_1(y - u_2) - C[F_1(y - u_2), F_2(u_2)] \\
&= \int_{-\infty}^{u_1} C_1[F_1(x), F_2(\min(y - x, u_2))] \; f_1(x) dx \\
&\quad + F_2(y - u_1) - C[F_1(u_1), F_2(y - u_1)] \\
&\quad + F_1(y - u_2) - C[F_1(y - u_2), F_2(u_2)].
\end{aligned} \tag{5.1}$$

The partial derivative of the copula distribution function, $C_1(u,v) = \partial_u C(u,v)$, is described in more detail in Appendix Chapter 14. Equation (5.1) is valid for random variables with a domain on the entire real line. Note that if X_1 has a domain on the non-negative portion of the real line and $u_2 > y$, then $F_1(y - u_2) = 0$, and similarly for X_2. This observation can simplify the calculation.

Under the Hood. See the *Online Supplement* for the verification of the excess of loss distribution function.

Example 5.3. Bivariate Excess of Loss Distribution. To show the flexibility of this set-up, we use different distributions for the two risks. Specifically, assume that X_1 has a gamma distribution with shape parameter 2 and scale parameter 2,000 and that X_2 has a Pareto distribution with shape parameter 3 and scale parameter 2,000. Their relationship is governed by a normal copula with parameter $\rho = 0.5$. The code uses $u_1 = 5,000$ and $u_2 = 1,500$.

Illustrative code shows how to determine the distribution function in two ways, one via integration following the formula in equation (5.1) and one via simulation. Figure 5.3 summarizes the excess of loss distribution function. Even though the two random variables are continuous, note the point of discreteness at $y = 6,500$ where the jump in the distribution function occurs.

Code. The *Online Supplement* provides Example 5.3 illustrative statistical code.

Quantiles. The `R` code shows how to determine quantiles (value at risk) for the excess of loss distribution in three ways:

- using the formula in equation (5.1) with the `R` `uniroot` function,
- via simulation, and
- as a constrained optimization approach that will be presented in Section 7.2.

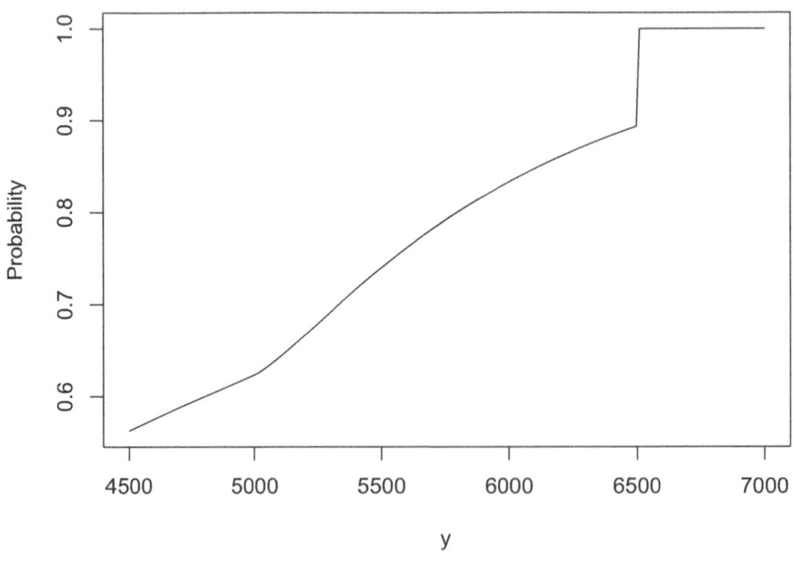

FIGURE 5.3
Excess of Loss Distribution Function. Based on upper limit parameters $u_1 = 5,000$ and $u_2 = 1,500$.

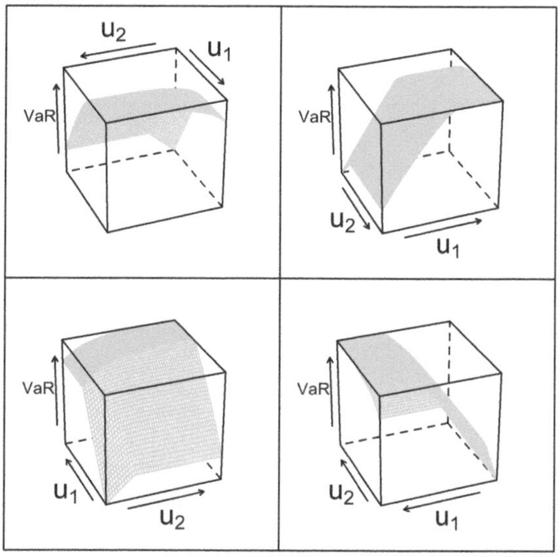

FIGURE 5.4
Value at Risk as a Function of Upper Limits. Four views of a three-dimensional plot of the value at risk VaR as a function of upper limits u_1 and u_2. These different viewpoints show that the VaR is less sensitive to the value of u_2 compared to u_1.

I use the distribution and parameter values from Example 5.3 and $\alpha = 0.85$. It turns out that the quantile is approximately 6,116. Figure 5.4 summarizes the quantile as a function of the two upper limits with a three-dimensional figure, shown from several perspectives.

5.2.2 Partial Derivatives of Distribution Functions

Partial Derivatives of the Distribution Function. I next develop partial derivatives of the distribution function. Based on equation (5.1), one can establish

$$\partial_{u_1} \Pr[S(u_1, u_2) \le y] = -f_2(y - u_1)\{1 - C_2[F_1(u_1), F_2(y - u_1)]\}. \qquad (5.2)$$

ⓘ *Under the Hood.* See the *Online Supplement* for the verification of the excess of loss distribution function partial derivative.

Note that this partial derivative is 0 for $u_1 > y$ and non-negative X_2. Intuitively, for large upper limits u_1, changes in u_1 do not affect the distribution function. Now, in the same way,

$$\partial_{u_2} \Pr[S(u_1, u_2) \le y] = -f_1(y - u_2)\{1 - C_2[F_2(u_2), F_1(y - u_2)]\}. \qquad (5.3)$$

This presentation assumes a copula symmetric in its arguments so that $C(u_1, u_2) = C(u_2, u_1)$. For symmetric copulas, one has

$$C_1(u_1, u_2) = \partial_{u_1} C(u_1, u_2) = \partial_{u_1} C(u_2, u_1) = C_2(u_2, u_1).$$

This assumption serves to ease some of the computational aspects, as follows.

Example 5.4. Partial Derivatives of Bivariate Excess of Loss Distribution Function. This is a continuation of Example 5.3. To illustrate, a block of code is available that

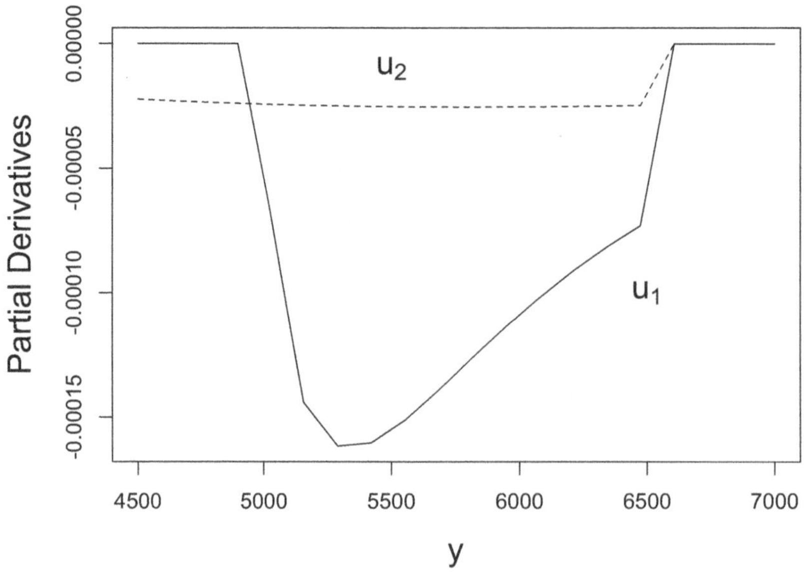

FIGURE 5.5
Partial Derivatives of an Excess of Loss Distribution Function. Based on upper limit parameters $u_1 = 5,000$ and $u_2 = 1,500$. The solid black curve provides the derivative with respect to u_1, the dashed curve is for u_2.

shows how to evaluate the partial derivatives. The resulting values are corroborated using numerical approximations from the R function `grad`. Using the distribution and parameter values from Example 5.3 and $\alpha = 0.85$, Figure 5.5 summarizes the behavior of these partial derivatives.

Code. The *Online Supplement* provides Example 5.4 illustrative statistical code.

Second-Order Partial Derivatives of the Distribution Function. Note that the partial derivatives in equations (5.2) and (5.3) are functions of a single upper limit, so the derivative with respect to both parameters is zero. One can also determine

$$\begin{aligned}\partial^2_{u_1} \Pr[S(u_1,u_2) \le y] &= -\partial_{u_1} \left[f_2(y-u_1)\left\{1 - C_2[F_1(u_1), F_2(y-u_1)]\right\}\right] \\ &= [f_2'(y-u_1)\left\{1 - C_2[F_1(u_1), F_2(y-u_1)]\right\} \\ &\quad + f_2(y-u_1)\{c[F_1(u_1), F_2(y-u_1)] f_1(u_1) \\ &\quad - C_{22}[F_1(u_1), F_2(y-u_1)] f_2(y-u_1)\}] \,.\end{aligned} \tag{5.4}$$

Switch the indices to get an expression for $\frac{\partial^2}{\partial u_2^2} \Pr[S(u_1,u_2) \le y]$.

Example 5.5. Second Partial Derivatives of Bivariate Excess of Loss Distribution Function. This is a continuation of Example 5.4. To illustrate, a block of code is available that shows how to evaluate the partial second derivatives. The resulting values are corroborated using numerical approximations from the R functions `grad` and `hessian`. Using the distribution and parameter values from Example 5.3 and $\alpha = 0.85$, Figure 5.6 shows the behavior of these partial second derivatives.

Code. The *Online Supplement* provides Example 5.5 illustrative statistical code.

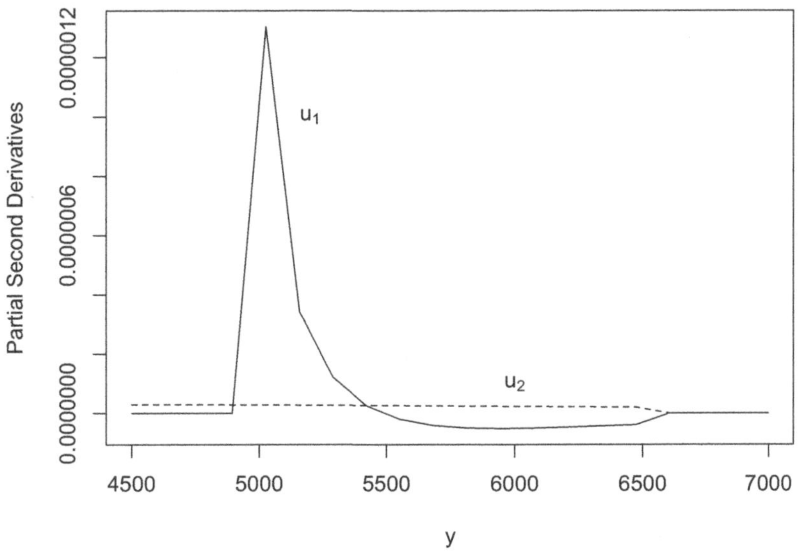

FIGURE 5.6
Partial Second Derivatives of an Excess of Loss Distribution Function. Based on upper limit parameters $u_1 = 5,000$ and $u_2 = 1,500$. The solid black curve provides the derivative with respect to u_1, the dashed curve is for u_2.

5.3 Risk Measure Sensitivities

As described in Section 2.1, the uncertainty of retained risks can be summarized using a risk measure. Further, Section 2.3.2 showed how to compute

$$\partial_\theta \ VaR_\alpha[g(X;\theta)] \quad \text{and} \quad \partial_\theta \ ES_\alpha[g(X;\theta)],$$

that is, the differential change in these risk measures per unit change in the parameter θ. The motivation is that it seems reasonable to want to know how such summary measures change in response to (small) changes in the risk retention parameters. The problems in Section 2.3.2 were rather specialized, with only a single random variable X and parameter θ. By making some mild additional assumptions to ensure smoothness, we are able to handle many additional situations. let us first introduce a general framework, and then specialize it to the two-risk excess of loss case.

5.3.1 Quantile Sensitivities

Specifically, now consider a generic random vector $\mathbf{X}$ and let $g(\mathbf{X};\boldsymbol{\theta})$ be a retention function that depends on outcomes in $\mathbf{X}$ and a vector of parameters $\boldsymbol{\theta}$. Let $\Pr[g(\mathbf{X};\boldsymbol{\theta}) \le y] = F(\boldsymbol{\theta};y)$ be the distribution function of the random variable $g(\mathbf{X};\boldsymbol{\theta})$ with corresponding density function $f(\boldsymbol{\theta};\cdot)$. For fixed $\boldsymbol{\theta}$, denote the inverse function at α to be $VaR_\alpha(\boldsymbol{\theta}) = VaR(\boldsymbol{\theta})$, the value at risk (suppressing the explicit dependence on α).

Assume sufficient local smoothness of $F(\boldsymbol{\theta};y)$ in both arguments y and $\boldsymbol{\theta}$ to permit taking partial derivatives with respect to these arguments. Then, one can use calculus to develop the *quantile sensitivity*,

$$\partial_{\boldsymbol{\theta}} VaR(\boldsymbol{\theta}) \quad = \frac{-1}{F_y[\boldsymbol{\theta};VaR(\boldsymbol{\theta})]} F_{\boldsymbol{\theta}}[\boldsymbol{\theta};VaR(\boldsymbol{\theta})]. \tag{5.5}$$

Here, I use $F_{\boldsymbol{\theta}}[\boldsymbol{\theta}; y] = \partial_{\boldsymbol{\theta}} F[\boldsymbol{\theta}; y]$ for the vector of partial derivatives with respect to $\boldsymbol{\theta}$ and similarly for F_y. Further, the matrix of second derivatives of the value at risk turns out to be

$$\begin{aligned} &\partial_{\boldsymbol{\theta}}\partial_{\boldsymbol{\theta}'} VaR(\boldsymbol{\theta}) \\ &= \frac{-1}{F_y[\boldsymbol{\theta}; VaR(\boldsymbol{\theta})]} \{F_{\boldsymbol{\theta}\boldsymbol{\theta}'}[\boldsymbol{\theta}; VaR(\boldsymbol{\theta})] + F_{yy}[\boldsymbol{\theta}; VaR(\boldsymbol{\theta})] \times \partial_{\boldsymbol{\theta}} VaR(\boldsymbol{\theta}) \partial_{\boldsymbol{\theta}'} VaR(\boldsymbol{\theta}) \\ &\quad + \partial_{\boldsymbol{\theta}} VaR(\boldsymbol{\theta}) \times F_{y\boldsymbol{\theta}'}[\boldsymbol{\theta}; VaR(\boldsymbol{\theta})] + F_{y\boldsymbol{\theta}}[\boldsymbol{\theta}; VaR(\boldsymbol{\theta})] \times \partial_{\boldsymbol{\theta}'} VaR(\boldsymbol{\theta})\}. \end{aligned} \tag{5.6}$$

Under the Hood. See the *Online Supplement* for the verification of the VaR partial derivative expressions.

Special Case. Two Risk Excess of Loss. In this case, the vector of parameters is $\boldsymbol{\theta} = (u_1, u_2)'$ and the retained risk function is the limited sum $S(u_1, u_2) = X_1 \wedge u_1 + X_2 \wedge u_2$. The distribution function, density, density derivative, and quantile are:

$$\begin{aligned} F(\boldsymbol{\theta}; y) &= F(u_1, u_2; y) = \Pr[S(u_1, u_2) \le y] \\ F_y(\boldsymbol{\theta}; y) &= \partial_y \; F(\boldsymbol{\theta}; y) = \partial_y \; F(u_1, u_2; y) \\ F_{yy}(\boldsymbol{\theta}; y) &= \partial_y^2 \; F(\boldsymbol{\theta}; y) = \partial_y^2 \; F(u_1, u_2; y) \\ VaR(\boldsymbol{\theta}) &= F_{\alpha}^{-1}(u_1, u_2). \end{aligned}$$

From equations (5.2) and (5.3), the vector of first derivatives with respect to parameters of the distribution function is

$$\begin{aligned} F_{\boldsymbol{\theta}}(\boldsymbol{\theta}; y) &= F_{u_1,u_2}(u_1, u_2; y) \\ &= \begin{pmatrix} -f_2(y-u_1)\{1 - C_2[F_1(u_1), F_2(y-u_1)]\} \\ -f_1(y-u_2)\{1 - C_2[F_2(u_2), F_1(y-u_2)]\} \end{pmatrix}. \end{aligned}$$

Taking another partial derivative, one has

$$\begin{aligned} F_{\boldsymbol{\theta}y}(\boldsymbol{\theta}; y) &= \partial_y F_{u_1,u_2}(u_1, u_2; y) \\ &= \begin{pmatrix} -[f_2'(y-u_1)\{1 - C_2[F_1(u_1), F_2(y-u_1)]\} \\ -f_2^2(y-u_1) C_{22}[F_1(u_1), F_2(y-u_1)]] \\ -[f_1'(y-u_2)\{1 - C_2[F_2(u_2), F_1(y-u_2)]\} \\ -f_1^2(y-u_2) C_{22}[F_2(u_2), F_1(y-u_2)]] \end{pmatrix}. \end{aligned}$$

The matrix of second derivatives with respect to parameters of the distribution function is

$$F_{\boldsymbol{\theta}\boldsymbol{\theta}'}(\boldsymbol{\theta}; y) = \begin{pmatrix} \partial_{u_1}^2 F(u_1, u_2; y) & 0 \\ 0 & \partial_{u_2}^2 F(u_1, u_2; y) \end{pmatrix},$$

where the diagonal elements are given in equation (5.4).

Code. The *Online Supplement* provides Example 5.6 illustrative statistical code.

Example 5.6. Bivariate Excess of Loss *VaR* Sensitivities. Let us continue to use the distribution and parameter values from Example 5.3 and $\alpha = 0.85$. As anticipated, from

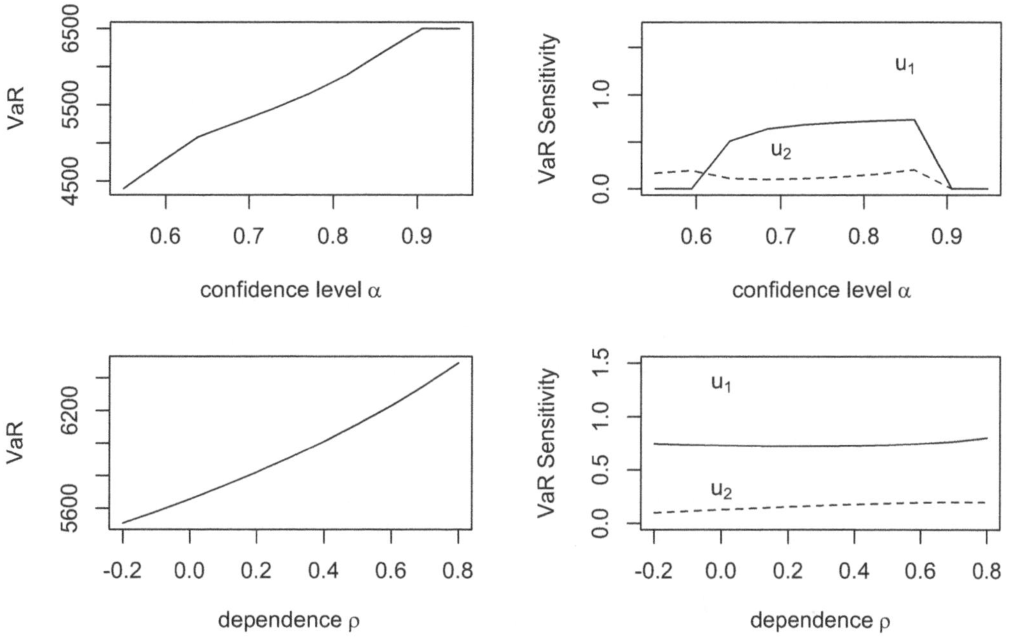

FIGURE 5.7
***VaR* and *VaR* Sensitivities for Excess of Loss.** Based on upper limit parameters $u_1 = 5,000$ and $u_2 = 1,500$. The solid black curve provides the derivative with respect to u_1, the dashed curve is for u_2. The upper panels are based on $\rho = 0.5$ and the bottom panels use $\alpha = 0.85$.

Figure 5.7 we see that the value at risk increases as both the confidence level α increases (top left panel) and as the dependence parameter ρ increases (bottom left panel). The two right-hand panels show that both VaR sensitivities are positive for all values of α and ρ.

5.3.2 Expected Shortfall Sensitivities

The development of the ES sensitivity employs an auxiliary function that will also be useful for us in other contexts. To define this function, consider a generic random variable Y having distribution function F_Y with a unique α quantile, $F^{-1}(\alpha)$. Now, define the function

$$ES1_F(z) = z + \frac{1}{1-\alpha} \left\{ \mathrm{E}[Y] - \mathrm{E}[Y \wedge z] \right\}.$$

From equation (2.3), one sees that the auxiliary function evaluated at the quantile is the expected shortfall, that is, $ES(\alpha) = ES1_F(VaR_\alpha)$.

After some calculations using this auxiliary function, one can show

$$\begin{aligned} \partial_{\boldsymbol{\theta}}\ ES_\alpha[g(\mathbf{X};\theta)] &= \frac{1}{1-\alpha} \left\{ \partial_{\boldsymbol{\theta}} \mathrm{E}[g(X;\boldsymbol{\theta})] - \partial_{\boldsymbol{\theta}} \mathrm{E}[g(X;\boldsymbol{\theta}) \wedge y]\big|_{y=VaR(\boldsymbol{\theta})} \right\} \\ &\quad + \left(1 - \frac{1}{1-\alpha} \{1 - F[\boldsymbol{\theta}; VaR(\boldsymbol{\theta})]\} \right) \times \partial_{\boldsymbol{\theta}} VaR(\boldsymbol{\theta}). \end{aligned} \tag{5.7}$$

Further,

$$\begin{aligned}\partial_{\boldsymbol\theta}\partial_{\boldsymbol\theta'}\ ES_\alpha[g(\mathbf{X};\theta)] &= \frac{1}{1-\alpha}\left\{\partial_{\boldsymbol\theta}\partial_{\boldsymbol\theta'}\mathrm{E}[g(X;\boldsymbol\theta)] - \partial_{\boldsymbol\theta}\partial_{\boldsymbol\theta'}\mathrm{E}[g(X;\boldsymbol\theta)\wedge y]|_{y=VaR(\boldsymbol\theta)}\right\}\\ &+\left(1-\frac{1}{1-\alpha}\{1-F[\boldsymbol\theta;VaR(\boldsymbol\theta)]\}\right)\times\partial_{\boldsymbol\theta}\partial_{\boldsymbol\theta'}VaR(\boldsymbol\theta)\\ &+\frac{1}{(1-\alpha)}F_y[\boldsymbol\theta,VaR(\boldsymbol\theta)]\times\partial_{\boldsymbol\theta}VaR(\boldsymbol\theta)\times\partial_{\boldsymbol\theta'}VaR(\boldsymbol\theta).\end{aligned} \tag{5.8}$$

Note that if $VaR(\boldsymbol\theta)$ is a point of continuity (no discrete jump in the distribution function), then $F[\boldsymbol\theta;VaR(\boldsymbol\theta)]=\alpha$. In this case, equations (5.7) and (5.8) become simpler. Specifically, the second term on the right-hand side of each equation becomes zero.

ⓘ *Under the Hood.* See the *Online Supplement* for the verification of the ES partial derivative expressions.

Special Case. Two Risk Excess of Loss. As before, the vector of parameters is $\boldsymbol\theta = (u_1,u_2)'$ and the retained risk function is $S(u_1,u_2) = X_1\wedge u_1 + X_2\wedge u_2$. In this case, we have

$$\partial_{\boldsymbol\theta}\mathrm{E}[g(X;\boldsymbol\theta)] = \begin{pmatrix}\partial_{u_1}\mathrm{E}[S(u_1,u_2)]\\ \partial_{u_2}\mathrm{E}[S(u_1,u_2)]\end{pmatrix} = \begin{pmatrix}1-F_1(u_1)\\ 1-F_2(u_2)\end{pmatrix}$$

and

$$\partial_{\boldsymbol\theta}\mathrm{E}[g(X;\boldsymbol\theta)\wedge y] = \begin{pmatrix}F_2(y-u_1)-C[F_1(u_1),F_2(y-u_1)]\\ F_1(y-u_2)-C[F_2(u_2),F_1(y-u_2)]\end{pmatrix}. \tag{5.9}$$

To see equation (5.9), we have the following.

ⓘ *Under the Hood.* See the *Online Supplement* for the verification of the ES excess of loss partial derivative expressions.

Taking derivatives again yields

$$\partial_{\boldsymbol\theta}\partial_{\boldsymbol\theta'}\ \mathrm{E}[g(X;\boldsymbol\theta)] = -\begin{pmatrix}f_1(u_1) & 0\\ 0 & f_2(u_2)\end{pmatrix}$$

and

$$\begin{aligned}&\partial_{\boldsymbol\theta}\partial_{\boldsymbol\theta'}\mathrm{E}[g(X;\boldsymbol\theta)\wedge y]\\ &=\begin{pmatrix}f_2(y-u_1)\{1-C_2[F_1(u_1),F_2(y-u_1)]\} & 0\\ 0 & f_1(y-u_2)\{1-C_2[F_2(u_2),F_1(y-u_2)]\}\end{pmatrix}.\end{aligned}$$

Example 5.7. Bivariate Excess of Loss - *ES* Sensitivities. We use the distribution and parameter values from Example 5.3 and $\alpha = 0.85$. These yield interesting results in Figure 5.8.

Looking Forward. Sensitivity analysis will be the main theme of Chapters 9 and 10 so the foundations laid in this section will bear fruit for us later on. From this section's detailed calculation of sensitivities, observe that the key aspect is the smoothness of parameters in potential outcomes of the retained risk and the retention parameters $\boldsymbol\theta$. This smoothness is important and will motivate the introduction of *kernel* methods when we introduce simulation methods in Chapter 7.

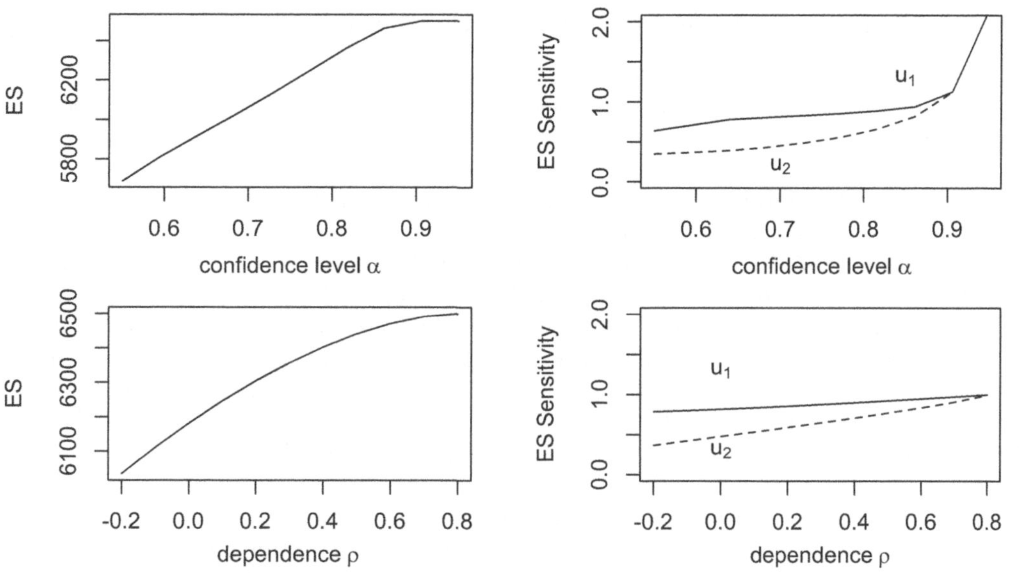

FIGURE 5.8
***ES* and *ES* Sensitivities for Excess of Loss.** Based on upper limit parameters $u_1 = 5,000$ and $u_2 = 1,500$. The solid black curve provides the derivative with respect to u_1, the dashed curve is for u_2. The two right-hand panels show that both ES sensitivities are positive for all values of α and ρ.

5.4 Visualizing Constrained Optimization

Using constrained optimization, the goal is to find those parameter values u_1, u_2 that minimizes an uncertainty measure of retained risk subject to the budget constraint that the risk transfer costs are limited by a given amount, RTC_{max}. With the equation (3.3) structure, one can write this as

$$\boxed{\begin{array}{cc} \text{minimize}_{u_1,u_2} & RM[S(u_1,u_2)] \\ \text{subject to} & RTC(u_1,u_2) \leq RTC_{max} \\ & u_1 \geq 0, \quad u_2 \geq 0. \end{array}} \tag{5.10}$$

This section introduces visualization techniques to help us understand the nuances of constrained optimization risk retention problems. As seen in Figure 5.4, it can be difficult to visualize a risk measure as a function of two upper limits. In addition, we want to get some insights by adding the budget constraint function that also could be represented as a three-dimensional figure (the constraint as a function of two upper limits). Instead, it is common in constrained optimization to utilize *contours*, a curve that is a function of two variables in which the function has a constant value.

5.4.1 Visualizing the Excess of Loss Problem

To explore visualization techniques, we start with ES minimization and then move on to VaR minimization. As will be demonstrated in Section 5.5, VaR minimization can be problematic.

Example 5.8. Visualizing Excess of Loss Constrained Optimization - *ES.* Consider the distributions specified in Example 5.3, using a positive correlation $\rho = 0.5$ and a confidence level $\alpha = 0.85$. Figure 5.9 shows contours of the expected shortfall risk measure ES and the risk transfer cost RTC for several choices of upper limits (u_1, u_2).

Specifically, the left-hand panel of Figure 5.9 shows the contours of the risk transfer cost RTC as solid curves. For this plot, darker shaded values indicate a larger value of RTC. The contours are constant RTC slices as one might take in a three-dimensional graph such as Figure 5.4. For example, if $RTC_{max} = 500$, then the feasibility region $\{(u_1, u_2) : RTC(u_1, u_2) \leq 500\}$ corresponds to the area to the right and above the curve corresponding to $RTC_{max} = 500$.

The right-hand panel retains these contours with constant values of the expected shortfall ES superimposed as dashed curves. For the optimization problem in Display (5.10), the objective is to find the smallest contour subject to being within this feasibility region. At the point of intersection where the smallest contour just meets the budget curve, the two curves are parallel.

To appreciate this visualization of the optimization problem, it is helpful to know where the optimal values lie when visualizing this graph. With $RTC_{max} = 1,500$, when minimizing

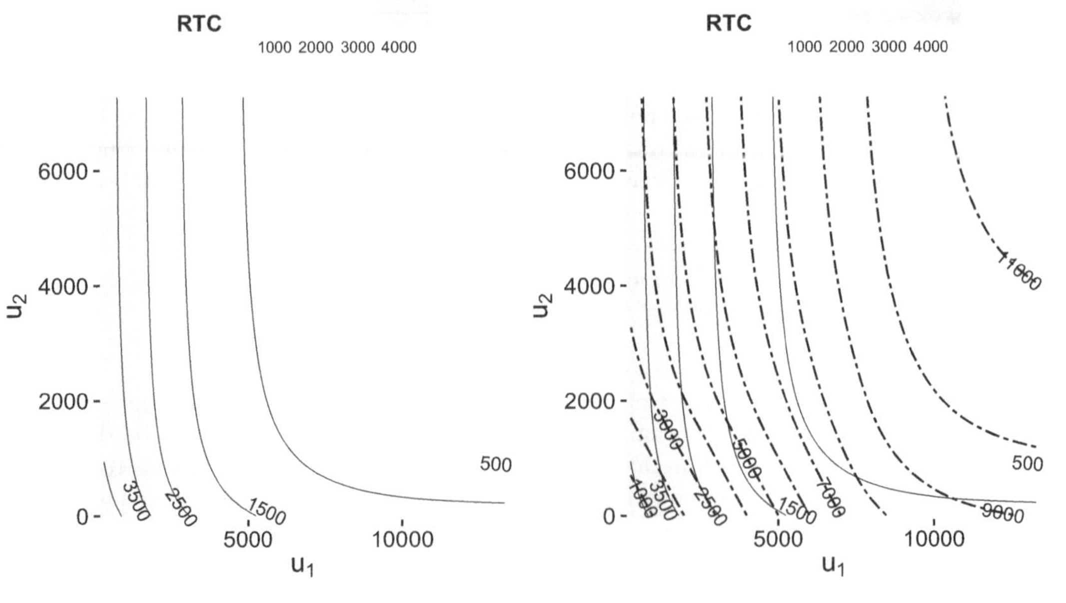

FIGURE 5.9
Expected Shortfall Contour Plot. The left-hand panel shows the contours of the risk transfer cost RTC as solid curves. The right-hand panel retains these contours with values of the expected shortfall ES superimposed as dashed curves.

ES, it turns out that the best values of the retention limits are $u_1^* = 3,379$ and $u_2^* = 3,387$ with expected shortfall $ES^* = 5,997$. This allows one to identify this point in Figure 5.9.

Example 5.9. Visualizing Excess of Loss Constrained Optimization - *VaR*. Continuing from Example 5.8, we now consider the value at risk VaR risk measure. Figure 5.10 shows the RTC as solid curves and the VaR as dashed curves. When the constraint is binding, the objective is to find the smallest contour subject to being at a given level of the risk transfer cost.

As before, it is helpful to know where the optimal values lie. It turns out that the best risk retention values are $u_1^* = 3,116$ and $u_2^* = 43,375$ (essentially infinity), resulting in a value at risk $VaR^* = 4,761$.

Similar calculations can be performed for other values of RTC_{max}. For example, if $RTC_{max} = 2,500$, it turns out that the best risk retention values are $u_1^* = 1,628$ and $u_2^* = 23,117$, resulting in a value at risk $VaR^* = 3,376$. As suggested by the graph, large changes in u_2 mean only small changes in the optimal values of u_1 and the VaR, suggesting that they are not sensitive to u_2.

From Figure 5.10, one sees that for this example we are able to determine the value of u_1 well but determining a value of u_2 is more problematic. When comparing this figure to the ES contour plot in Figure 5.10, it seems like the distinction between ES and RTC contours is more pronounced than the distinction between VaR and RTC contours, suggesting that it is easier to determine the optimal values for the ES problem.

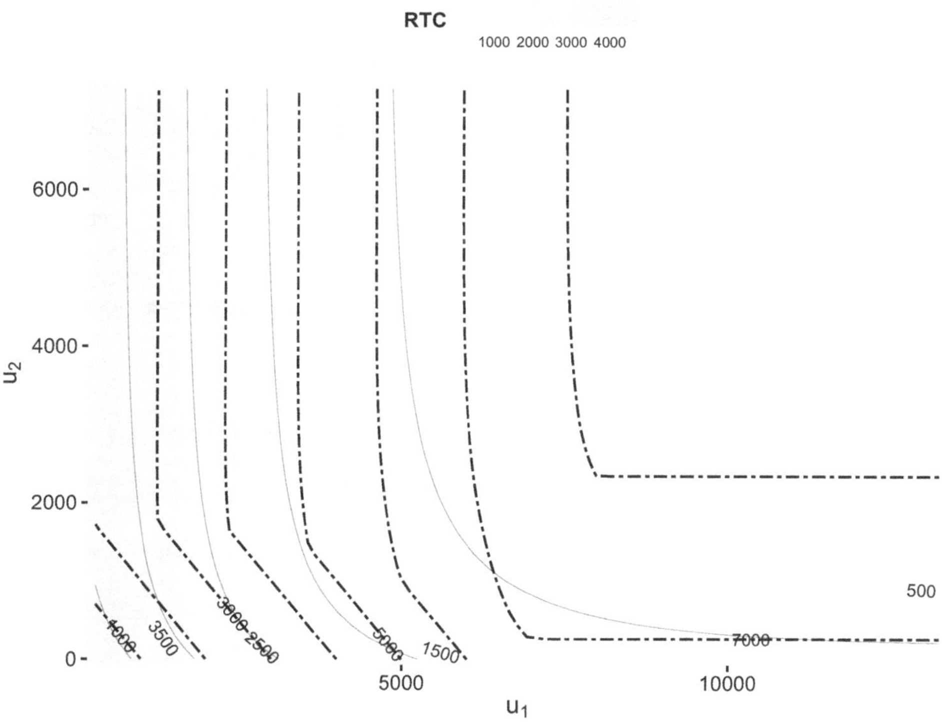

FIGURE 5.10
Value at Risk Contour Plot. This figure shows the contours of the risk transfer cost RTC as solid curves and value at risk VaR superimposed as dashed curves.

5.4.2 Using Distribution Functions

When the risk measure is the value at risk (VaR), an equivalent problem formulation is

$$\boxed{\begin{array}{ll} \text{minimize}_{u_1,u_2,x} & x \\ \text{subject to} & \Pr[S(u_1,u_2) \le x] \ge \alpha \\ & RTC(u_1,u_2) \le RTC_{max} \\ & u_1 \ge 0, \quad u_2 \ge 0. \end{array}} \tag{5.11}$$

Intuitively, finding the smallest x value that minimizes the probability condition is equivalent to the quantile as described in Section 2.1. In this section, we simply provide code that solves each problem and show the equivalence between the two solutions. Section 7.2 will develop this general approach in a broader context, explaining its strengths and limitations compared to the basic approach, such as in Display (5.10).

The minimization problem in Display (5.11) is complicated to visualize because now there are three decision variables, x, u_1, and u_2, over which one seeks to evaluate the probability of the retained risk $\Pr[S(u_1,u_2) \le x]$ and the risk transfer cost RTC. Figure 5.11 shows different aspects of the problem; the top two panels hold the value of u_2 as constant and the bottom two hold u_1 constant. The upper left-hand panel is a contour plot of the probability of retained risk with the solid curves representing the contours (the background

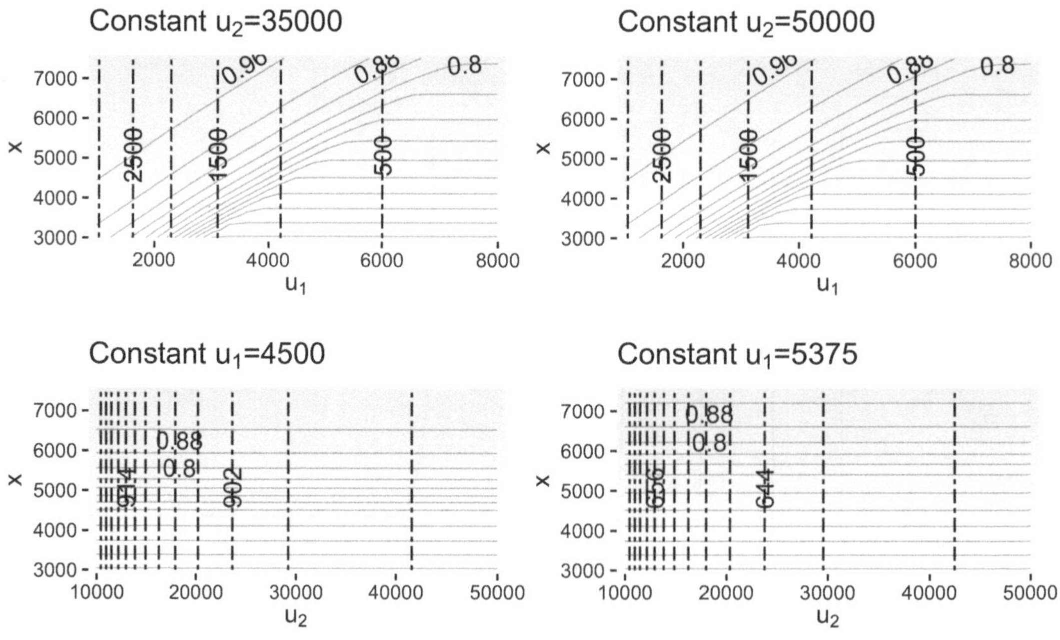

FIGURE 5.11
Distribution Function Contour Plot. All panels show the contours of the probability of retained risk as solid curves and the contours of the risk transfer cost RTC as dashed curves. The top two panels hold u_2 fixed, the bottom two hold u_1 fixed. In part, these plots demonstrate that the optimal solution is robust to the choice of u_2.

shading is for values of x). As before, the dashed curves represent contours of the RTC. For this panel, because we hold $u_2 = 35,000$ fixed, the contour of RTC is simply a constant vertical line. The value of $u_1 = 6,000$ corresponds to a risk transfer cost of $RTC = 500$. When $x = 7,000$, this vertical line is close to the contour corresponding to a probability value of 0.85. As we see below, these values turn out to give near optimal values.

There is little difference when comparing the top two panels of Figure 5.11. This is because the surface is relatively flat over different values of u_2, meaning that changes in this variable have relatively little effect on the problem solution. The bottom two panels of Figure 5.11 each hold u_1 constant. Here, in addition to vertical lines for the contours of RTC, we see near horizontal lines for the contours of $\Pr[S(u_1, u_2) \leq x]$. This is additional evidence that changes in u_2 have relatively little effect on the problem solution.

To help interpret the figures, note that the optimal values of the problem in Display (5.11) are equivalent to those in Display (5.10). Interestingly the code for this alternative formulation (as in Display (5.11)) is much faster than for the code that optimizes directly the VaR risk measure. Intuitively, one can see in Figure 5.11 the sharp distinction among the contour curves. This is in contrast to the earlier Figure 5.9 where the near parallel contours suggested difficulties in the algorithms arriving at optimal solutions.

5.5 Restricting the Feasible Region

As in Display (5.10), we seek to solve the problem

$$\begin{array}{ll} \text{minimize}_{u_1,u_2} & RM[S(u_1,u_2)] \\ \text{subject to} & RTC(u_1,u_2) \leq RTC_{max} \\ & u_1 \geq 0, \quad u_2 \geq 0. \end{array}$$

The feasible region is within the quadrant where values of u_1 and u_2 are non-negative. Moreover, there is an additional restriction in the budget constraint, $RTC(u_1, u_2) \leq RTC_{max}$. As will be seen, for small values of RTC_{max} this restriction can be considerable. By recognizing in advance the restrictions on the feasible region, optimization algorithms will have a smaller region to search over which to optimize, potentially improving their convergence.

As a tactic for reducing the feasible region, one might look to solutions where one selects parameter values so that transfer costs are at the maximum, that is, values of u_1 and u_2 so that $RTC(u_1, u_2) = RTC_{max}$. In addition, analysts are sometimes simply given an equality constraint, "this is what you spend." You will recall that this choice results in an **active**, or *binding*, constraint.

By restricting ourselves to feasible sets where the constraint is binding, we have reduced the number of "free" decision variables by one. This can be especially useful in the case of two risks with upper limits where there are only two decision variables; with a binding constraint, we have reduced the dimension of the problem to a single variable. That is, if a value of u_1 is known, then subject to mild regularity conditions, we can solve for the value of u_2 through the equation $RTC(u_1, u_2) = RTC_{max}$. However, this tactic comes with caveats; it is also well known in the optimization literature (see for example, Nocedal and Wright (2006), Section 15.3) that this approach can introduce some subtle errors into the problem formulation. As summarized by Nocedal and Wright (2006), "Elimination techniques must be used with care, however, as they may alter the problem or introduce ill conditioning."

5.5.1 Budget Restriction

As introduced in Section 3.2, the risk transfer cost RTC is the expense associated with offloading risks. You can think of limiting the RTC by a fixed amount, RTC_{max}, as our budget restriction. As with all constrained optimization problems, imposing a constraint can restrict the feasible set of candidate parameter values.

To develop intuition, we make two simplifying assumptions. First, let us focus on risk transfer costs that are *additive* so that we may write $RTC(u_1, u_2) = RC_1(u_1) + RC_2(u_2)$, where $RC_j(u)$ is the risk transfer cost for the jth risk, $j = 1, 2$. As will be discussed in Section 7.5.1, this is a very natural assumption to make with separate risk transfer agreements.

Second, consider "fair" risk costs of the form $RC_j(u) = \int_u^{\infty}[1 - F_j(x)]dx$ for $j = 1, 2$, as in equation (3.4). It is easy to extend this fair risk cost using multiplicative constants to allow for administrative expenses. Alternatively, risk managers may prefer to use measures such as quantiles for risk costs instead of expectations.

Guidelines for Setting the Maximal Risk Transfer Cost

We first look to bounds on the RTC function to guide setting the maximal cost RTC_{max}. For example, if RTC_{max} is greater than the upper bound of the function RTC then the constraint does not impose any limitations on the minimization problem and one can ignore the constraint. In the same way, if RTC_{max} is less than the lower bound of the RTC function, then there are no feasible parameter values for the optimization problem.

With additive fair risk transfer costs, we have

$$\begin{aligned} RTC[u_1, u_2] &= RC_1(u_1) + RC_2(u_2) \\ &= \int_{u_1}^{\infty} [1 - F_1(x)]\ dx + \int_{u_2}^{\infty} [1 - F_2(x)]\ dx. \end{aligned} \tag{5.12}$$

This RTC is bounded by

$$\begin{aligned} 0 = RTC[\infty, \infty] &\leq RTC[u_1, u_2] \\ &\leq RTC[0, 0] = \mathrm{E}(X_1) + \mathrm{E}(X_2). \end{aligned}$$

Thus, we only consider values of maximal cost RTC_{max} such that $0 < RTC_{max} < \mathrm{E}(X_1) + \mathrm{E}(X_2)$, so that the constraint has some meaning.

In addition, sometimes one wants to know about bounds on the RTC function when there are restrictions on one of the parameters. For example, suppose that $u_2 = 0$, meaning that all of the risk X_2 is transferred (nothing is retained). Then, the bounds on the RTC function over different choices of u_1 are

$$\mathrm{E}(X_2) = RTC[\infty, 0] \leq RTC[u_1, 0] \leq RTC[0, 0] = \mathrm{E}(X_1) + \mathrm{E}(X_2).$$

Thus, when $u_2 = 0$, a selection of $RTC_{max} < \mathrm{E}(X_2)$ means that the feasible set is null so that the problem is not well defined. Table 5.1 summarizes this and additional bounds.

As a consequence, suppose that one sets the maximal transfer cost so that $RTC_{max} < \min\{\mathrm{E}(X_1), \mathrm{E}(X_2)\}$. Then, on the feasible set where $RTC(u_1, u_2) \leq RTC_{max}$, both u_1 and u_2 are positive.

Limitations on Risk Retention Parameters Imposed by Constraints

For the feasible set, we require $u_1 \geq 0, u_2 \geq 0$ and $0 \leq RC_1(u_1) + RC_2(u_2) \leq RTC_{max}$. This means that $RC_j(u_j) \leq RTC_{max}$ that we can also write as $u_j \geq RC_j^{-1}[RTC_{max}]$, for $j = 1, 2$. Here, RC_j^{-1} is the inverse of the cost function RC_j that may be defined as

TABLE 5.1
Bounds on the Fair Risk Transfer Cost Function.

Known Parameter Values	Lower Bound	Upper Bound
$u_1 = 0$	$\mathrm{E}(X_1)$	$\mathrm{E}(X_1) + \mathrm{E}(X_2)$
$u_1 = \infty$	0	$\mathrm{E}(X_2)$
$u_2 = 0$	$\mathrm{E}(X_2)$	$\mathrm{E}(X_1) + \mathrm{E}(X_2)$
$u_2 = \infty$	0	$\mathrm{E}(X_1)$

$RC_j^{-1}(t) = \inf\{u : RC_j(u) \leq t\}$. Thus, if the maximal risk transfer cost is sufficiently small, then we can get some information about the lower bounds for the risk retention parameters u_1, u_2. This is illustrated in the following example.

Example 5.10. Parameter Limits for Gamma and Pareto Distributions. Let X_1 have a gamma distribution with shape parameter 2 and scale parameter 2,000 and let X_2 have a Pareto distribution with shape parameter 3 and scale parameter 2,000. Their relationship is governed by a normal copula with parameter $\rho = 0.5$. For demonstration purposes, we assume a maximal risk transfer cost is 0.30 times the sum of expected losses. With $\mathrm{E}(X_1) = 4{,}000$, $\mathrm{E}(X_2) = 1{,}000$, this means that $RTC_{max} = 1{,}500$.

For this example, it turns out that the lower bound for u_1 is $RC_1^{-1}(1500) = 3{,}113$ and the lower bound for u_2 is 0. Because the maximal risk transfer cost exceeds the expected risk, $RTC_{max} > \mathrm{E}(X_2)$, it seems reasonable that there is no additional information about the lower bound for u_2.

5.5.2 Active Constraint Feasible Region

It is natural to test limiting values so that one approaches the point where a constraint becomes *active* in the sense that $RTC(u_1, u_2) = RTC_{max}$. How does the choice of RTC_{max} affect the set of risk retention parameters that achieve an active constraint? These parameter bounds could also provide useful starting values for the broader constrained optimization problem (where we do not impose an active constraint in advance).

Let us now consider limitations on the feasible set imposed by the active budget constraint $RTC(u_1, u_2) = RTC_{max}$. With $RC_1(u_1) + RC_2(u_2) = RTC_{max}$, we define

$$AT_{1,RC}(u_1) = RC_2^{-1}[RTC_{max} - RC_1(u_1)] = u_2, \tag{5.13}$$

a *trade-off function* among parameter values at the active constraint. The requirement that $0 \leq u_2 \leq \infty$ implies constraints on u_1 in that

$$\max\left[0, RTC_{max} - RC_2(0)\right] \leq RC_1(u_1) \leq RTC_{max}. \tag{5.14}$$

ⓘ *Under the Hood.* See the *Online Supplement* for the verification of the bounds on active constraints.

To see this result, consider the following minimal scenario.

Example 5.11. Fair Transfer Costs, Uniform Distributions. Assume that X_1, X_2 are both uniformly distributed on [0,1]. From equation (5.12), we have $RC_j(u) = (1-u)^2/2$ and so $RC_j^{-1}(y) = 1 - \sqrt{2y}$. With this and equation (5.14), we have

$$\max\left[0, RTC_{max} - \frac{1}{2}\right] \leq \frac{(1-u_1)^2}{2} \leq RTC_{max}.$$

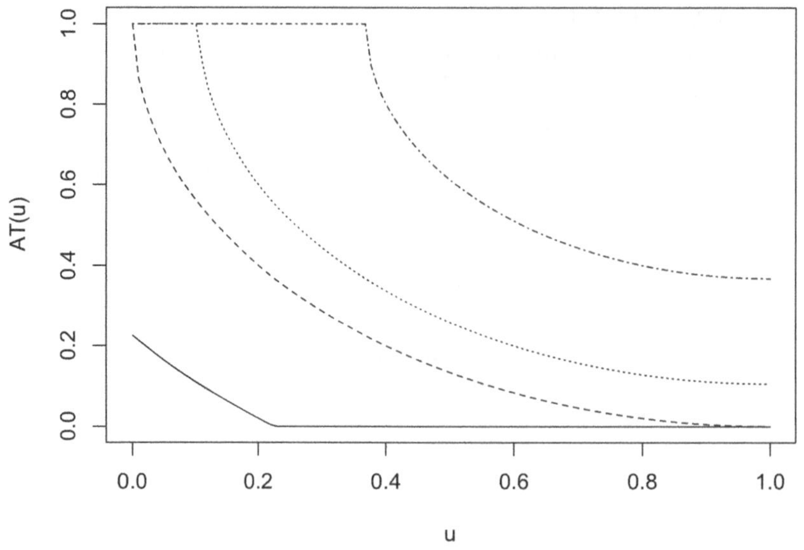

FIGURE 5.12
Plot of Trade-off Function Between Two Upper Limits for an Active Constraint. The black solid is for maximum risk transfer cost RTC_{max} 0.8, the dashed is for 0.5, the dotted for 0.4, and the dot-dash for 0.2.

This is equivalent to

$$\max[0, 1-\sqrt{2RTC_{max}}\ ] \leq u_1 \leq 1-\sqrt{\max[0,(2RTC_{max}-1)]} \quad .$$

Values of u_1 outside of the boundaries may represent legitimate risk parameters but cannot achieve an active constraint.

For valid values of u_1, from equation (5.13), we have

$$AT_{1,RC}(u_1) = 1-\sqrt{2RTC_{max}-(1-u_1)^2}\ ,$$

for the trade-off function. Figure 5.12 shows this trade-off function with these bounds. Note that values of u_1 or u_2 at zero signify total risk transfer and values at one signify total risk retention.

Code. The *Online Supplement* provides Example 5.11 illustrative statistical code.

Limitations on Risk Retention Parameters Imposed by an Active Constraint

For an active constraint, (u_1, u_2) are such that $RC_1(u_1)+RC_2(u_2) = RTC_{max}$ and so there is only one free parameter, bring us back to the discussion in Section 5.5.1 with a known parameter. Formally, given u_1, one can determine u_2 through the active trade-off function in equation (5.13). In the same way, we can define a second active trade-off function

$$AT_{2,RC}(u_2) = RC_1^{-1}[RTC_{max} - RC_2(u_2)] = u_1.$$

that provides an expression to determine u_1 based on knowledge of u_2.

From equation (5.14), we can define two feasible sets:

$$\begin{array}{l} FS_1 = \{u_1 \geq 0 : \max\{0, RTC_{max} - RC_2(0)\} \leq RC_1(u_1) \leq RTC_{max}\} \\ FS_2 = \{u_2 \geq 0 : \max\{0, RTC_{max} - RC_1(0)\} \leq RC_2(u_2) \leq RTC_{max}\} \end{array} \quad .$$

TABLE 5.2
Bounds and Risk Measures.

	Feasible Set 1	**Feasible Set 2**
u_1 Based on Lower u in the FS	3113	5989
u_2 Based on Lower u in the FS	2717751	0
VaR Based on Lower u in the FS	4231	5989
ES Based on Lower u in the FS	5809	5989
u_1 Based on Upper u in the FS	5989	3113
u_2 Based on Upper u in the FS	0	100000000
VaR Based on Upper u in the FS	5989	4231
ES Based on Upper u in the FS	5989	5809

- Assuming that the constraint is binding or active, we can also get an upper bound on each parameter (that may be infinity).
  - The upper bound for u_1 comes from the inequality $\max\{0, RTC_{max} - RC_2(0)\} \leq RC_1(u_1)$ and similarly for u_2.
- Note that these feasible sets depend on the maximal risk transfer cost, RTC_{max}.

Example 5.12. Active Constraint Parameter Limitations. Table 5.2 provides these bounds using the assumptions of Example 5.10. Using equation (5.14) and FS_1, we can determine a lower bound imposed by the active constraint limitation on u_1 as $u_{1,lower} = RC_1^{-1}(RTC_{max}) = 3,113$. With the value of $u_{1,lower}$, we can compute the corresponding value for $u_2 = AT_{1,RC}(u_{1,lower}) = 2,717,751$ (that is essentially infinity), as well as the $VaR = 4,757$ and $ES = 6,674$.

Because $RTC_{max} > \mathrm{E}(X_2) = RC_2(0)$, we also have a finite upper bound on u_1. This is $u_{1,upper} = RC_1^{-1}[RTC_{max} - RC_2(0)] = 5,989$. To summarize, from the table we see that the set of bounds on u_1 for the first feasible set FS_1 is [3113, 5989]. In the same way, the set of bounds on u_2 for the second feasible set FS_2 is [0.00, 100000000], meaning that essentially no bounds are available for this parameter.

5.5.3 Active Constraint Objective Function

By working with feasible sets limited by the active constraint, we can replace the problem summarized in Display (5.10) with

$$\boxed{\begin{array}{ll} \text{minimize}_{u_1} & RM[S(u_1, AT_{1,RC}(u_1))], \\ \text{subject to} & \max\{0, RTC_{max} - RC_2(0)\} \leq RC_1(u_1) \leq RTC_{max} \\ & u_1 \geq 0. \end{array}} \qquad (5.15)$$

This uses the active trade-off function $AT_{1,RC}(u)$ defined in equation (5.13). Although certainly less intuitive, it is only a one-parameter problem. We could also express Display (5.15) in terms of the second trade-off function $AT_{2,RC}$ and feasible set FS_2. Although in principle they have the same solutions, the choice can matter in terms of numerical stability.

This section demonstrates, as suggested by Nocedal and Wright (2006), that the objective function varying over this restricted region can exhibit unanticipated behavior.

To start, we now return to Example 5.11 where the risks have identical distributions. In this case, the two functions are the same. The point of this illustration is to emphasize the importance of the dependence between risks.

Example 5.13. Visualizing the Objective Function, Uniform Distributions. Extending Example 5.11, we now assume the maximal risk transfer cost is $RTC_{max} = 0.2$. With this, the parameter bounds become

$$\begin{aligned} 0.3675 &\approx 1 - \sqrt{0.4} \\ &= \max[0, 1 - \sqrt{2RTC_{max}}\,] \leq u_1 \leq 1 - \sqrt{\max[0, (2RTC_{max} - 1)]} = 1. \end{aligned}$$

From $RC(u) = (1-u)^2/2$, one can easily verify that the inverse function is $RC^{-1}(y) = 1 - \sqrt{2y}$. Thus, with equation (5.13), we have the active trade-off function to be

$$AT_{RC}(u) = RC^{-1}[RTC_{max} - RC(u)] = 1 - \sqrt{0.4 - (1-u)^2}.$$

Figure 5.13 summarizes results; it is a **remarkable figure**. This figure suggests that the optimization problem depends in a critical way on the dependence, reinforcing observations made in Example 5.2.

- The left-hand panel suggests that the VaR does not depend on u_1 at the solid black line corresponding to the case of independence. The dotted curve is for a negative Spearman correlation -0.5, the dashed curve is for a positive correlation 0.5 (I use confidence level $\alpha = 0.7$ for this graph). For negatively dependent risks marked by the dashed curve, an analyst could use a standard optimization technique, such as seeking a point where the gradient is zero, to determine the best value of u to minimize the VaR. However, if one uses the same technique for positive dependent risks remarked by the dotted curve, the "best" value of u

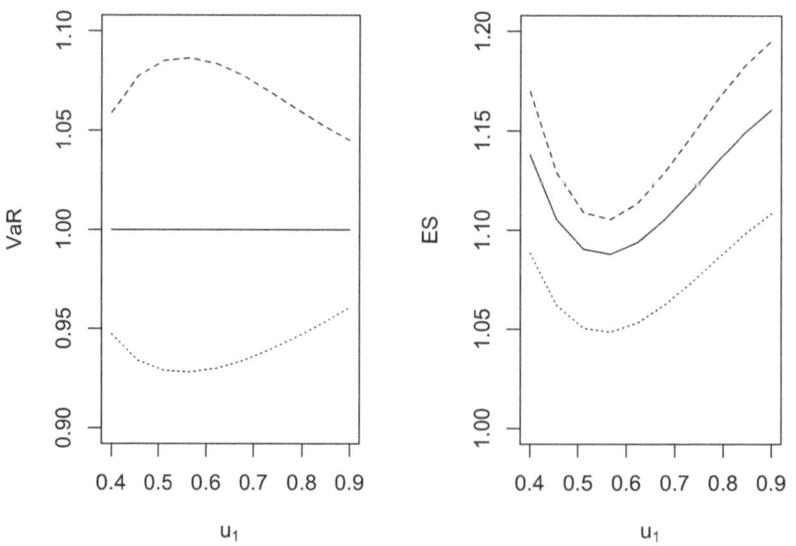

FIGURE 5.13
Plot of Risk Measures at the Active Constraint for Uniform Distributions. The left-hand panel gives the value at risk at 70 percent confidence level, and the right-hand one gives the corresponding ES. For these panels, the solid black line corresponds to the case of a zero Spearman correlation 0.0. The dotted curve is for a negative Spearman correlation -0.5 and the dashed curve is for a strong positive Spearman correlation 0.5.

results in the **maximum** VaR. The solid black line is flat, suggesting that the analyst would find this optimization problem to be challenging.

- The right-hand panel shows the relation between the upper limit and the ES. This figure is more comforting in that, regardless of the dependence, one could use a standard optimization technique to determine its minimal value. Less comforting is the fact that one could get a very different optimal point, depending on whether one uses the VaR or ES as the risk measure.

This example underscores the importance of dependence between risks, a topic that we take up in Chapter 12.

Example 5.14. Visualizing the Objective Function with an Active Constraint. The observations from Example 5.11 are not unique to the uniform distribution. To verify this, we return to the set-up in Example 5.12. Figure 5.14 shows the value at risk and the expected shortfall for u_1 in the interval $[u_{1,lower}, u_{1,upper}]$.

From the left-hand panel of Figure 5.14, we see that $u_1 = u_{1,lower}$ minimizes the value at risk. A numerical optimization routine that searches for a stationary point (with a zero derivative) will have difficulty reaching this conclusion. In contrast, from the right-hand panel, minimization of the expected shortfall can readily be accomplished with numerical optimization routines. The associated code demonstrates how to use the `R` function `optimize`. It verifies that, minimizing VaR, the best value of u_1 is 3,113 with value at risk $VaR^* = 4{,}757$ and expected shortfall $ES^* = 6{,}673$. Further, minimizing ES, the best value of u_1 is 4,257 with value at risk $VaR^* = 5{,}038$ and expected shortfall $ES^* = 5{,}038$.

Code. The *Online Supplement* provides Example 5.14 illustrative statistical code.

This work assumes a fixed value of RTC_{max}. It is reasonable to ask whether the results vary qualitatively if one changes the maximal risk transfer cost. Figure 5.15 shows the VaR and the ES over different levels of RTC_{max}. The left-hand panel tells us that we are likely to encounter problems with the VaR regardless of the value of RTC_{max}. The right-hand panel tells us that we are likely to have little difficulty with the ES regardless of the value of RTC_{max}.

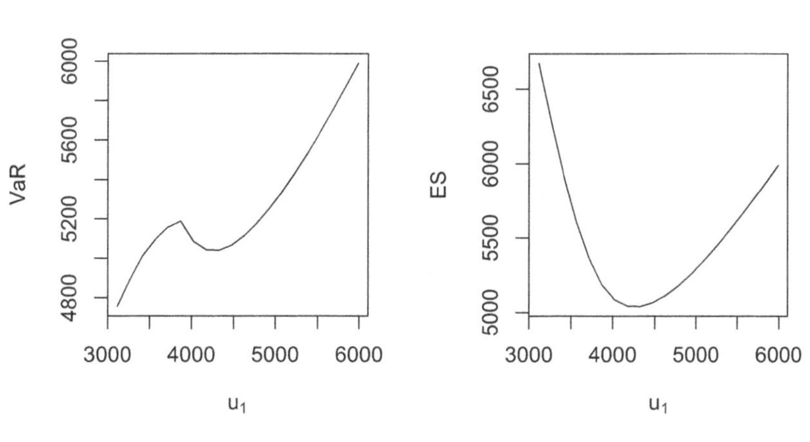

FIGURE 5.14
Plot of *VaR* and *ES* with an Active Constraint. The left-hand panel reports the VaR, the right-hand panel reports the ES.

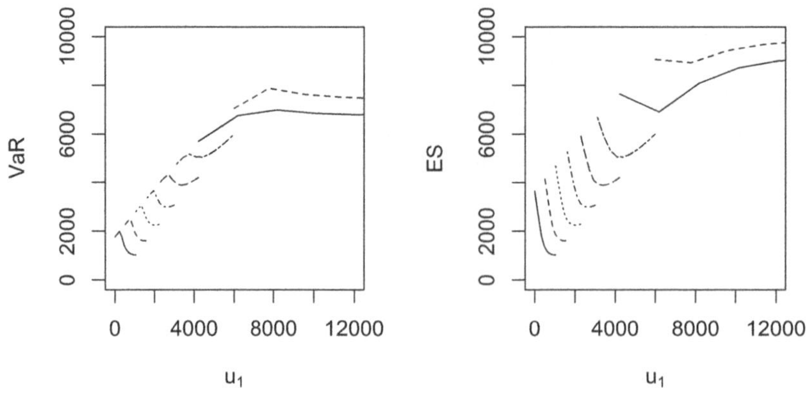

FIGURE 5.15
Plot of *VaR* and *ES* at the Active Constraint with Varying RTC_{max}. The left-hand panel reports the VaR, the right-hand panel reports the ES. Each curve corresponds to a different value of RTC_{max}.

5.6 Supplemental Materials

5.6.1 Further Resources and Readings

The statistics literature has a long history on sensitivity of quantiles to unusual or aberrant data, cf. Serfling (1980). In contrast, the management science literature uses quantile sensitivities, as in Hong (2009), to understand how a quantile changes when a parameter changes. To interpret the quantile sensitivity, Hong (2009) showed how to write it as a conditional expectation. Using notation from Jiang and Fu (2015), who provide less stringent conditions for the relationship to hold, we have

$$\mathrm{E}\left[\partial_\theta \; g(X;\theta)|g(X;\theta) = x\right] = -\frac{\partial_\theta F_{g(\mathbf{X};\theta)}(x)}{\partial_x F_{g(\mathbf{X};\theta)}(x)}.$$

In management science, these results have been utilized as part of the general simulation literature on evaluating derivatives, cf. Fu (2008).

The Section 5.3 risk measure sensitivities were portrayed in terms of risk retention parameters $\boldsymbol{\theta}$ as these are the key variables of interest for portfolio risk retention. As a follow up, Chapter 9 will consider so-called "auxiliary" variables such as the mean of each risk and will provide motivation to calculate sensitivities with respect to these variables. No additional work will be needed for these auxiliary variables; the calculation techniques introduced in this chapter will hold for this new application.

The $ES1$ function was first put forth by Rockafellar and Uryasev (2002).

Difficulties with value at risk optimization are well known in the finance literature. For example, Alexander (2013) argues that the use of VaR minimization in the asset portfolio problem can lead to perverse outcomes and argues for the use of expected shortfall.

See Lee (2023) for an extensive investigation of risk retention policies utilizing active constraints.

5.6.2 Exercises

See the *Online Supplement* for selected exercise solutions.

Section 5.1 Exercises

Exercise 5.1. Lack of Convexity for Value at Risk. The sample code in Example 5.1 provided an example of the lack of convexity for the excess of loss contract based on the value at risk measure using simulation techniques. Replicate this example using the deterministic methods of evaluating the value at risk summarized in Section 5.2.

Exercise 5.2. Lack of Convexity for Expected Shortfall. The sample code in Example 5.1 provided an example of the lack of convexity for the excess of loss contract based on the value at risk measure using simulation techniques. Continue using simulation techniques but now replicate this example using the expected shortfall instead of value at risk and summarize results as in Figure 5.16.

Section 5.2 Exercises

Exercise 5.3. Special Case. Excess of Loss Distribution Function for Independent Uniformly Distributed Losses. Let us get some familiarity with the results in Section 5.2 by assuming that both risks have uniform (on (0,1)) distributions and are independent. Assume $u_2 \leq u_1$.

a. Show that

$$\Pr[S(u_1,u_2) \leq y] = \begin{cases} y^2/2 \times I(y \leq 1) + \{1-(2-y)^2/2\} \times I(1 \leq y \leq 2) & \text{if } y < u_2 < u_1 \\ u_2^2/2 + y - u_2 & \text{if } u_2 < y < u_1 \\ u_1u_2 - \frac{1}{2}(u_1+u_2-y)^2 & \text{if } u_2 < u_1 < y \\ \quad + (y-u_1)(1-u_1) + (y-u_2)(1-u_2). & \end{cases}$$

b. For $u_1 = 0.8$, $u_2 = 0.4$, provide code to display the function as in Figure 5.17.

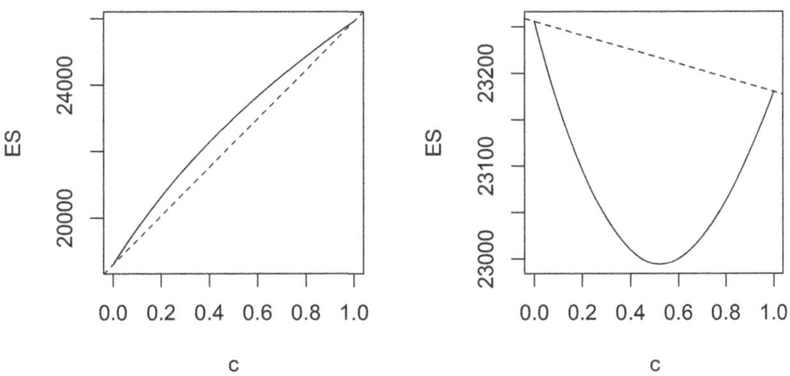

FIGURE 5.16
Exercise 5.2. Excess of Loss Expected Shortfall.

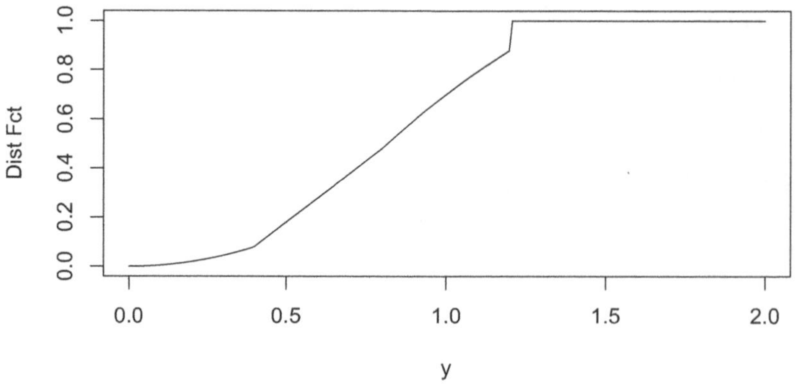

FIGURE 5.17
Exercise 5.3. Excess of Loss Distribution Function.

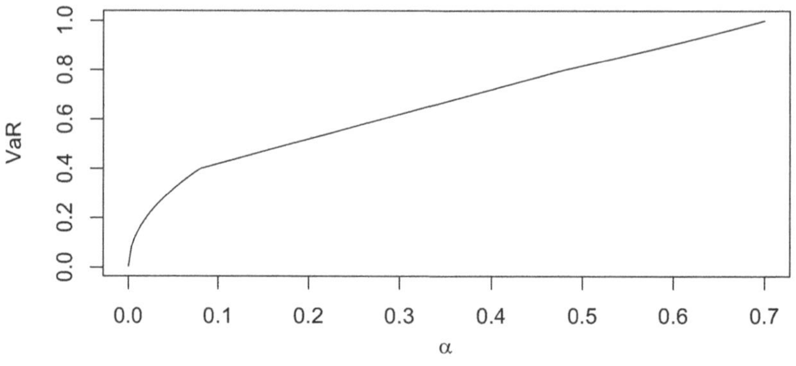

FIGURE 5.18
Exercise 5.4. Excess of Loss Value at Risk versus Level of Confidence (α).

Exercise 5.4. Special Case. Excess of Loss Value at Risk for Independent Uniformly Distributed Losses. Use the assumptions of Exercise 5.3, provide code to replicate the graph of the value at risk versus the level of confidence α in Figure 5.18.

Section 5.3 Exercises

Exercise 5.5. Risk Sensitivity for the Sum of Two Risks. This is a continuation of Exercises 2.3, 2.4, and 2.6. The risk sensitivities presented in equations (5.5) and (5.7) appear to be mathematically challenging so it might be helpful to provide some numerical context. To this end, consider the sum of two risks and the baseline retained risk $g(S) = g(S; d = 100, c = 0.8, u = 30000)$ with confidence level $\alpha = 0.95$.

a. Determine the quantile sensitivity that appears in equation (5.5). Because of the complexity of the quantile, use the `R` function grad() from the package `numDeriv` to compute numerical approximations of the derivatives.

b. Determine the ES sensitivity that appears in equation (5.7). *Hint*: For the partial derivative of the limited expectations, compute $\mathrm{E}[g(\mathbf{X}; \boldsymbol{\theta}) \wedge z]$ as a function of argument z and parameters $\boldsymbol{\theta}$ and then calculate a numerical derivative.

For the limited expected value in the ES sensitivity with equation (5.7), use the form

$$\mathrm{E}[g(\mathbf{X};\boldsymbol{\theta}) \wedge z] = \int_0^z \left[1 - F_{g(\mathbf{X};\boldsymbol{\theta})}(y)\right] dy.$$

Exercise 5.6. Single Variable Upper Limit - VaR. Consider a single random variable X with an upper limit so that $\theta = u$ and $g(X;\theta) = X \wedge u$. Show that the results from Table 2.2 are *not* consistent with those from equation (5.5). Infer from this that smoothness conditions on the retention parameter $\theta = u$ are not satisfied.

Exercise 5.7. Single Variable Upper Limit - ES. Consider a single random variable X with an upper limit so that $\theta = u$ and $g(X;\theta) = X \wedge u$. Although this set-up is similar to Exercise 5.6, show that one can use equation (5.7) to determine

$$\partial_u ES_\alpha(X \wedge u) = \begin{cases} 1 & \text{if } u \leq F_\alpha^{-1} \\ \frac{1-F(u)}{1-\alpha} & \text{if } u > F_\alpha^{-1}. \end{cases}$$

This corroborates the results from Table 2.2.

Exercise 5.8. Special Case. Excess of Loss Value at Risk for Independent Uniformly Distributed Losses. Use the assumptions of Exercise 5.3, determine the first and second derivatives of the value at risk with respect to the first retention parameter, u_1. Describe what this says about the convexity of the value at risk function.

Section 5.5 Exercises

Exercise 5.9. Constrained Optimization with an Equality Constraint. Consider the trade-off function, defined in equation (5.13).

a. Show that the derivative of the trade-off function can be expressed as $-RC_1'(u)/RC_2'[AT_{RC}(u)]$.

b. In the case of fair transfer costs, show that the derivative is $-\{1 - F_1(u)\}/\{1 - F_2[AT_{RC}(u)]\}$.

Exercise 5.10. Derivatives of the Risk Measure at the Equality Constraint. To ease the notation a bit, define $u_2^* = AT_{RC}(u)$ for the second upper limit (that is a function of the first upper limit u). Show that

$$\begin{aligned} \partial_u VaR[u, u_2^*] = & \frac{1}{F_y[u, u_2^*, y]} \{f_2(y-u)\{1 - C_2[F_1(u), F_2(y-u)]\} \\ & + f_1(y-u_2^*)\{1 - C_2[F_2(u_2^*), F_1(y-u_2^*)]\} AT_{1RC}(u)\}, \end{aligned} \tag{5.16}$$

at $y = VaR[u, u_2^*]$.

Exercise 5.11. Special Case. Identical Distributions. Consider the assumptions of Exercise 5.10 and further suppose that we have identical distributions so that $F_1 = F_2 = F$ and similarly for the densities. Give a simpler expression for equation (5.16) and argue that $u = RC^{-1}(RTC_{max}/2)$ is a stationary point for the VaR. Note this result does not depend on the dependence and relate this to Figure 5.13.

Exercise 5.12. Special Case. Independent Distributions. Consider the assumptions of Exercise 5.10 and now assume independent risks. Provide sufficient mild conditions so that a stationary point of the VaR can be found at u such that $f_2(y-u) = f_1(y-u_2^*)$, where $y = VaR[u, u_2^*]$.

6

Managing Portfolios of Insurable Risks

Chapter Preview. This book is designed to help technical analysts support the risk management function. As suggested in Chapter 1, the risk management function is broad and goes well beyond the scope of this book. Risk managers must first identify potential risks, think about how to mitigate them, and decide whether they are insurable. This book focuses on just one piece of this - the insurable portion. Nonetheless, it is helpful for analysts to become familiar with the broad scope of the risk function, and that is the purpose of this chapter.

To describe the broad scope of the risk function, Section 6.1 begins with fundamentals of risk control and transfer. Next, Section 6.2 summarizes methods available to the manager for transferring risk. These include methods for full retention, for full transfer, and approaches that are in between full retention and transfer. Section 6.3 delineates the types of coverages available from insurers. As examples, Sections 6.3.2 and 6.3.3 showcase studies that demonstrate how these different methods can be implemented. Section 6.4 looks to the future by exploring technology-driven tools being used in the insurance sector known as *InsurTech* as well as another tool, *robo-advising*, that is currently used in the financial sector and may someday migrate to the insurance sector.

As an alternative to traditional risk transfer approaches, Section 6.5 summarizes methods for decentralized insurance that includes a more detailed discussion of risk-sharing rules in Section 6.5.2. This builds upon Section 4.5 where we introduced general theoretical structures for exchanging risks. In contrast, this chapter delves into practical risk management methods used by various organizations. Although much of the recent literature on decentralized insurance and peer-to-peer exchanges is more geared for individuals, the main focus of this book is on risks faced by general commercial firms (not just insurers).

6.1 Risk Control and Transfer Fundamentals

Before deciding on a strategy for handling one or a set of risks, a risk manager's first task is risk identification, assessing which risks are worthy of attention. Risk identification is industry and firm specific. The risks faced by a car wash franchise in Madison, Wisconsin are very different from those faced by an actuarial consulting firm in Canberra, Australia. As a consequence, this text does not further discuss the suite of potential risks that a risk manager must first identify except to acknowledge that this is an important step in the process. Readers may look to any of a number of excellent resources that introduce the risk management process, some of which are referenced in Section 6.6.

Having identified a set of risks, we know from Chapter 1 that some are insurable, some partially insurable, and some uninsurable. Stepping back a bit, this section considers risks in two parts, first in terms of risk control activities and then in terms of retention and risk transfer.

DOI: 10.1201/9781003469568-6

6.1.1 Risk Control Activities

Managers engage in risk control activities to **mitigate,** or lessen, the potential impact of risks. These activities serve to shape the risk distribution and therefore impact the decisions for financial risk transfers such as insurance. Risk managers have at their disposal a number of ways in which they can manage, and somewhat *control*, risks facing a person or firm. These include:

- risk avoidance,
- loss prevention and/or reduction, and
- risk control transfer.

Risk avoidance. The first decision is whether or not to take on a risk. A company considers risky projects in the face of competing opportunities: its capital position, management strategy, regulatory constraints, and many other aspects (such as administrative costs, acquisition expenses, runoff expenses, taxes, asset management fees, and profit). While risk avoidance is typically proactive, it may also occur after a loss, e.g., abandonment. Although the criteria for taking on (or abandoning) a risk are important, due to its industry-specific nature we will not dwell on this decision. In contrast, the emphasis here is on managing risks that a firm already has.

Nonetheless, "risk avoidance" is of particular interest when the firm is an insurance company. In this case, the company may elect to not take on risks through its *underwriting* process. For an insurer, risk control begins with underwriting, the process whereby an insurer decides on whether to accept or reject a risk, and then classifies a prospective policyholder into one of several established groups known as a "rating class," thereby establishing a price range for an agreed upon insurance coverage. The uncertainty associated with this process is that premiums will be inadequate to cover insurance claims, both initially and over the life of the agreement. Additionally, there is a risk that several policies may submit claims simultaneously due to a large-scale calamity. For risk transfers, recall that reinsurers insure insurance companies (discussed in Section 4.1). We also note that the banking industry has this same type of underwriting ability to avoid unacceptable risks. However, because of the specialized nature of this industry, we do not seek to develop banking rules in this book.

Loss prevention or reduction. In general, *loss prevention* activities reduce the frequency of losses. In comparison, *loss reduction* activities are geared to lessen the amount of damage if a loss occurs and so reduce the severity. To be concrete, let us illustrate with some examples of loss prevention activities:

- Dams and water resource management can mitigate problems due to flooding.
- Shoveling, salting, and heated walkways can mitigate problems of icy sidewalks.
- Fire resistant construction of structures are less susceptible to fire.
- Security systems can prevent unauthorized access to data.
- Training programs can reduce the likelihood of injury.

For comparison purposes, here are a few examples of loss reduction activities:

- Installation of a sprinkler system. A fire needs to activate the system, thus not affecting frequency.
- Salvage. Even a loss for which an insured is compensated may be of some worth. For example, a totally damaged car may be sold for scrap. As another example, a damaged but repairable piece of equipment may be sold on a secondary market.

- Subrogration. After paying a claim, the insurer inherits the insured's right to recoup losses from a responsible third party. Any money reclaimed from another responsible party reduces the net loss.

Risk control transfer. This type of risk control causes some entity, other than the firm experiencing the loss, to bear some or all of the risk's financial burden. What is being transferred?

1. The property or activity responsible for the risk may be transferred. As an example, a general contractor concerned about risk costs may transfer activities to a subcontractor (for a fixed price).
2. The liability, but not the property/activity, may be transferred. For example, a retailer may assume responsibility for product damage that occurs after the product leaves the manufacturer (even if the manufacturer would otherwise be responsible). The contracts that implement such transfers are known as "exculpatory".

In a risk control transfer, the "transferee" (the party accepting the risk) excuses the "transferor" (the party transferring the risk) from liability. The transferor's exposure is eliminated.

A risk control transfer shifts the property or activity itself to another entity; it can also eliminate or reduce the transferor's responsibility for losses to the transferee. This is in contrast to a risk *financing* transfer such as insurance, discussed in the next section, that will focus on providing external funds that will pay for losses that occur, not a transfer of liability.

6.1.2 Retention and Risk Transfer

Risk financing methods are used to provide resources to the risk owner that reimburse the cost of a loss. These methods can be categorized into two main types: retention and risk transfer.

- **Retention.** The source of funds for loss reimbursement is the organization, including borrowed funds that the organization must repay.
- **Risk transfer.** The source of funds for loss reimbursement is an external party such as an insurer. This method is also known as *risk financing transfer* to distinguish it from *risk control transfer*.

A company's risk financing program uses a combination of retention and risk transfer, even on a specific risk. Complete transfer is atypical, as is full retention. On the one hand, many exposures to loss are so financially inconsequential that they can be safely retained and paid out of a company's normal cash flow. On the other hand, an organization may not purchase any insurance or have other formal transfer agreements in place but transfer can still occur due to legal limits on liability, through limited liability inherent in the corporate form of the organization, or through bankruptcy law. Virtually every firm uses a combination of retention and risk transfer.

For many organizations, insurance pricing is based on the principle that, over time, they will bear the full cost of losses, except for rare catastrophes. Nonetheless, the decision of retention versus transfer can be affected by tax laws, generally accepted accounting principles, and civil law (that may treat retained and transferred losses differently).

Here are some broad considerations affecting the choice between retention and risk transfer.

1. *Firm Capacity.* One element is the capacity of the organization to bear the loss. Without required capacity, the risk manager may be forced to seek alternatives to retention.
2. *Degree of Control.* The organization's degree of control over the risk affects the attractiveness of retention programs. The greater the control, the more attractive is retention. This is primarily due to moral hazard. Insurance has the potential to weaken incentives to prevent or reduce loss. In general, retention increases the incentives of an organization to establish and maintain loss prevention and mitigation activities.
3. *Expense Loadings.* Other things being equal, a higher risk transfer loading fee (for expenses, cost of capital, and profits), means more retention.
4. *Services.* Insurers can also provide valuable services to organizations, decreasing the attractiveness of retention. Loss adjustment and loss-control services are good examples. For contrast, some organizations believe that many of the services provided by insurers can be done better internally.
5. *Taxes.* In general, tax laws often favor insurers over policyholders, making insurance more attractive.
6. *Other.* There may be legal, economic, and public policy limitations on the transfer of risk.

For timing, both retention and risk transfer may be *contemporaneous*, *prospective*, or *retrospective*.

- Contemporaneous retention is simply paying for losses as they occur. It occurs without any advanced funding and is sometimes known as "pay-as-you-go." For example, the peer-to-peer exchanges introduced in Section 4.5 are commonly pay-as-you-go. Section 6.5 discusses these exchanges in further detail.
- Prospective retention involves accumulating funds in anticipation of future losses. This is similar to insurance in that one pays now for future losses. Specific examples include:
  - Liability or earmarked accounts that can absorb fluctuations in volatility, and
  - Earmarked asset accounts. For example, a risk manager for a municipality could hold a trust fund account to pay for losses and avoid tax increases when losses occur.
- Retrospective retention involves payments to reflect past losses. To illustrate, consider the planned use of borrowing to finance potential losses, with repayments coming from future revenues. Retrospective agreements are common for large commercial firms with an abundance of capital.

6.2 Corporate Risk Financing Methods

Complete transfer and pure retention represent opposite ends of a continuum; between these two ends lie a variety of methods combining transfer and retention.

- The transfer end of continuum contains guaranteed cost insurance, experience-rated plans, and retrospectively rated plans.

- The retention end of the continuum includes self-insurance, captive insurers, and finite or risk insurance plans.
- Methods in between these two ends include consumer proprietaries and cooperatives, self-insurance pools, risk retention groups, and administrative service/stop-loss plans.

6.2.1 Methods Emphasizing Risk Transfer

Insurance represents a classic method of risk transfer. Broadly, insurance is a transfer of uncertain risks for a certain price and can be viewed as a pre-loss financing method for future contingent claims. Within this broad classification, there are several features that affect the degree of risk transfer.

Guaranteed Cost Insurance. Here, the transferee, an insurer, agrees to reimburse the transferor, a policyholder, for losses incurred in exchange for a premium that is fixed at the time of the contract. Guaranteed cost insurance is on the end of the continuum where risk and uncertainty are fully transferred. This approach is widely used in personal insurance and may be the only one available to small and medium-sized organizations.

Experience-Rated Insurance is also known as "prospective experience-rating". For this type of arrangement, a firm's losses are used to update the insurer's understanding of the policyholder's risk profile; this information is used to formulate premiums for the next period. It is not an adjustment to current period premium.

Experience-rated plans are typically available to organizations with some degree of statistical credibility. They are widely used in workers' compensation and general liability insurance. Through experience rating, the policyholder has some incentive to control losses through its risk management efforts. This direct control over a risk that in turn influences the premium makes this qualitatively different than the guaranteed cost approach.

Retrospective Rating is also known as "retrospective experience-rating". Here, the premium is not known in advance but is determined after the policy period and is a function of losses incurred in the period. This resembles a "high-deductible" plan, with the exception that losses less than the deductible are paid by the insurer but then charged back to the policyholder as part of the "premium." To complicate matters, retrospective rating may be used for coverages that take many years to develop (e.g., workers' compensation) and so several adjustments to the "premium" may be made over time, involving different financial statements.

Prospective experience rating is more responsive to an organization's losses than guaranteed cost insurance and retrospective rating is even more so. Retrospective rating is typically available only to large organizations.

6.2.2 Methods Emphasizing Retention

Self-insurance, in which the organization takes responsibility for losses from its risks, represents the most basic form of retention. Some writers modify this definition by imposing an additional constraint that a "self-insured" firm has special earmarked funds accumulated in advance to pay for potential losses.

Earmarking funds for potential losses may be viewed as a disadvantage to a firm; the opportunity cost is the difference between the return on alternative available investments in the firm and the lower return on the liquid self-insurance fund. However, even earmarked funds might not suffice to cover potential losses. The self-insurance fund might prove insufficient to cover potential losses which means that the firm would have to raise additional funds at the time of a major loss. If this difference were relatively large, such losses could have an adverse impact on the market price of a publicly-traded firm's stock. Nevertheless, self-insurance is viewed as a viable option for large firms when the number of exposure units is sufficiently large for the risk manager to predict losses within some reasonable bounds, cf. Brockett et al. (1986).

Captive Insurers. A captive is an insurance company wholly owned and established by one or more non-insurance companies to insure the risks of its owners. Captives are a form of self-insurance whereby the insurer is owned wholly by the insured. The vast majority of Fortune 500 companies have captive subsidiaries, cf. NAIC (2020). Once established the captive operates like any commercial insurance company and is subject to an insurer's regulatory standards including reporting, capital, and reserve requirements.

Several factors motivate captive use, including control over terms and conditions of the coverage being offered, cost efficiency, strategically leveraging (re)insurance market dynamics, and managing retained risk within a complex organization in a structured way, cf. Aon Risk Solutions (2019). Moral hazard and adverse selection present no issues with this arrangement because both insurer and insured are part of the same organization.

Most commonly, captives insure only the risk of its parent, known as a single parent or "pure" captive. In addition to single-parent captives, there are group/association captives, rent-a-captive, risk retention groups, agency captives, branch captives, senior or diversified captive, protected cell captive and producer owned reinsurance companies (PORCs), cf. State of Vermont (2020).

In the U.S., *fronting* is a technique for using a captive insurer to provide coverage when there is a requirement that insurance be provided by an insurer licensed in the state. A licensed insurer, the "front," writes coverage for the parent company but does not retain responsibility, it reinsures the coverage with the captive.

Finite or Risk Insurance Plans. Essentially, these variations resemble banking arrangements in function. In this case, the "bank" is a separate entity that arranges and manages the plan for a self-insured entity. The simplest form of a banking arrangement could be a line of credit that an organization may draw upon in the event of unexpected losses. More elaborate schemes involve issuing debt instruments to fund potential losses. In these plans, the risk is spread across time rather than across exposures.

A common feature of finite insurance plans is the creation of a managed account of money. The firm makes premium payments into the account which becomes the principal source for financing losses. A contract is set, e.g., 3-5 years, and premiums are paid. The finite insurance provider may offer administrative services, may structure some excess insurance to cap losses, and may provide additional guarantees, e.g., interest rate guarantees on the account. Other banking services, such as cash management or line of credit, may also be provided. For the largest organizations, in principle there is little difference between a banking arrangement and a retrospectively rated insurance plan.

The finite risk principle can also be applied to *reinsurance*, cf. Insurance Information Institute (2023a)).

6.2.3 Methods Intermediate between Risk Transfer and Retention

A wide variety of risk financing methods exist between methods emphasizing risk transfer and those emphasizing retention, as summarized below.

Consumer Proprietaries/Cooperatives. A consumer proprietary is a type of stock insurance company and is a private insurance pool where the stock is owned by its policyholders, similar to mutual insurers. For example, X.L. Insurance Co. Ltd and A.C.E. Insurance Co. Ltd have become major providers of large liability policies that cover only losses in excess of a given amount. The consumer cooperative is an older form of insurance. For example, Protection and Indemnity clubs are shipowner mutual insurers providing liability insurance to its members.

On the one hand, we can think of these organizations as providing something similar to a guaranteed cost insurance. On the other hand, due to the typically small number of organizations forming the pool, access to capital is limited and it may be that member organizations are actually retaining more risk than they would with a contract from an external insurer.

Self-Insurance Pools. These are arrangements whereby participating entities mutually agree to provide insurance coverage for each other. For instance, a dozen city governments may agree to finance workers' compensation by pooling member contributions to pay claims and administrative expenses. The Wisconsin Property Fund introduced in Section 1.2.2 and further described in Section 6.3.3 is an example of this type of arrangement.

Occasionally, pools are incorporated as insurance companies. More often, they are formed under an intergovernmental cooperative agreement. By not incorporating as an insurer, they are not subject to insurance regulation and avoid certain insurance taxes. Are government sponsored pools more efficient than the commercial insurance market? As a matter of practice, their functions are indistinguishable from one another. Although there are a few pools in Europe and Australia, most are in the U.S.

Risk Retention Groups. A risk retention group (RRG) is a (limited liability) corporation whose primary purpose is assuming and pooling liability exposures of its members. Examples include:

- Groups offering environmental pollution liability coverage to firms in the electronics industry.
- Educators legal liability protection for colleges and universities.
- General and product liability coverage for livestock and pet food producers.
- RRGs write more medical malpractice coverage than any other line of business.

The U.S. Liability Risk Retention Act of 1986 created new opportunities for the creation of pools. Under this act, organizations are permitted to band together to set up RRGs for retaining liability exposure (other than workers' compensation) and purchasing groups that can purchase liability insurance on a group basis for their members.

An RRG closely resembles a mutual insurer. In the U.S., it only needs to be licensed in one state and so avoids many expenses and delays than if it had to be licensed in each state in which it operates (the norm for insurers in the U.S.). More information about risk retention groups can be found through the National Association of Risk Retention (2023) or the National Association of Insurance Commissioners (2023b).

Purchasing Groups. These groups were also created by the same 1986 act. Like an RRG, a purchasing group must be composed of persons/entities with similar liability exposures by virtue of common premises or operations. Unlike RRGs, a purchasing group does not even have to be a corporation. For example, a trade association could create a purchasing group by passing a resolution.

Administrative Services Only (ASO) / Minimum Premium Plans or Stop-Loss Plans. ASO plans offer administrative services for claims processing. Under an ASO only plan, the plan sponsor is responsible for the financing. Sometimes, the ASO provider is an insurer.

Under a stop-loss plan, also known as minimum premium plan, the insurer provides the claim settlement service with the plan sponsor paying the claims themselves. The insurer guarantees a maximum cost (stop-loss) within a specified timeframe. For example, this maximum cost may be in terms of a monthly health claim cost per employee. This maximum cost is similar to the maximum premium in a retrospective rating plan.

6.3 Commercial Insurance Coverages

What types of insurance are available to cover risks faced by businesses?

We started in Section 1.1 by citing a survey of risk managers that listed potential risks faced by businesses as well as by looking to OECD's summary of the international marketplace for a sense of insurance coverages purchased globally. This section delineates insurance coverages available to protect businesses against insurable risks. I begin with more structured contracts suitable for small and medium-sized firms and then discuss more flexible arrangements desirable for larger firms.

6.3.1 BOPs and Commercial Insurance

So that firms need not purchase separate policies for different risks, it is common for insurers to offer a policy that covers several risks simultaneously. By *bundling* risks together into a single contract, such policies offer a convenient and cost-effective option for businesses that need a set of core coverages. These core coverages typically include general liability insurance, property insurance, and business interruption insurance. In the United States, such contracts are known as *business owners policies* (BOPs). In Europe, it is common to refer to them as a "package policies" or "combined policies" and, in Australia, as "business package policies" or "multi-peril insurance."

In all jurisdictions, these policies bundle several essential coverages for small and medium-sized businesses into a single package. They are generally designed to provide standardized coverage packages meaning that the options for customizing risk retention may be limited. That is, the coverage limits and deductibles may be predetermined by the insurance provider based on the standard package offered. Even though these policies are designed to provide affordable and convenient coverage for a wide range of small businesses, the risk retention options may be less customizable compared to larger, more tailored commercial insurance policies.

A BOP is not limited to a single set of core coverages. It is common to add coverage for risks due to commercial auto, cyber liability, directors and officers, employment practices liability, and so forth, depending on the needs of the business.

Larger firms require more flexibility than afforded by a BOP. *Commercial insurance* typically refers to a broad category of insurance policies designed to protect businesses from various risks. Furthermore, it includes different types of coverage, such as general liability, property, workers' compensation, and more, which can be tailored to the specific needs of a business.

To illustrate the need for flexibility, the following Section 6.3.2 presents the case of a large organization (a university) that has insurance needs comparable to a large commercial organization. For contrast, Section 6.3.3 describes an insurance pool of government entities, the Wisconsin Property Fund. Although not a commercial insurer, the fund functions similarly to one in that it provides standard policies which cover several risks for members of the pool. I feature both entities in this book because data are available that allows us to calibrate optimal insurable portfolios in the same way that a commercial insurer or risk advisor could do for its clients.

6.3.2 Australian National University Case

To give a concrete example of broad nature of the risk management function and commercial insurance coverage for several risks, this section returns to the Australian National University (ANU) case that was introduced in Section 1.2.1. ANU has extensive risk control activities in place focusing on risk avoidance, loss prevention or reduction, and risk control transfer, consistent with the general summary in Section 6.1. These activities are designed to mitigate the financial impact that a risk can have on the organization. Anticipating and reducing the potential impact of risks are key responsibilities of risk managers. You can learn more about ANU's risk philosophy in an overview of ANU Risk Management Policy Statement, ANU (2022b).

Nonetheless, even with the best risk management processes, losses do occur and risk financing methods are needed to provide resources for recovering the cost of a loss. As described in Section 6.2, these methods fall into two broad classes: retention and risk transfer. ANU takes a layered approach to financing risks:

- At the lowest level, individual business units are responsible for handling small adverse outcomes. Business units may include departments, schools and colleges, and administrative units such as financial management.
- For small and intermediate outcomes, ANU has created a *Self-Insurance Reserve (SIR) Pool.*
- For potentially larger adverse outcomes, ANU transfers risks to external insurance companies (for a price).
- A type of risk sitting outside of this layered structure is workers' compensation. This has a separate history and claims experience that is briefly summarized below.

Overview of Insurance Programs

The amount of risk retention and transfer depends upon the risk type. To understand the variety of risk types and the relationship among the different layers, Table 6.1 classifies various risk types. For example, for the building and contents of a property, individual business units are responsible for losses up to 15,000 (Australian dollars or AUD). Losses in excess of this go to the SIR pool. The pool retains losses but is reimbursed by insurers for losses in excess of 5,000,000. As another example, for cyber claims, individual business units have no responsibility and the pool manages all losses. But, losses in excess of 250,000 are reimbursed by insurers.

Table 6.1 summarizes 15 coverages and premiums paid in year 2020-2021. Data in Table 6.1 comes from information provided by Gallagher, a brokerage firm retained by ANU. For confidentiality reasons, Table 6.1 excludes the following coverages:

TABLE 6.1
ANU Insurance Premiums, 2020-2021.

Class of Insurance	SIR Deductible	Insurer	Deductible	Limit	Premium
Property	15,000	London Syndicate and Others	5,000,000	1,000,000,000	23,564,759
General and Products (G & P) Liability	5,000	Newline	100,000	20,000,000	340,000
G & P Umbrella Liability		Liberty	20,000,000	50,000,000	27,500
G & P 1st Excess Liability		QBE	50,000,000	100,000,000	27,500
G & P 2nd Excess Liability		Chubb	100,000,000	150,000,000	17,500
G & P 3rd Excess Liability		CGU	150,000,000	200,000,000	16,000
G & P 4th Excess Liability		Zurich	200,000,000	250,000,000	20,000
Cyber	NA	London	250,000	2,000,000	75,721
Crime	NA	AIG	100,000	20,000,000	100,000
Employment Practices Liability	5,000	AIG	100,000	2,000,000	84,000
Expatriate	NA	Chubb			11,676
Group Personal Accident	NA	Chubb	Various	As per schedule	104,920
Corporate Travel	NA	Chubb	Various	As per schedule	75,000
Professional Indemnity	NA	Newline	100,000	\$20m/\$40m	130,000
Medical Malpractice	NA	Newline	100,000	\$20m/\$40m	
Clinical Trial	NA	Newline	2,500	\$20m	
Statutory Liability	NA	Berkley Insurance Australia (SUA)	\$1,000/\$15,000	1,000,000	8,360
Motor Vehicle	NA	Vero	1,000	As per schedule	84,700
Marine Cargo	NA	Richard Oliver (QBE)	5,000	5,000,000	6,127
Marine Hull	NA	Richard Oliver (QBE)	150	5,000,000	11,552
TOTAL					24,407,255

- Directors & Officers
- Excess Directors & Officers
- Directors & Officers Supplementary Legal Expenses.

The insurance program is important to the overall financial health of ANU. According to the 2020 Annual Report (page 96), total expenditures for ANU in 2020 were 1,315 (in millions). Thus, the 24,407,255 AUD in insurance premiums constitutes approximately 1.86% of total university expenditures.

As is evident from Table 6.1, the property risk is by far the most important, accounting for about 95% of the premium. Notably, the property deductible is currently 5,000,000 AUD, representing a large uninsured risk. The second most important risk type is General and Products Liability, representing 448,500 AUD in premiums or about 2% of the total. The other 13 coverages sum to 692,056 AUD that represents about 3% of the total.

Generally, different risks are associated with different insurers and so have different contracts. However, sometimes multiple risks fall under a single insurer. For example, Chubb takes on the expatriate, group personal accident, corporate travel, and a layer of the G & P Liability. In addition, sometimes multiple insurers get involved in a single risk. This is implicit for property risk and explicit for the general and products liability. For the G & P Liability risk, one can see a layered approach for the coverages.

More detailed descriptions of these coverages are available in Frees and Butt (2022).

Self-Insurance Reserve Pool

Major universities such as ANU allocate budgetary responsibilities by layers, beginning from smaller units such as colleges and departments. In addition, ANU owns or is affiliated with several other organizations such as ANU Enterprise Pty Ltd, ANU Union, and so forth. (ANU (2018) provides a list of "named insureds" organizations that participate.) To avoid each financially responsible unit purchasing insurance according to their own needs, in 1994 ANU organized the so-called "Self-Insurance Reserve" or SIR for short. According to the ANU (2022a), the SIR pool is "for insurance losses that except for the policy excess or coverage limit, would otherwise be covered under the commercial insurance program." Through this mechanism, the university facilitates economies of scale and reduces costs of insurance purchases.

On the one hand, this mechanism affects incentives of units within ANU and so is important. See the ANU (2018) for more information about the structure and operations of the SIR pool. On the other hand, because the SIR pool is owned and operated by ANU, there is no external transfer of risk. In the context of risk financing, this level of detail could be ignored.

ANU Workers' Compensation

Workers' compensation provides cash and medical benefits to workers who are injured or become ill in the course of their employment and provides cash benefits to the survivors of workers killed on the job.

This risk has several features that warrant consideration of this coverage separate from other financial risks. These features include:

- Workers' compensation benefits are typically *paid out over several years.* These benefits help to lower the cost and impact of work-related injury and illness through replacing income, paying medical and rehabilitation expenses, and for permanent impairment payments, all of which can occur over a long period of time. This is in contrast to most lines of general insurance where payments for claims are settled relatively quickly. It is true that claim settlement for some liability coverages can take a long time but these are exceptions to the general rule. You might think of an analogy to life insurances; in life insurance, the death benefit is paid rather quickly after a claim has been filed whereas for life annuity benefits, payments typically occur over the course of many years. In the same way, workers' compensation benefits can occur over many years, depending on the recovery of the injured worker.

- In many countries including Australia, workers' compensation coverage is *mandatory* for businesses. For other lines of general insurance, risk transfer such as insurance is simply good business practice yet is not required. Another exception to this rule is medical malpractice which is also typically mandatory. For mandatory coverages, there are typically additional regulatory scrutiny and administrative requirements compared to coverages that are not required.
- Although coverage is required, in many jurisdictions regulators provide options for self-funding, or *self-insuring,* this risk. Self-insurers often use outside firms known as *third-party administrators* to settle workers' compensation claims, in part because of the administrative requirements imposed by regulatory scrutiny.

Given the context described in Section 6.2.2, it is not surprising that ANU essentially self-insures this risk. ANU is a large organization with financial capacity to absorb deviations in experience. Moreover, most workers' compensation claims occur with high frequency and low severity relative to other risk classes, making it easier to manage. Further, to provide protection against very large unexpected claims, ANU subscribes to an insurance policy with a retention level of 750,000 per event. As at 31 December 2019 no claims have, or are expected to, exceed this retention level.

To get a sense of the size of this obligation, the 2020 ANU annual report (page 97) lists the current liability provision for workers' compensation to be 2,588 (thousands of AUD) and the non-current liability provision to be 21,638 (thousands of AUD); the latter is a provision for future anticipated claims. This is further evidence that workers' compensation claims are paid out over many years and is one reason that we do not advocate using the same liability portfolio methods as with the other risks that ANU faces.

For this risk, further information is available through the website ANU (2023).

6.3.3 Wisconsin Property Fund Case

The ANU case describes an organization that manages risks inherent in its operations. In contrast, the Wisconsin property fund, introduced in Section 1.2.2, is a self-insurance pool formed under an intergovernmental cooperative agreement. This is an example of an organization that takes on risks in a manner similar to an insurance company.

The Wisconsin Local Government Property Insurance Fund (LGPIF) was established to provide property insurance for local government entities that include counties, cities, towns, villages, school districts, and library boards. The fund insures local government property such as government buildings, schools, libraries, and motor vehicles. State government buildings are not covered; the LGPIF is for local government entities that have separate budgetary responsibilities and who need insurance to moderate the budget effects of uncertain insurable events. Claims for state government buildings are charged to another state fund that essentially self-insures its properties. Coverage for state properties and local government units was established by the state in 1911 when it was difficult for these units of government to obtain reasonably priced coverage from the private sector. In 1979, this separate fund was established to provide insurance coverage to government properties not owned by the State of Wisconsin. The fund was dissolved in 2017, ostensibly because the private marketplace was sufficiently robust to provide the needed coverage at reasonable prices.

During its existence, the LGPIF acted as a stand-alone insurance company, charging premiums to each local government entity (policyholder) and paying claims when appropriate. As described in Section 1.2.2, the Property Fund tracks six coverage groups that can be treated as individual lines of businesses. Although the LGPIF was not permitted to deny coverage for local government entities, these entities could work with the open market to secure coverage. Thus, the LGPIF acted as a "residual" market to a certain extent, meaning that other sources of market data may not reflect its experience.

When considering the balance between risk retention and risk transfer, it is natural to consider three perspectives:

- A typical fund member that is interested in minimizing the uncertainty of its retained risk given a pre-determined maximum level of risk transfer contribution to the fund.
- The fund itself, or the pool of retained risks from all fund members. From this perspective, the fund may seek to transfer a portion of its retained risk to an external party such as an insurer/reinsurer.
- The reinsurer. Here, the perspective is that, for a price, a reinsurer has agreed to take on all risks except a designated portion that typically includes small routine claims.

Section 8.1 will consider the details underpinning each risk retention problem. Interestingly, each perspective represents a different optimization problem that naturally leads to different, and sometimes conflicting, retention policies. However, note that when the fund pool considers retention limits for a typical member it will balance that member's retained risk with others in the portfolio. In the same way, when the fund transfers risk to a reinsurer, the reinsurer will also consider other risks that it retains. Thus, although the three perspectives appear to be in direct conflict, conflicting interests are mitigated by the presence of other risks not directly included in each optimization problem.

6.4 Robo-Advising and InsurTech

Technical analysis of the balance between risk retention and risk transfer constitutes a type of technological innovation in the insurance sector, a field that has become known as *InsurTech.* To see how this fits, this section provides an overview of InsurTech, beginning with an introduction to *FinTech* while emphasizing a specific type, *robo-advising.*

6.4.1 Introduction to Robo-Advising

What is Robo-Advising?

Methods for solving the Section 3.4 asset allocation problem continue to evolve and remain relevant in part because they form the backbone of *robo-advising.* A robo-advisor (RA) is an automated digital financial system that provides financial advice with limited human intervention. These digital wealth management platforms provide investment solutions primarily focused on passive and cost-efficient instruments, while efficiently managing these allocations using rebalancing mechanisms. Robo-advisors are designed to incorporate inputs received from the investor.

The robo-advising industry started in 2007 by new entrants who saw it as an opportunity to bring advice to customer segments too small for traditional financial advisors. A core innovation of robo-advisors is the automation of wealth management processes that traditionally have been manual; robo-advising refers to attempts made to automate any part of the investment management process.

According to Morningstar Report (2022), industry-wide assets were around 440 billion USD at year-end 2021. Although large, it made up only a small fraction of the 27.9 trillion USD U.S. retail market. Like other aspects of the digital economy, future generations of clients are anticipated to welcome robo-advising and fuel its future growth.

How does an RA work?

As described in Grealish and Kolm (2021), robo-advising platforms begin collecting client information that includes age, net worth, investment goals, risk capacity and risk tolerance. The investment goals may include generating income, saving for retirement, planning for large future expenditures (such as the purchase of a house), and establishing a financial safety net for emergencies.

Next comes implementation of the investment strategy that can involve selecting a set of investment options such as exchange-traded funds (ETFs) followed by portfolio optimization. (An ETF is a bundle of securities that one buys or sells on a stock exchange.) Portfolio optimization may be based on the mean-variance asset allocation strategy or extended versions using alternative risk measures such as expected shortfall. Some robo-advisors incorporate additional features into their models such as transaction costs, tax strategies, and subjective viewpoints.

Robo-advisors are also used in the ongoing management of the investment strategy. Portfolios typically require rebalancing for shifts in market movements, unforeseen changes in cashflows, or opportunities for "tax-loss harvesting." (Tax-loss harvesting means that one sells investments that are down, replaces them with similar investments, and use the losses to offset other realized taxable investment gains. The net result is a savings in taxes paid.) Some robo-advisors also provide updated automated advice as the investor approaches retirement or other major lifestyle changing events that affect asset consumption patterns.

Benefits of an RA

An important benefit of RAs is the lower costs attained through the use of technology. As summarized in D'Acunto et al. (2019) and Beketov et al. (2018), robo-advisors have several strengths when compared to traditional financial advisors.

- They offer various options to control, customize, and construct investment portfolios from multiple devices (such as smartphones and laptops). The technology can simplify and speed up contact with clients. RAs also have transparent workflow and monitoring systems, requiring a low minimum investment.
- Robo-advisors enjoy greater simplicity and efficiency in implementing strategies due to built-in automated algorithms.
- Robo-advisors typically use replicable algorithms that utilize advanced quantitative methods of portfolio management and optimization. As one consequence, RAs do not suffer from potential biases that human advisors may have. The interaction between human advisors and clients often resembles a sales transaction in which the advisor has an incentive to maximize personal incentives that may differ from an investor's interests.

6.4.2 What is FinTech?

Robo-advising is the delivery and execution of financial advice through automated algorithms on digital platforms. Thus, it falls squarely into the sector known as *FinTech*, an abbreviation for financial technology. This is a term used to describe any technology that delivers financial services digitally; it may include online banking, mobile payments, and cryptocurrency. As described in Benamraoui and Abdulkader (2020), other types of FinTech include:

- *Trading* is a popular area for machine-learning/artificial intelligence (AI) applications. Considering the growing speed and complexity of trades, AI techniques are becoming an essential part of trading practice. These techniques have the capacity to analyze large amounts of data to generate trading signals. Algorithms can be trained to automatically execute trades based on these signals, which has given rise to the industry of algorithmic (or "algo") trading. In addition, AI techniques can reduce transaction costs by automatically analyzing the market and subsequently identifying the best time, size, and venue for trades.
- *Digital lending* is the use of digital platforms to originate and renew loans in order to deliver faster and more efficient decisions. Online technology could be used to improve traditional banking relationships but also be used for non-traditional mechanisms, such as crowdfunding and peer-to-peer (P2P) lending. The concept of crowdfunding continues to evolve since it emerged to fund charities and facilitates improved access for businesses to investors. Crowdfunding capitalizes on social networks and online platforms by drawing on small contributions to fund a project. In contrast, P2P lending has achieved this by removing the middleman and connecting borrowers and lenders directly to one another. P2P lending usually takes the form of business loans or lines of credit. Investors typically provide funding assets through a lending platform.
- *Blockchain.* The blockchain technology can be described as a decentralized ledger that keeps a permanent record of digital transactions while also ensuring the purity of digital information. Blockchains overcome the need for third-party organizations or central administrators; only the participants in the blockchain control the network without any intermediaries. Blockchains have enabled organizations to automate their transactions to improve the flow of information between the organization and its external stakeholders.

6.4.3 What about InsurTech?

Many of these financial innovations can be applied directly, or adapted to, the insurance sector. *InsurTech* is a term used to describe any technology that delivers insurance services digitally, as in Cortis et al. (2019). The evolution of InsurTech is based in part on newly available *data* and can be described in terms of the *insurance activities* that it influences.

Data

As with FinTech, the explosion of interest in InsurTech has been partially driven by the increasing volume and breadth of data. Three types of data which have already made important impacts on pricing and underwriting insurance activities are:

- Mobile devices and wearables can generate massive amounts of data from sensor-equipped mobile devices and wearables. Early marketplace entrants were those of a wrist-borne nature but now wearables include fashion items such as jewelry, clothing and shoes. As these devices become more affordable to the general public, insurers benefit from this surge of available data to improve their pricing models for life and health insurance.

- These wearables can capture biometric information such as those from physical activity (e.g., number of steps, time spent sitting, miles cycled), cardiovascular measures (heart rate, heart rate variability, ECG, blood pressure), sleep data (quantity and quality), body temperature, galvanic skin response, blood sugar and even pollution exposure.

- Auto telematic devices can pick up diverse driving metrics such as location, time of day, mileage, driving frequency, behavior around hazardous zones, speed, rates of acceleration and braking habits.
- Home sensors and the Internet of Things (IoT). The network of physical devices embedded with sensors and connectivity, allowing the transmission and communication of data, forms the Internet of Things (IoT). Applications range from smart home devices (e.g., smoke alarms, thermostats and fridges) to environmental monitoring (e.g., examining air and water quality) and have permeated the market.

These variables not only impact pricing and underwriting but also allow consumers to assess their own outcomes on a timely basis. Insurance agreements can be established to encourage better consumer behavior (e.g., slow down speedy driving) and thus provide real benefits to society.

Examples of big data include text, audio, images, and video files that are known as unstructured data. The insurance industry as a whole is actively experimenting and developing approaches for incorporating these massive sources of data into their insurance activities. Table 6.2 summarizes new types of data sources that lead to new data.

Insurance Activities

Underwriting Process. The underwriting process is undergoing extensive transformations due to the introduction of AI algorithms. The business lines most readily digitized are those that have smaller potential losses and can be processed readily such as travel, event, and car insurance. Even when the process requires human intervention, AI can aid in almost every scenario, allowing underwriters to pull in external data in order to strengthen their

TABLE 6.2
Analytic Trends.

Data Sources	Algorithms
Mobile devices and wearables	Statistical learning
Auto telematics	Artificial intelligence
Home sensors (Internet of Things)	Structural models
Drones, micro satellites	
Data	**Software**
Big data (text, speech, image, video)	Text analysis, semantics
Behavioral data (including social media)	Voice recognition
Credit, trading, financial data	Image recognition
	Video recognition

Source: Stephen Mildenhall, Personal Communication. From Frees and Gao, 2020.

decision-making processes. It is likely that all underwriters will be using digital technologies to support their underwriting decisions over time. Some analysts foresee that underwriting as we know it today will eventually cease to exist for most personal and small-business products across life and property and casualty insurance.

Claims Management Process. As with the underwriting process, AI is making a major impact on claims management associated with smaller, more frequently occurring claims such as auto insurance. Using AI image analysis and automated estimation algorithms, claims estimators can quickly resolve claims more accurately. For example, the AI can process a photo of vehicle damage and provide an estimate of the cost based on factors such as the size and orientation of a dent. Noncomplex claims are approved within seconds while complicated claims are handed over to humans in the claims department.

In addition to claims settlement, insurers may employ AI fraud investigation systems that identify potentially fraudulent claims. Such systems can investigate social media accounts, criminal records, property and vehicle history and other documentation submitted with the claim, thus enabling more time for the claims department to analyze results and close claims in a timely manner. Similar services that weed out possible fraudulent policyholders at underwriting stage are available in the market.

Distribution and Customer Interaction. Distribution is the process of delivering products or services into a target market. Like underwriting and claims management, AI has already impacted distribution channels for lines of business with smaller, more frequently occurring claims such as auto insurance. For these lines, the traditional quotation and policy issuance process is time-consuming and error prone. As alternatives, some InsurTech platforms use AI to streamline the process and automatically generate a quote, helping underwriters and brokers save time and issue policies faster.

In addition, insurers increasingly utilize chatbots to communicate with existing and potential customers. A chatbot is a digital service capable of holding natural-sounding conversions with individuals to accomplish a particular task, such as answering consumers' questions, providing guidance, checking billing information, and addressing common inquiries and transactions.

Another potential disruptor for insurance distribution channels is peer-to-peer (P2P) insurance. P2P insurance is a risk-sharing network where a group of people pools their premiums together to insure against a risk. Modern implementations of P2P are technology enabled. Because it also offers a fundamentally different approach to risk sharing, we discuss this innovation in a later section.

6.4.4 Managing Insurable Risks Portfolios

Currently, most InsurTech activities can be classified as facilitating the pricing, underwriting, claims management, and distribution and customer interaction processes of insurers. Moreover, blockchain and P2P insurance have the potential to encourage innovative risk-sharing models.

However, one notable absence is the development of any tools comparable to robo-advisor in the insurance sector. As described earlier, from a technical viewpoint a robo-advisor can be thought of as a solution to the asset allocation problem, coupled with client information for a tailored solution (not just a frontier) that incorporates a digital platform to execute the financial advice through automated algorithms on digital platforms.

This book proposes a framework for constructing risk frontiers for insurable risk portfolios, paving the way for potential digital applications in the future. Conceptually, it is not difficult to imagine gathering additional information about firm attributes, a risk manager's risk tolerance, and other preferences for the development of a digital platform to execute the financial advice through automated algorithms. As with the early development of robo-advising, it is likely that this solution would initially be most useful for small and moderate size firms.

6.5 Decentralized Insurance

Building upon the "risk owner - risk taker" framework, Section 4.5 introduced the concept of risk sharing among multiple parties, historically motivated by insurance and reinsurance companies exchanging risks. To a lesser extent, this framework also applies to risk-sharing among like-minded businesses, such as the pools and risk retention groups described in Section 6.2.3. This section broadens this discussion by extending risk-sharing concepts to individuals that is coming to be known as *decentralized insurance.* Our focus is on a peer-to-peer (P2P) business model, a specific type of decentralized insurance.

6.5.1 Decentralized Insurance and Reciprocal Insurance Exchanges

Sharing Economy. With the explosion of the sharing economy and internet technology, many industries have adopted decentralized peer-to-peer concepts. For example, Uber, AirBnb, and eBay, are wildly successful start-ups that brought P2P business models to their respective taxi, hotel, and products industries. These firms succeed not by buying fleets of taxis, chains of hotels, or warehouses of goods, but rather by connecting consumers to people like them, their "peers," who can provide the desired good or service.

Reciprocal Insurance Exchanges. The sharing of uncertain risks among like-minded individuals or firms is not new but rather reflects the origins of mutual insurance dating back centuries, cf., National Association of Insurance Commissioners (2023a) and Feng (2023). This sharing principle is reflected directly in the mission of mutual insurers where members care for each other's financial needs in the event of misfortune. As will be seen, decentralized insurance frameworks most closely resemble traditional *reciprocal insurance exchanges* cf. Rego and Carvalho (2020). A reciprocal insurance exchange is an organization where policyholders directly exchange insurance contracts with one another. Unlike a mutual insurance company, it is an unincorporated association, not a traditional corporation. In the United States, the association administrator is known as an attorney-in-fact and policyholders are called subscribers. Although reciprocals are far less common than mutual and stock insurers, some major insurance providers are organized as reciprocals, notably USAA and Farmers Insurance. Reciprocal insurance exchanges are less common in Europe and Asia compared to the United States, but they do exist in several countries. Lloyd's of London, a risk exchange that provides marine and specialty insurance for diverse risks, is a well-known example.

Decentralized Insurance. Traditionally, insurance companies have operated as central authority figures for groups of policyholders. They enter into contracts with policyholders to take on a portfolio of risks, collect premiums, pay claims, and set aside reserve funds for future obligations. This structure contributes to the reputation of insurance as a service industry with high expenses. The left-hand panel of Figure 6.1 is a schematic of this business model, with policyholder risks scattered around the insurer at the center.

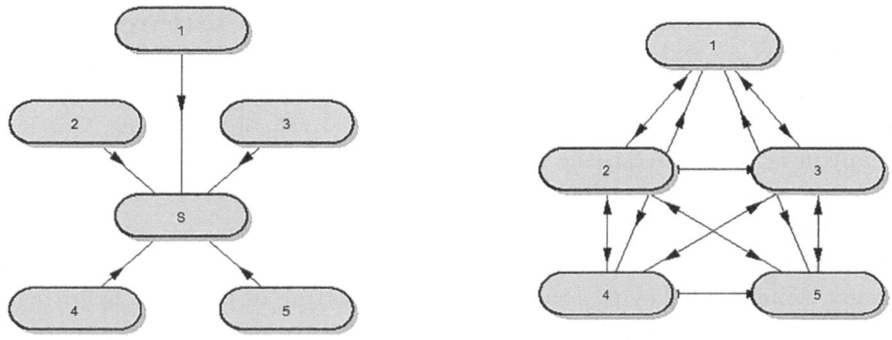

FIGURE 6.1
Centralized versus Decentralized Approaches to Risk Sharing. The left-panel provides a schematic for a centralized approach, the right-hand panel is for a decentralized approach.

The right-hand panel of Figure 6.1 is a schematic of a decentralized network where the central authority figure is absent. In decentralized insurance, the role of market intermediaries is reduced or eliminated. It is sometimes also referred to as *disintermediated insurance*, as such practices bypass or cut out the central insurer role as a middleman. Just as Uber does not maintain a fleet of taxis, the administrator of a decentralized insurance network does not accumulate premiums and claim obligations which results in much lower overhead for central administrative expenses than a traditional insurer.

Figure 6.1 also suggests that administration of a decentralized insurance model can be more complex than a traditional insurer model. This complexity has been addressed by recently developed technology that permits participants to pay premiums/contributions, make claims, and receive claim payments readily. Moreover, some systems utilize blockchain technology for premium and claim payments coupled with bitcoins to further reduce transaction costs. According to the National Association of Insurance Commissioners (2023a), Teambrella claims to be the first P2P insurer using this model based on bitcoin.

Traditional Insurance versus Decentralized Models. What are the strengths and limitations of traditional insurance delivery systems when compared to the newer decentralized models? According to Feng (2023), traditional models offer three important advantages:

- *Standardization of policies.* Insurers use common policies and practices, reducing administrative costs.
- *Single authority figure.* Each policyholder deals only with an insurer which can reduce confusion and anxiety for consumers.
- *Economies of scale.* Insurance is an expensive service industry and insurers can spread costs and achieve economic efficiencies by taking on many risks.

In addition, traditional models have significant limitations:

- *Insurance premiums are high relative to the actual cost of benefits.* This is the most significant limitation in terms of motivating the introduction of modern decentralized platforms. Insurance is a high expense and service industry. Some services, such as claims adjudication, and some expenses, such as compliance with government regulations, are necessary. However, others such as agent commissions, costs of solvency capital, overhead, and so forth, are due to the central authority figure structure and are thought to be mitigated with a decentralized system.

- *Concentration of market power.* Because insurers need a requisite amount of capital to operate, this necessarily restricts entrance to the marketplace. Moreover, in many jurisdictions, relatively few insurers have a significant portion of the market share, giving the impression that they have undue influence on insurance policy decisions.
- *Lack of trust in the insurance industry.* Fraud is a major concern for insurers, many claimants believing that a slight misreporting of facts can help their case for payments significantly while not really hurting the financial situation of a major insurer. In contrast, particularly for a small group of family members, friends, or individuals with common interests who combine their premiums to insure against risks, it is thought that a decentralized model can result in a lower incidence of fraudulent claims. Fraud can be addressed through a variation known as a "broker model" to be introduced in Section 6.5.2.

Models without Insurers. Some decentralized systems adopt a pure mutual approach, where risks are shared among participants without any transfer to an insurance company. These models are developed entirely outside the traditional insurance sector and sometimes known as "self-governing models," or "no-insurance-company models." Like reciprocal insurance exchanges, these systems have no separate legal personality but only organize indemnity agreements among participants.

For instance, China's "mutual aid platforms" illustrate this approach. As one example, Xianghubao (a subsidiary of Ant Financial) was introduced in October 2018 and in less than a year had amassed more than 100 million users. It was organized to provide benefits that are payable upon the diagnosis of critical illness (such as cancer). When a member joins the platform, that person is obligated to split the cost of mutual aid with others at the end of each month. Thus, the business model does not involve an insurer. Xianghubao was closed in early 2022 due to regulatory changes, industry challenges, and strategic decisions by the parent company. See, for example, Abdikerimova and Feng (2022) and Feng (2023) for further discussions of this interesting case.

Models with Insurers. Many innovative technology exchanges have been organized by insurers or insurtech firms serving the insurance industry, so it is natural for these models to include the presence of an insurer. Moreover, for exchanges among individuals, it is natural to include insurers in order to limit potential obligations of participants by transferring tail risks to insurers.

According to the classification proposed by Rego and Carvalho (2020), the dominant type of P2P insurance is the P2P *broker model.* The main distinctive features of this model are as follows.

- Participants form groups (mainly online). A portion of their premiums goes into a common fund, while the rest gets paid to an insurance or reinsurance company.
- If the common fund is insufficient to pay for the claims then the insurance carrier pays the excess over retained premiums.
- Conversely, if the pool has few claims then the surplus is given back to the participants or to a cause the pool members care about.

The risk and insurance literature cites Friendsurance, a German startup launched in 2010, as the first to enter the market as a P2P model.

6.5.2 Risk-Sharing Rules

We have already seen an introduction in Section 4.5 to some risk-sharing rules that apply to decentralized insurance.

Risk-Sharing Rules without Insurers. As in Section 4.5.1, linear exchanges enable direct risk transfer among participants, bypassing central collection and insurers. Furthermore, the conditional mean-sharing rules of Section 4.5.3 facilitate economically efficient risk sharing within a centralized pool.

Recall that a central system such as mutual insurance not only collects risks centrally but also accumulates surplus capital over time, which can be used to pay dividends to policyholders, reduce premiums, or invest in growth. In contrast, decentralized systems like P2P may still centralize risk collection, but typically function on a "no surplus, no deficit" principle. This means collected premiums solely cover claims and operating costs, with any excess returned to policyholders.

For some additional options, recall the notation that n agents have potential risks $X_1, \ldots, X_n$ and we wish to determine risk-sharing rules $g_i(\cdot)$ such that $Y_i = g_i(X_1, \ldots, X_n)$ is the ith agent's share after entering into a risk-sharing arrangement. The pooled risk is defined to be $S = X_1 + \ldots + X_n$. With centrally distributed networks, the ith agent's share is $Y_i = g_i(S)$. Some basic requirements of risk-sharing rules are:

1. $g_i(s)$ is strictly increasing in s for each i, and
2. the sum over all agents preserves the total, that is, $\sum_i g_i(s) = s$.

It is easy to see that the following two special cases have these properties:

- conditional mean risk-sharing rule, where $g_{1,i}(s) = \mathrm{E}[X_i|S = s]$, and
- proportional risk-sharing rule, where $g_{2,i}(s) = \frac{\mathrm{E}(X_i)}{\mathrm{E}(S)} \times s$.

In addition, a *quantile risk-sharing* rule has been proposed recently in Denuit et al. (2022) which is described as follows. Let F_i be the distribution function of X_i. For simplicity, assume that the function F_i is continuous so that the inverse is well-defined and unique. Extensions to cases with discontinuities are not difficult and appear in Denuit et al. (2022).

With the sum $S = X_1 + \cdots + X_n$, define a comonotonic version $S^c(V) = F_1^{-1}(V) + \cdots + F_n^{-1}(V)$, with a uniform random variable V. Let $F_{S^c}(s) = \Pr[S^c(V) \leq s]$ be the corresponding distribution function. With this notation, we define a quantile risk-sharing rule as

$$g_{3,i}(s) = F_i^{-1}(p_s^c),$$

where $p_s^c = F_{S^c}(s)$ is the distribution function evaluated at s. One can check that this rule satisfies the two basic requirements for risk-sharing rules.

Under the Hood. See the *Online Supplement* for the verification of basic properties of the quantile risk-sharing rule.

Risk-Sharing Rules with Insurers. In risk-sharing arrangements, *counter-party risk* is a potential problem. That is, the amounts $Y_i = g_i(S)$ to be paid at the end of the period remain unknown until then, and some participants may be unable, or unwilling to pay their contributions. Also, unless participants are ready to pay expensive contributions, the risk-bearing capacity of the community is limited. In the "broker model," the organizer of the P2P insurance system acts as an intermediary and buys insurance protection on behalf of the participants whose liability is limited to a lower layer of the total losses.

The ex-post contributions in the self-governing model are replaced in the broker model with a combination of

- an ex-ante provision π_i, depending on the distribution of both X_i and S and
- an ex-post cash back CB_i, in case of favorable experience.

Pool and Participant Retention Levels. Specifically, we suppose that there exists an upper limit, or stop-loss, cover with retention level u_P. With this cover, pool participants (in total) retain responsibility for claims smaller than u_P whereas the insurer has responsibility for excess claims, $(S - u_P)_+$. Define $u_i = g_i(u_P)$, which turns out to be the retention level for the ith participant. From the second property of the risk-sharing rule, we note that $\sum_i u_i = \sum_i g_i(u_P) = u_P$.

Insurance Contribution. Suppose that the price of the stop-loss protection is of the form $\pi^{SL} = (1 + \theta_{SL})\mathrm{E}[(S - u_P)_+]$. Then, one can define the share of the ith participant to be

$$\pi_i^{SL} = (1 + \theta_{SL})\mathrm{E}\{[g_i(S) - u_i]_+\}.$$

Because $\pi^{SL} = \sum_i \pi_i^{SL}$, the sum of participant contributions covers the stop-loss protection.

Sharing Contribution. The loss $\min(S, u_P)$ is retained by the pool. The share for agent i is $\min(Y_i, u_i)$. It is not difficult to check that the sum of individual contributions equals the loss retained by the pool.

Surplus and Cash Back. One way to think about this responsibility is that, in advance, each member contributes u_i. Because this sums to $\sum_i u_i = u_P$, the sum of participant contributions is sufficient to cover retained pool losses. Thus, the total contribution, in advance, is $\pi_i = \pi_i^{SL} + u_i$. This means that, in the event that $S < u_P$, there will be a surplus. If there is a surplus, then it can be shared among participants or given to a charity.

The surplus that can be attributed to participant i is CB_i, the "cash back" which is defined as $CB_i = (u_i - Y_i)_+$. It is not difficult to check that $(u_P - S)_+ = \sum_i CB_i$, so the sum of the cash backs for all agents equals the amount available from the pool.

ⓘ *Under the Hood.* See the *Online Supplement* for the verification of basic properties of the cash back.

6.6 Supplemental Materials

6.6.1 Further Resources and Readings

Actuaries and other financial analysts will enjoy learning more about the fundamentals of risk control and transfer as outlined in Section 6.1. To this end, see some classic references including Willams et al. (1995), Huebner et al. (1996), Harrington and Niehaus (1999), and Skipper and Kwon (2007).

If you would like to learn more about the commercial insurance coverages, see Insurance Information Institute (2023b), NAIC (2023), Metz and Valentine (2023), and Serpa and Krishnan (2017).

You can learn more about reciprocals in Mayers and Smith (2000) or via popular outlets such as Yates (2023).

Decentralized insurance has received substantial attention recently, both from industry and academia. For a regulator's perspective, see National Association of Insurance Commissioners (2023a). In addition to the references cited in Section 6.5, see the works of Michel Denuit and co-authors as in Denuit (2020), Denuit and Robert (2021b), among others, for the foundations of this rapidly developing literature. From the perspective of this text, we look forward to the time when this stream of thought takes on the case of multivariate risks.

7

Constructing Multivariate Portfolios

Chapter Preview. The constrained optimization approach can be used to determine risk retention allocations that are intuitively appealing and have sound and rigorous foundations. However, a difficulty is that this approach can be computationally demanding. Even in today's world of fast desktop and cloud-based servers, the single evaluation of a risk measure can be computationally challenging. Constrained optimization techniques rely on many such evaluations, making this approach computationally prohibitive for even a small number of risks.

To circumvent these difficulties, this chapter utilizes a simulation strategy outlined in Section 7.1. In many cases, simulation allows analysts to quickly evaluate complex functions. However, there is a cost to this; the simulation approach introduces approximation errors when numerically evaluating functions. This cost is partially mitigated in that the size of the error can be controlled, primarily by increasing the number of replications. Section 7.6 provides additional insights for controlling simulation uncertainty in risk retention problems.

A basic strategy can be improved upon by treating our objective functions, the risk measures, as outputs of a minimization process as outlined in Section 7.2. In addition, Section 7.3 utilizes *kernel* estimation methods to smooth the simulated objective function. Both of these techniques will speed up the convergence of our algorithms in real time.

Section 7.4 shows how to construct an insurable risk portfolio for an excess of loss contract. Not only is excess of loss a foundational contract, it is our first non-linear retention function applied to many (multivariate) risks. As such, analysis of this contract opens the pathway to an entire host of non-linear retention agreements, some of which are introduced in Section 7.5.

7.1 Simulation Strategy

Simulation, as employed here, is also known as the *Monte Carlo* technique, cf. Metropolis and Ulam (1949). For the simulation stage, one specifies a large number, R, of replications. Then, for each replicate $r = 1, \ldots, R$:

- Create a realization of dependent multivariate uniform random variates $(V_{1r}, V_{2r}, \ldots, V_{pr})$. For example, one way to do this is via the `R` package `copula` using the function `rCopula`.
- Create a realization of dependent multivariate losses $(X_{1r}, X_{2r}, \ldots, X_{pr})$, denoted by the vector $\mathbf{X}_r$.
  - One way to do this is to use the inverse of each distribution function, $X_{jr} = F_j^{-1}(V_{jr})$, $j = 1, \ldots, p$.
  - For example, if the jth variate has a Pareto distribution, this can be done via the `R` package `actuar` using the function `qpareto`.

DOI: 10.1201/9781003469568-7

- Create a realization of that portion retained by the risk owner, $g(\mathbf{X}_r;\boldsymbol{\theta})$ (for a given set of risk retention parameters $\boldsymbol{\theta}$).

Using the entire simulation distribution, one then creates a function to approximate the risk measure, say, $RM_R(\mathbf{X};\boldsymbol{\theta})$. For example, if the risk measure is the value at risk, then in `R` one could use the `quantile` function.

For risk transfer functions, the initial focus is on quota share and excess of loss policies. In addition, also featured is a combination policy, simply to demonstrate the flexibility of the simulation strategy:

- Quota share: $g_{QS}(\mathbf{X};\boldsymbol{\theta}) = \sum_{j=1}^{p} \theta_j X_j$
- Excess of loss: $g_{XL}(\mathbf{X};\boldsymbol{\theta}) = \sum_{j=1}^{p} (\theta_j \wedge X_j)$
- Combination policy: $g_{Combo}(\mathbf{X};\boldsymbol{\theta}) = \sum_{j \in Set_{QS}} \theta_j X_j + \sum_{j \in Set_{XL}} (\theta_j \wedge X_j)$, where Set_{QS}, Set_{XL} denotes the sets of risks included in the quota share and excess of loss policies.

Using the same approach, one could create a simulated approximation of the risk transfer cost, say, $RTC_R(\boldsymbol{\theta})$. To illustrate, assuming fair risk transfer costs,

$$
\begin{aligned}
RTC_R(\boldsymbol{\theta}) &= \frac{1}{R}\sum_{r=1}^{R}\left\{\sum_{j=1}^{p} X_{rj} - g(\mathbf{X}_r;\boldsymbol{\theta})\right\} \\
&= \mathrm{E}_R\left\{\sum_{j=1}^{p} X_j - g(\mathbf{X};\boldsymbol{\theta})\right\},
\end{aligned}
$$

where $\mathrm{E}_R(\cdot)$ is the empirical (or simulation) expectation. For some problems we will see that numerically calculating the risk transfer cost is relatively straight-forward and there is no need to introduce a simulation-based approximation.

In addition to a budget restriction ($RTC_R(\boldsymbol{\theta}) \leq RTC_{max}$), it will also be common to constrain the risk retention parameters. For example, most problems will require the parameters to be non-negative. In addition, it is common for quota share problems to restrict parameters to be bounded above by one; this prevents the risk owner from transferring more than a realized loss.

To be concrete, define $\mathbf{1}_j$ to be a vector of zeros with a one in the jth position. Then, the requirement $\theta_j \geq 0$ is equivalent to $-\mathbf{1}_j'\boldsymbol{\theta} \leq 0$. In the same way, the requirement $\theta_j \leq 1$ is equivalent to $\mathbf{1}_j'\boldsymbol{\theta} \leq 1$. Because these restrictions are linear in the parameters, we can summarize them as $\mathbf{P}\boldsymbol{\theta} \leq \mathbf{p}_0$. As we use m for the number of constraints, the dimensions of $\mathbf{P}$ and $\mathbf{p}_0$ are $(m-1)\times p$ and $(m-1)\times 1$, respectively. (The budget restriction is the mth constraint.)

With a risk measure, transfer cost function, and parameter restrictions in place, we then solve the simulation version of Display (3.3) given as

$$
\boxed{
\begin{array}{lc}
\text{minimize}_{\boldsymbol{\theta}} & RM_R[g(\mathbf{X};\boldsymbol{\theta})] \\
\text{subject to} & RTC_R(\boldsymbol{\theta}) \leq RTC_{max} \\
 & \mathbf{P}\boldsymbol{\theta} \leq \mathbf{p}_0 \ .
\end{array}
}
\qquad (7.1)
$$

In the operations research and management science literature, the combination of simulation and optimization techniques is known as *simulation optimization*, cf. Fu (2015a).

Example 7.1. Optimal Retention Policies for ANU Risks. To see how this works, let us revisit the *ANU Risk Data* from Section 6.3.2 with $p = 14$ risks (omitting the property risk). Simulation of ANU risks is realistic and thus a bit complex. The risk parameters and the generation of the random variables will be described beginning in Appendix Section 8.3.4.

For the retention policies, let us consider an excess of loss, quota share, and a combination policy as described above. For the quota share component of the combination policy, I selected three risks, Group Personal Accident, Corporate Travel, and Motor Vehicle; the other risks are on an excess of loss basis. These three risks have relatively high frequency (compared to the other risks) and so would not typically be the subject of a single excess of loss contract; Section 7.5.3 describes alternatives.

For simplicity, fair pricing is used to determine the risk transfer cost and I use $RTC_{max} = 433$ (in thousands of AUD) for the maximal risk transfer cost budgeted. As will be seen in later examples in this chapter, this is a typical value on a frontier of costs. For the other restrictions, all risk retention parameters are non-negative and quota share parameters are bounded above by one. Thus, for optimizing the excess of loss contract, there are $m = 15$ parameter restrictions (one for the budget plus one for each of 14 parameters), for the quota share there are 29 restrictions ($= 1+2\times 14$), and 18 for the combination policy ($= 1+14+3$).

The optimization was coded using the package `alabama` without taking additional derivatives, as described in Section 4.4.2. Subsequent sections in this chapter will describe methods for improving this optimization procedure although the results show that even this rudimentary approach is successful for this application.

The objective is to minimize the standard deviation of retained risks. This traditional risk measure was chosen for simplicity and more realistic measures will be considered in the following section. Table 7.1 summarizes the optimization results for the three retention functions. The results show that the excess of loss and combination policies achieved a lower standard deviation of retained risks than the quota share policy. Each optimization run was based on minimization over 14 risk retention (decision) parameters although only the first, second, and sixth are displayed in the table (θ_1, θ_2, and θ_6, for General and Property Liability, Cyber, and Group Personal Accident, respectively). For the combination policy, I used the optimal values from the corresponding excess of loss and quota share policies as starting values. It turned out that these were very close to the resulting optimum.

With specific values of the risk retention parameters, one can compute any risk measure desirable, such as VaR and ES. Table 7.1 also shows the value at risk and expected shortfall at the optimal risk retention parameters, computed using level of confidence $\alpha = 0.80$. These values will serve as comparisons with subsequent examples.

TABLE 7.1
ANU Optimization Results.

	RTC	***VaR***	***ES***	**Std Dev**	θ_1^*	θ_2^*	θ_6^*
Excess of Loss	433	1443.9	2492.1	1024.7	1419.422	1005.995	59.132
Quota Share	433	1269.1	2674.7	1312.8	0.119	0.946	1.000
Combination	433	1443.9	2492.1	1024.7	1419.422	1005.995	1.000

Note: Objective function is standard deviation of retained risks.

7.2 Risk Measures as Minimizers

Although the representation in Display (7.1) is intuitively appealing, the evaluation of associated risk measures can be computationally challenging. As a tool to assist with this computational challenge, this section demonstrates how to write two risk measures, the VaR and the ES, as solutions of constrained optimization problems.

7.2.1 Value at Risk

First recall from equation (2.1) that we can write the value at risk as a quantile, $VaR[g(X;\boldsymbol{\theta})] = F^{-1}_{g(X;\boldsymbol{\theta})}(\alpha)$, with confidence level α. Moreover, as suggested in Section 5.4.2, we can also express the value at risk as the solution of a constrained minimization problem,

$$\boxed{\begin{array}{ll} \text{minimize}_{z_0} & z_0 \\ \text{subject to} & F_{g(X;\boldsymbol{\theta})}(z_0) \geq \alpha. \end{array}} \tag{7.2}$$

ⓘ *Under the Hood*. See the *Online Supplement* for the verification of the VaR as a solution of Display (7.2).

The value at risk $F^{-1}_{g(X;\boldsymbol{\theta})}(\alpha)$ provides a solution to the problem in Display (7.2) for fixed values of $\boldsymbol{\theta}$. We can now take the minimum of all such solutions over different values of $\boldsymbol{\theta}$ while restricting consideration to a feasible set. Thus, subject to mild continuity requirements, our problem of interest

$$\boxed{\begin{array}{ll} \text{minimize}_{\boldsymbol{\theta}} & VaR_{\alpha}[g(X;\boldsymbol{\theta})] \\ \text{subject to} & RTC(\boldsymbol{\theta}) \leq RTC_{max} \end{array}}$$

is equivalent to

$$\boxed{\begin{array}{ll} \text{minimize}_{z_0,\boldsymbol{\theta}} & z_0 \\ \text{subject to} & F_{g(X;\boldsymbol{\theta})}(z_0) \geq \alpha \\ & RTC(\boldsymbol{\theta}) \leq RTC_{max}. \end{array}} \tag{7.3}$$

Using this approach, one no longer has to compute the quantile directly, only the distribution function of retained risks. This formulation is known as *chance-constrained programming*, cf. Henrion (2024).

Example 7.2. Bivariate Excess of Loss Using VaR Optimization. We return to Example 5.8 that considers two risks that depend on one another through a Gaussian copula with parameter $\rho = 0.5$. The first risk is X_1, with a gamma distribution having shape parameter 2 and scale parameter 2,000, and the second is X_2, with a Pareto distribution having shape parameter 3 and scale parameter 2,000. A level of confidence $\alpha = 0.85$ is used. For comparison purposes, recall that for this example we were able to use deterministic methods to compute the optimal upper limits $u_1 = 3{,}113$ and $u_2 = 364{,}924$ with value at risk $VaR = 4{,}757$ and expected shortfall $ES = 6{,}673$.

From this joint distribution, $R = 50,000$ losses were simulated. The appended `R` code shows two approaches. The first follows Display (7.1), using the quantile as the risk measure (as before, we use $RTC_{max} = 1,500$ for the budget constraint). This does well but, as we learned in Chapter 5, it is important to start near the optimal values. The second approach is based on Display (7.3), using a "usual" simulated distribution function. This approach did not fare well in this experiment; an improvement is described in the next section.

7.2.2 Expected Shortfall

As with the value at risk in Display (7.2), we can also express the expected shortfall ES as the solution of a minimization problem. To this end, consider again a generic random variable Y having distribution function F with a unique α quantile, $F^{-1}(\alpha)$. Recall in Section 5.3.2 we defined the function

$$ES1_F(z_0) = z_0 + \frac{1}{1-\alpha}\mathrm{E}[(Y - z_0)_+].$$

With this, it is easy to check that the solution of the unconstrained minimization problem

$$\boxed{\text{minimize}_{z_0} \qquad ES1_F(z_0)}$$

is $F^{-1}(\alpha)$. Moreover, the minimum value is $ES1_F[F^{-1}(\alpha)] = ES$.

Under the Hood. See the *Online Supplement* for the solution of the ES minimization problem.

To extend this to the risk retention problem, we first use the retained risk variable $g(\mathbf{X};\boldsymbol{\theta})$ in place of Y. Next, we take the minimum of all such solutions over different values of $\boldsymbol{\theta}$ while restricting consideration to a feasible set. Here is a summary of the simulation version of the expected shortfall retention problem:

$$\boxed{\begin{array}{ll} \text{minimize}_{z_0,\boldsymbol{\theta}} & z_0 + \frac{1}{(1-\alpha)}\mathrm{E}_R\{[g(\mathbf{X};\boldsymbol{\theta}) - z_0]_+\} \\ \text{subject to} & RTC_R(\boldsymbol{\theta}) \le RTC_{max} \\ & \mathbf{P}\boldsymbol{\theta} \le \mathbf{p}_0 \quad . \end{array}} \qquad (7.4)$$

Example 7.3. Bivariate Excess of Loss Using ES Optimization. This is a continuation of Example 7.2. As before, we use $R = 50,000$ simulated bivariate risks. The associated `R` code shows two approaches. The first follows Display (7.1), using the expected shortfall as the risk measure. The second is based on Display (7.4). For comparison purposes, recall from the discussion in Example 7.2 that we were able to use deterministic methods to compute the optimal upper limits $u_1 = 4{,}257$ and $u_2 = 781$ with value at risk $VaR = 5{,}038$ and expected shortfall $ES = 5{,}038$.

Both approaches work well for this demonstration although the first did better than the second. This may be that the first approach, from Display (7.1) with the expected shortfall as the risk measure, optimizes over two parameters in contrast to the three parameters (x, u_1, u_2) in the second. Further, through experimentation, we learned that starting values are more important than the number of simulations.

7.2.3 Quota Share Problem

The quota share contract, described in Section 4.3.2, provides a useful starting point for understanding risk retention strategies because of its simplicity relative to other contracts.

This agreement is essentially identical to the asset allocations, the difference being that the latter typically requires that the risk retention parameters sum to one. In particular, you will recall from Section 4.4.1 that these methods are convex when one uses the variance or expected shortfall to summarize the uncertainty of retained risks.

In addition, it turns out that the decision to use variance or expected shortfall does not matter for a class of problems based on *elliptical* distributions. An elliptical distribution is a generalization of the concept of a symmetric distribution such as a normal or $t-$ distribution; a technical description can be found in Appendix Section 14.2.1. Theoretically, it can be shown, for quota sharing of *elliptical* risks, that any scale and translation invariant risk measure traces out the same frontier as one determined by minimizing the variance, see, for example, Embrechts et al. (2005), Proposition 6.13. For a more explicit type of result, Xiao and Valdez (2015) show that

$$ES(\boldsymbol{\theta}'\mathbf{X}) = k \times \sqrt{\boldsymbol{\theta}'\boldsymbol{\Sigma}\boldsymbol{\theta}} - \boldsymbol{\theta}'\mathrm{E}(\mathbf{X}),$$

where k is a constant that depends on the distribution of $\mathbf{X}$ and the level of confidence α but not the decision variables in $\boldsymbol{\theta}$. Thus, the two problems,

$$\begin{array}{rll} ES \text{ problem:} & \text{minimize}_{\boldsymbol{\theta}} & ES(\boldsymbol{\theta}'\mathbf{X}) - \delta_1\boldsymbol{\theta}'\mathrm{E}(\mathbf{X}) \\ \text{standard deviation problem:} & \text{minimize}_{\boldsymbol{\theta}} & \sqrt{\boldsymbol{\theta}'\boldsymbol{\Sigma}\boldsymbol{\theta}} - \delta_2\boldsymbol{\theta}'\mathrm{E}(\mathbf{X}), \end{array}$$

are equivalent with the only difference being in the penalty terms δ_1 and δ_2.

Example 7.4. Quota Sharing of ANU Risks. To see how this works, let us revisit the *ANU Risk Data* from Example 7.1 where we already calculated the optimal quota share policy using the standard deviation to measure the uncertainty of retained risks. That example used the general purpose `alabama` constrained optimization package in `R`. As an alternative, this example computes the optimal policies using CVXR, a package that utilizes the convexity, for both the variance and the expected shortfall risk measures.

Table 7.2 summarizes optimization results minimizing the variance and the expected shortfall although only the first three quota share coefficients are printed, for brevity. When the optimal coefficients are computed by optimizing the variance measure, it is straightforward to evaluate the expected shortfall, and similarly for optimization with the expected shortfall. These results indicate a close proximity between the two problems, suggesting that risk managers need not be too concerned about the choice of which risk measure to use.

An elliptical distribution is a type of multivariate symmetric distribution that includes the normal and t-distributions as special cases, see, Section 14.2.1. Risks encountered in risk retention problems such as the ANU case are unlikely to have an elliptical distribution (in the ANU case, many risks have a non-elliptical Tweedie distribution). Nonetheless, as seen in the empirical work in Example 7.4, parameter optimal values can be surprisingly similar

TABLE 7.2
Optimal Quota Sharing of ANU Risks.

	RTC	*VaR*	*ES*	**Std Dev**	θ_1^*	θ_2^*	θ_3^*
Variance	433	1997.4	3788.6	1308.2	0.071	1.000	0.977
Expected Shortfall	433	1992.9	3764.9	1314.8	0.138	0.939	0.886

for quota sharing problems. Presumably, this is because the family of elliptical distributions is rich and can approximate many distributions of interest in risk retention problems.

Going forward our emphasis is on the value at risk and expected shortfall to measure the uncertainty of portfolios of risks. As quantile-based risk measures, they are preferred in that they focus on the tails of the distribution in contrast to the variance/standard deviation which emphasizes deviations from the middle. To be fair, we note that when using the variance as the risk measure, one only needs to specify the first two moments which includes the mean, variance, and covariances of the risks. In contrast, for the value at risk and the expected shortfall, more information about the joint distribution is needed.

7.3 Kernel Smoothing

The minimization procedure in Display (7.1) can be improved by **smoothing** the simulated estimates of the distribution using well established methods from statistics.

7.3.1 Nonparametric Smooth Estimation

To provide a quick overview of the kernel smoothing idea, let us think of a simple situation where we have a set of R i.i.d. replications, $\{Y_1, \ldots, Y_R\}$, of a single random quantity. The usual (nonparametric) estimator of the distribution function, $\Pr(Y \leq y) = F_Y(y)$, is

$$\hat{F}_Y(y) = \frac{1}{R} \sum_{r=1}^{R} I(Y_r \leq y) = \mathrm{E}_R\{I(Y \leq y)\},$$

where you will recall that E_R is the empirical (or simulation) expectation. It is the "usual" estimator because it has several nice properties, including being unbiased. To estimate the corresponding probability density function, $f_Y(y)$, it is common to use a *kernel density estimator* of the form

$$f_{Rk}(y) = \frac{1}{R\, b} \sum_{r=1}^{R} k\left(\frac{y - Y_r}{b}\right),$$

where $k(\cdot)$ is a probability density function centered about 0 and b is a constant known as the *bandwidth* parameter. Motivated by this estimator, one can define an alternative distribution function estimator of the form

$$\begin{aligned} F_{Rk}(y) &= \int^{y} f_{Rk}(x)dx = \frac{1}{R} \sum_{r=1}^{R} K\left(\frac{y - Y_r}{b}\right) \\ &= \mathrm{E}_R\left\{K\left(\frac{y - Y_r}{b}\right)\right\}, \end{aligned}$$

where $K(y) = \int^{y} k(z)dz$ is the distribution function corresponding to density $k(\cdot)$. When compared to the usual estimator, it is known that the estimator F_{Rk} is slightly biased but has a lower variance. So, for routine applications, there is a trade-off in terms of bias and variance and certainly the estimator $\hat{F}_Y(\cdot)$ is more commonly applied.

We can use these smoothed simulated distributions to estimate other quantities. For example, for a generic function $h(\cdot)$, we can approximate $\mathrm{E}[h(Y)]$ using

$$
\begin{aligned}
\mathrm{E}_{Rk}[h(Y)] &= \int h(y) f_{Rk}(y) dy \\
&= \int \left\{ \frac{1}{R} \sum_{r=1}^{R} h[Y_r + bz] \right\} k(z) dz \\
&= \int \mathrm{E}_R \left\{ h[Y_r + bz] \right\} k(z) dz.
\end{aligned} \tag{7.5}
$$

Note that the evaluation of $\mathrm{E}_{Rk}[h(Y)]$ in equation (7.5) requires only a single one dimensional integration, a tractable task.

For our applications, we routinely use the retained risk $Y = g(\mathbf{X}; \boldsymbol{\theta})$ - our interest is in the distribution function as we vary (and try to optimize over) the parameters in $\boldsymbol{\theta}$. A drawback of the usual estimator is that $\hat{F}_Y$ employs indicator (or "check") functions of the form $I[g(\mathbf{X}_r; \boldsymbol{\theta}) \leq y]$ that are **discontinuous** in the parameters (even if the function g is smooth). In contrast, we can use a smooth kernel function $K(\cdot)$ such as the cumulative normal probability function in the alternative estimator F_{Rk} that does not share this drawback. You can learn more about the mathematics underpinning the kernel smoothing in Appendix Section 10.5.3.

Example 7.5. Bivariate Excess of Loss Using VaR Optimization and Smoothing. This is a continuation of Example 7.2, using the second approach based on the smooth simulated distribution function. Specifically, we consider

$$
\begin{array}{|ll|}
\hline
\text{minimize}_{z_0, \boldsymbol{\theta}} & z_0 \\
\text{subject to} & F_{Rk}(x) \geq \alpha \\
 & RTC(\boldsymbol{\theta}) \leq RTC_{max}. \\
\hline
\end{array}
$$

In the appended sample code, we use the `kcde` function from the `R` library `ks`. An advantage of using this package is that it automatically determines an appropriate bandwidth b. This approach fares better than the approach based on the usual estimator investigated previously. However, the value at risk is still not an easy problem due to the need to start near the optimal values, as noted in Chapter 5.

7.3.2 Expected Shortfall and its Derivatives

Using equation (7.5), we may evaluate the function needed for expected shortfall minimization as

$$
ES1_{Rk}(z_0, \boldsymbol{\theta}) = z_0 + \frac{1}{1-\alpha} \int \mathrm{E}_R \left\{ [g(\mathbf{X}; \boldsymbol{\theta}) + bz - z_0]_+ \right\} k(z) dz. \tag{7.6}
$$

Expected Shortfall Derivatives. This function is smooth in the decision variables $(z_0, \boldsymbol{\theta})$, a property that we can use to speed up the convergence of the algorithm. Recall our Section 4.4 discussion on using derivatives to speed convergence.

Specifically, we can take partial derivatives of $ES1_{Rk}$ with respect to the decision variables, to get

$$
\partial_{z_0} \ ES1_{Rk}(z_0, \boldsymbol{\theta}) = 1 - \frac{1}{1-\alpha} \mathrm{E}_R \left\{ 1 - K \left(\frac{z_0 - g(\mathbf{X}; \boldsymbol{\theta})}{b} \right) \right\} \tag{7.7}
$$

and

$$\partial_{\boldsymbol{\theta}}\ ES1_{Rk}(z_0, \boldsymbol{\theta}) = \frac{1}{1-\alpha}\mathrm{E}_R\left\{\left[1 - K\left(\frac{z_0 - g(\mathbf{X};\boldsymbol{\theta})}{b}\right)\right] \partial_{\boldsymbol{\theta}} g(\mathbf{X};\boldsymbol{\theta})\right\}. \tag{7.8}$$

Excess of Loss Special Case. To provide intuition, we consider the special case of (multivariate) excess of loss. Here, the set of risk retention parameters is $\boldsymbol{\theta} = (u_1, \ldots, u_p)'$. In addition, the risk retention function is $g(\mathbf{X};\boldsymbol{\theta}) = X_1 \wedge u_1 + \cdots + X_p \wedge u_p = S(\mathbf{u})$. For the first parameter,

$$\begin{aligned}\partial_{\theta_1}\ g(\mathbf{X};\boldsymbol{\theta}) &= \partial_{u_1}\left(X_1 \wedge u_1 + \cdots + X_p \wedge u_p\right) \\ &= \partial_{u_1}\left(X_1 \wedge u_1\right) = I(X_1 > u_1).\end{aligned}$$

Replicating this for other parameters and organizing the results as a vector yields

$$\partial_{\boldsymbol{\theta}}\ g(\mathbf{X};\boldsymbol{\theta}) = \begin{pmatrix} I(X_1 > u_1) \\ \vdots \\ I(X_p > u_p) \end{pmatrix}.$$

Exercise 7.6 provides guidance on verifying these results. As suggested in Section 4.4.3, the following sections use these partial derivatives extensively as part of the minimization process.

7.4 Constructing a Portfolio Frontier

When we focused on the quota share problem in Section 7.2.3, due to the linearity of quota sharing, the optimization problem in Display (7.4) was convex and so could be solved very efficiently. However, other risk retention problems are nonlinear (such as an upper limit) and so we utilize more general constrained optimization solvers. Although these general solvers can be applied to a broader set of problems, they have limitations: they are generally less computationally efficient (slower), provide only local optimization, and they rely on starting values to begin their recursive procedures.

This section shows how to handle nonlinear contracts by focusing on excess of loss. Specifically, we now consider p risks $X_1, \ldots, X_p$ with upper limits $\mathbf{u} = (u_1, \ldots, u_p)$ so that the retained risk is

$$S(\mathbf{u}) = X_1 \wedge u_1 + \cdots + X_p \wedge u_p.$$

The simulation version of the expected shortfall (multivariate) excess of loss retention problem is:

$$\boxed{\begin{array}{ll}\text{minimize}_{z_0,\mathbf{u}} & ES1_{Rk}(z_0, \mathbf{u}) = z_0 + \mathrm{E}_{Rk}\{[S(\mathbf{u}) - z_0]_+\} \\ \text{subject to} & RTC_R(\mathbf{u}) \le RTC_{max} \\ & u_1 \ge 0, \ldots, u_p \ge 0.\end{array}} \tag{7.9}$$

Assuming fair risk transfer costs, we may write $RTC_R(\mathbf{u}) = RC_{R1}(u_1) + \cdots + RC_{Rp}(u_p)$, where $RC_{Rj}(u) = \mathrm{E}_R\left\{(X_j - u)_+\right\}$.

Example 7.6. Constructing a Portfolio Frontier for ANU Risks. To see how this works, we return to Example 7.4 on the *ANU Risk Data* with $p = 14$ risks except now we consider excess of loss, not quota sharing. As before, use $\alpha = 0.80$ for the level of confidence.

TABLE 7.3
Optimal Excess of Loss of ANU Risks (RTC_{max} = 433).

Param	Opt Value	Param	Opt Value	Opt Prop	Param	Opt Value	Opt Prop
RTC	433	u_1	1116.1	0.92	u_8	1109.3	0.967
VaR	1529.3	u_2	1113.8	0.969	u_9	1584.4	1
ES	2476.9	u_3	1114.4	0.967	u_{10}	1132.5	1
StdDev	1031.5	u_4	1119.7	0.975	u_{11}	694.6	1
		u_5	784.3	1	u_{12}	885.2	1
		u_6	150.8	1	u_{13}	642.4	1
		u_7	309.9	1	u_{14}	787.8	1

(Chapter 8 will investigate the effects of varying α.) As in Example 7.1, starting values are determined using the α quantile of each risk. Later, in Appendix Section 7.7.3, I show how to compute starting values quickly by using the variance as an approximation to the risk measure of interest, the expected shortfall.

The optimization of Display (7.9) uses the `R` package `alabama` and the results are summarized in Table 7.3. From this table, first note that the realized value of the risk transfer cost RTC at the optimal equals the budgeted amount, $RTC_{max} = 433$. Thus, the constraint is binding which, for our applications, is often a good signal that the algorithm has converged.

In addition to the optimal upper limits u^*, the table also provides corresponding proportions. So, for example, for the first risk the optimal upper limit is $1,116.1$ (in thousands of AUD). With the simulated distribution having $R = 100,000$ replicates, this corresponds to the 92nd percentile. Further, note that risks $5, 6, 7, 9, \ldots, 14$ are at the 100th percentile. For risk managers, we interpret this to mean that, given a budget of 433 (thousands), the algorithm essentially recommends self-insuring these risks (taking $u = \infty$) and using the budget to provide protection for the other risks.

7.4.1 A Frontier of Trade-offs

Range of RTC_{max}. Even for risk managers that have a good sense of the budget that they have available for risk protection, there is still considerable interest in understanding alternatives over a range of budget possibilities. At one end, a value of $RTC_{max} = 0$ corresponds to no risk transfer, meaning that $u_j = \infty$ for all j for the excess of loss contract. At the other end of the spectrum, we can develop intuition by assuming fair costing and expectations for risk transfers. This means that $\mathrm{E}(X_1 + \cdots + X_p)$ is the cost of full insurance or full transfer; for excess of loss, this yields $u_j = 0$ for all j.

To illustrate, for the ANU Example 7.6, from the simulated distribution, it is easy to calculate the cost of full transfer as 1,443.37. For a range of maximal risk transfer costs, I start with 95% of the full transfer cost and then proceed in deciles down to the 5% level for a total of 11 points. The levels of 100% and 0% are omitted because theory tells us values of risk retention levels at those points. In your own work, you may well choose different ranges of the maximal risk transfer cost but these choices seem sensible for the purposes of this book.

Starting values. Working with this range of maximal risk transfer costs helps immensely with the issue of starting values for the algorithm. This is because when starting with a very high level of RTC_{max}, close to full risk transfer cost, the upper limits will be close to zero (they are zero when RTC_{max} is at 100% of expectations). Thus, I use some small

TABLE 7.4
ANU Excess of Loss ES Optimization Frontier.

RTC_{max}	RTC	VaR	ES	Std Dev	u_1	u_2	u_3
1371.2	1370.1	75.3	79.0	3.6	0.0	0.0	0.0
1299.0	1293.1	151.4	153.6	2.0	1.1	0.0	0.0
1154.7	1145.1	307.1	309.6	9.6	0.0	0.0	0.0
1010.4	1010.4	457.2	471.8	32.2	4.6	3.0	3.0
866.0	866.0	661.4	874.7	205.9	102.6	98.6	97.1
721.7	721.7	896.0	1350.5	435.5	299.7	284.9	277.2
577.3	577.3	1133.4	1869.7	707.7	597.3	587.5	583.0
433.0	433.0	1529.4	2476.9	1031.4	1116.2	1114.3	1114.4
288.7	288.7	1573.2	3197.1	1416.8	2604.1	1711.5	1749.5
144.3	144.3	1573.2	3918.8	1929.9	5890.8	2648.8	2730.7
72.2	72.2	1573.2	4279.7	2279.5	9773.0	3437.9	3603.5

values of the upper limits as starting values for the algorithm. Proceeding to the next level of RTC_{max} (say, moving from the 95% to the 85% level), I can use optimal values from the prior level as starting values for the current level, and so on.

Frontier. Table 7.4 summarizes the results of 11 different optimization problems, one for each value of RTC_{max}. Not surprisingly, it shows that as the budget, RTC_{max}, decreases, the uncertainty of retained risk, measured by the ES, increases. As noted earlier, although the optimization is done with respect to the ES measure, once the optimal upper limits are obtained, it is straightforward to also determine other risk measures. Hence, in Table 7.4 one can provide the VaR and the standard deviation, *StdDev*, as alternative risk summary measures. Table 7.4 also shows the actual risk transfer cost, RTC, that in most cases equals the maximum and so the constraint is binding. The exceptions are in the cases where the value of RTC_{max} is large (the first three rows). Here, optimal upper limits are near the boundary of zero; it is more difficult to achieve convergence in these problems via simulation.

From Table 7.4, we see that the summary measures, VaR, ES, and the standard deviation (*Std Dev*), all increase as the maximal risk transfer cost RTC_{max} decreases, as anticipated. Further, this is also true of most of the retention parameters u_j. In particular, the table indicates when individual upper limits u_j are equal or close to zero, the case of no retention or full insurance. This case is of particular interest to risk managers.

Code. The supplements provide a set of code can be used for either excess of loss or quota share, utilizing kernel smoothing and gradients to speed convergence.

Visualizing the Frontier

Figure 7.1 shows the optimal value of the uncertainty measure versus the risk transfer cost. For each panel, one clearly sees the negative relationship between uncertainty and transfer costs. For the middle panel, these are the optimal values of the expected shortfall. Thus, you can interpret this as an optimal "frontier" - given a risk transfer cost, this is the best (smallest) value of uncertainty that can be achieved. Uncertainty frontiers are familiar constructs to risk managers in an asset allocation context and so this interpretation is appealing.

Because we are minimizing expected shortfall, the left and right panels of Figure 7.1 are not technically "frontiers." However, as with the middle panel, they demonstrate the trade-off between costs of transferring risks and the uncertainty of retained risks. Some risk

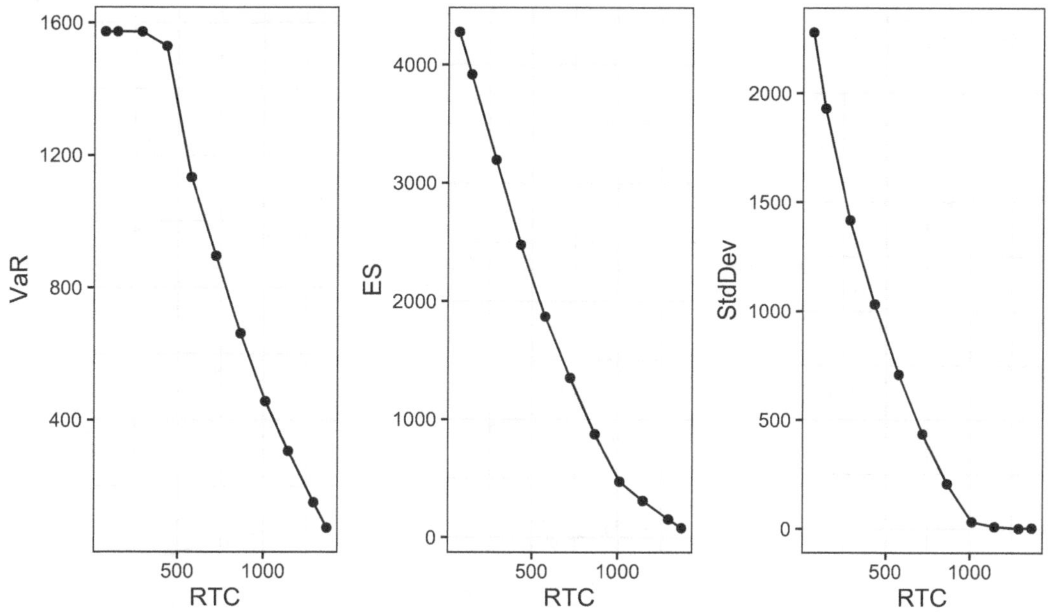

FIGURE 7.1
ANU Excess of Loss Optimization Results—ES Criterion.

managers are comfortable with the value at risk as a measure of uncertainty, whereas others prefer the more traditional standard deviation. Depending on the intended audience, it can be valuable to show that one reaches qualitatively similar conclusions regardless of the risk measure utilized. Section 8.2.3 will revisit the topic of the choice of the risk measure.

Alternative Frontiers. Naturally, analysts are not constrained to portraying measures of uncertainty over a range of risk transfer costs. Alternatively, or in addition, one could display measures of uncertainty versus almost any model parameter or summary measure, depending on the point of emphasis. As illustrations, Chapter 8 will show alternative frontiers that vary by the confidence level α and the range value at risk parameter β. Chapter 9 will show alternatives by levels of dependence. The convention adopted in this book, utilizing risk transfer costs, is motivated by analogies to the Markowitz investment portfolio optimization and so is likely to be familiar to insurance risk managers.

Risk Retention Coefficients

It can be difficult to interpret the practical implications of an uncertainty measure, even a standard deviation. In contrast, upper limits are parameters that appear in contracts and so are familiar to risk managers. Figure 7.2 shows how each upper limit changes according to the level of the risk transfer cost. For example, from the upper right panel of Figure 7.2 and the corresponding u_3 column in Table 7.4, we see that the upper limit for the crime coverage decreases slowly as the risk transfer increases. Even when $RTC = 868$ (which is about 60% of the full transfer amount 1,443.37), the optimal upper limit is only 97.1, suggesting that most of this risk should be transferred (e.g., insured) when a moderate transfer budget is available.

In Figure 7.2, the dashed line represents the 99th percentile of the risk displayed in each panel. When the recommended upper limit exceeds that benchmark, we typically interpret

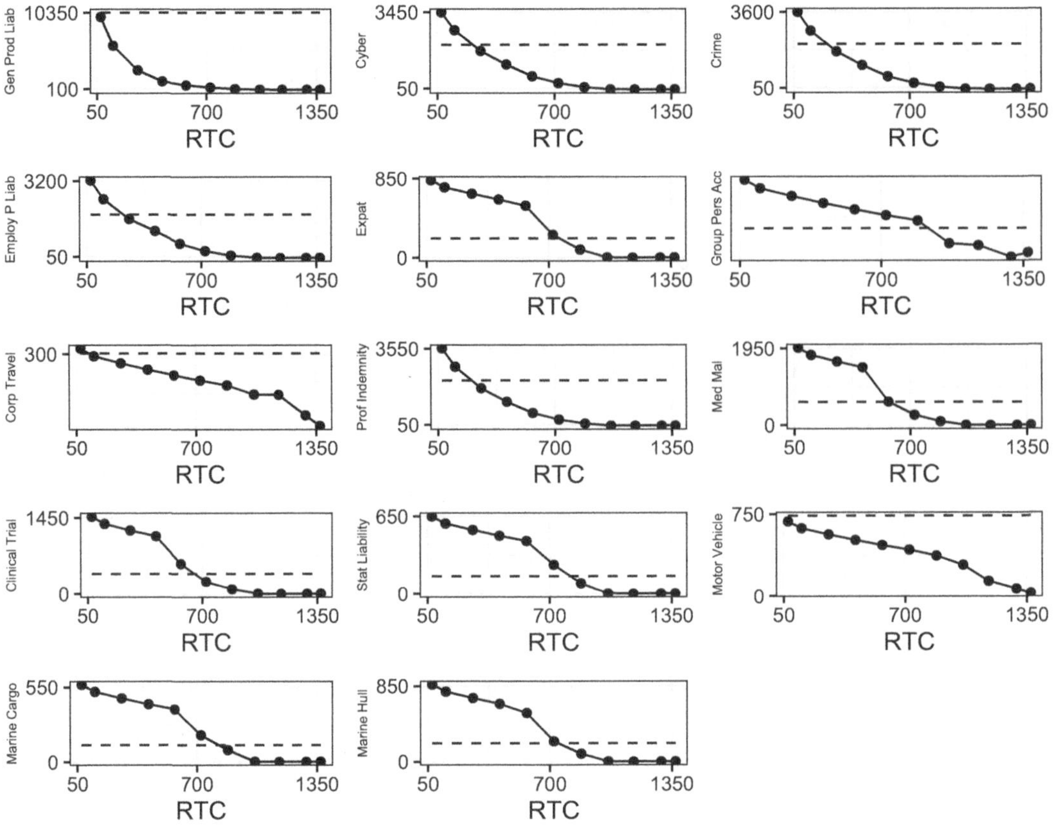

FIGURE 7.2
ANU Optimal Upper Limits versus Risk Transfer Costs, by Risk Type.

this to mean that all of this risk is being retained, corresponding to $u_j = \infty$. At this level, when a finite risk retention level is used, it is typically not motivated by empirical modeling such as the framework we are pursuing but rather considerations external to the modeling.

7.4.2 Comparing Different Objective Functions

We learned in Section 7.2.3 that minimizing the expected shortfall objective function is equivalent to minimizing the standard deviation for the quota share problem, at least for elliptical risks. How different are the optimal parameter estimates for the excess of loss problem?

To show the similarity among the results, Figure 7.3 compares values of $StdDev$, ES, and two of the retention parameters, u_1 and u_{10}, for the ANU data. Here, the optimization is done twice, first by minimizing the variance/standard deviation function in Section 7.1 and second by minimizing the expected shortfall in Section 7.4.1. In Figure 7.3, each plotting point represents a different level of RTC_{max}.

The top row of Figure 7.3 compares the two risk measures obtained using the different objective functions. It shows that the values of the risk measures are largely comparable. In addition, the bottom row compares values of two retention parameters, the first and the tenth, obtained from the two minimization criteria. Values of the first parameter obtained by

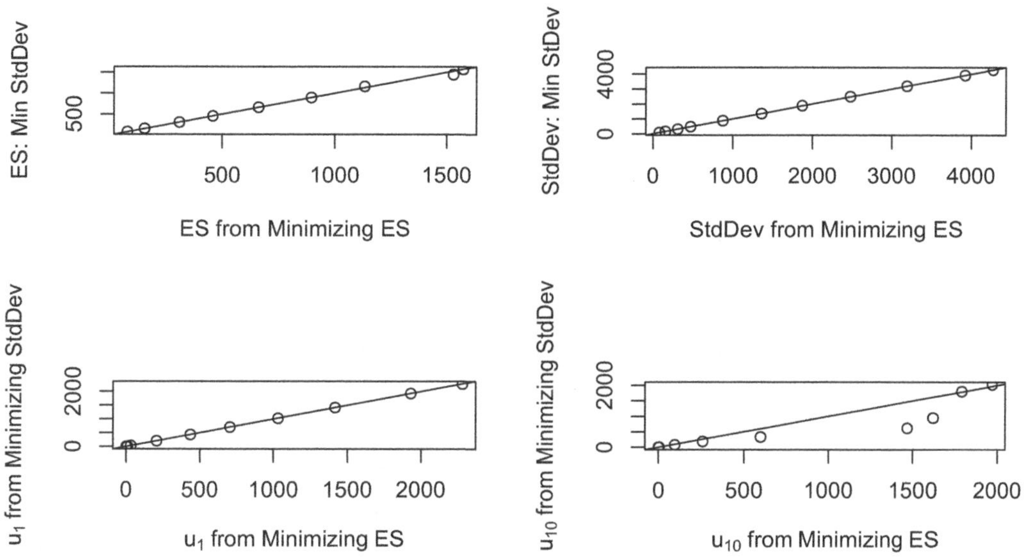

FIGURE 7.3
Comparison of Risk Measures and Retention Parameters by Optimization Method.

TABLE 7.5
Correlation among Methods for Determining Retention Parameters.

Param	u_1	u_2	u_3	u_4	u_5	u_6	u_7
Correlation	1	0.999	0.97	0.976	0.98	0.954	0.997
Param	u_8	u_9	u_{10}	u_{11}	u_{12}	u_{13}	u_{14}
Correlation	0.999	0.976	0.927	0.945	0.984	1	0.948

different objective functions are largely equivalent whereas those from the tenth parameter displays markedly different values for some levels of RTC_{max}.

One can summarize the relationships given in the Figure 7.3 panels using correlation statistics, such as shown in Table 7.5. Here, a correlation close to one suggests a linear relationship, that is, a close relationship for each retention parameter constructed using different objective functions. From the table, we see that although there are some differences when using different objective functions, there is substantial agreement for each of the 14 risks. I follow up with this in Appendix Section 7.7.3 where standard deviation minimization is employed to provide starting values for expected shortfall minimization.

7.5 General Risk Retention Functions

Section 7.4 described simulation strategies to optimize our first nonlinear risk transfer agreement, the excess of loss. Having done so, similar strategies can be applied to a host of other risk transfer agreements. To distinguish among alternative agreements, this section describes general risk retention functions that are of interest to risk managers. In this book, we have used the notation $g(\mathbf{X};\boldsymbol{\theta})$ to represent a risk owner's retained risk after a risk transfer agreement $g(\cdot;\cdot)$ based on multivariate risks $\mathbf{X} = (X_1, \ldots, X_p)$ and a given set of risk retention parameters $\boldsymbol{\theta}$. This section describes different categories of risk transfer agreements.

7.5.1 Separate Contracts for each Risk

One way to think about a firm's risk retention policy is to postulate a different contract for each risk so that the amount retained for the jth risk is of the form $g_j(X_j;\boldsymbol{\theta}_j)$. Many cases of interest also assume an additive structure among retained risks so that the total retained risk can be expressed as

$$g(\mathbf{X};\boldsymbol{\theta}) = \sum_{j=1}^{p} g_j(X_j;\boldsymbol{\theta}_j). \tag{7.10}$$

As examples, we have already considered the quota share agreement with $\theta_j = c_j$ and $g_j(X_j;c_j) = c_j \cdot X_j$ as well as the excess of loss agreement with $\theta_j = u_j$ and $g_j(X_j;u_j) = X_j \wedge u_j$. The extension in equation (7.10) allows some of the contracts to be excess of loss and others to be quota share, as in Example 7.1. Also recall that with Chapter 2 based on univariate risks ($p = 1$), in equation (2.6) we considered a risk retention function with three risk retention parameters, $\boldsymbol{\theta} = (d, c, u)$ where d is the deductible, c is the coinsurance amount, and u is the upper limit of retention. Some risks may have deductibles, coinsurance, and upper limits.

Working with additive structures simplifies the numerical convergence of our optimization procedures. Moreover, as we will see in Chapter 11, an additive structure will simplify the analysis of conditions for when risk transfers achieve optimal values.

Note that the contracts represented by the risk transfer agreements $g_j(\cdot)$ may or may not be with different vendors. If a vendor takes on a set of risks, other risks are irrelevant from that vendor's perspective and may be treated as independent. Nonetheless, from the perspective of the risk owner, the risks themselves may still be dependent.

7.5.2 Combining Risks

Naturally, there will also be many instances when it is not possible to consider different risks additively. Rather than attempt to enumerate the infinite number of possible nonlinear combinations, this section describes selected examples of interest.

Example 7.7. Combined Deductibles from Bundled Insurance Contracts. In personal insurance, combined deductibles are common when insurers "bundle" together several types of insurance that had historically been offered separately, see, for example, Dong et al. (2022). For example, in travel insurance X_1 might represent the loss due to luggage damage and X_2 is the loss due to personal injury. Then, a type of insurance policy covers

$$Y_{insurer} = \left[(X_1 - d_1)_+ + (X_2 - d_2)_+ - d_{combined}\right]_+ .$$

In this expression, d_1 is the deductible for luggage claims and $(X_1 - d_1)_+$ is the luggage excess. In the same way, d_2 is the deductible for personal injury and $(X_2 - d_2)_+$ is the personal injury excess. The insurer is responsible for the excess over a combined deductible, $d_{combined}$, over these two separate excesses. Note that these are "deductibles" from the perspective of the insurer but can also be viewed as upper limits from an individual's perspective.

Example 7.8. Global Protections Coverage. As described in Section 4.1.4, agreements between an insurer and a reinsurer can cover multivariate risks in a nonlinear way. To illustrate, in that section we introduced an agreement where the reinsurer's portion of the total risk is

$$Y_{reinsurer} = \min\left(\sum_{j=1}^{p} \min[(X_j - u_{1,j})_+, u_{2,j}], AAL\right).$$

Here, $u_{1,j}$ represents a reinsurer's deductible, $u_{2,j}$ represents is upper limit for the jth risk, and AAL stands for the annual aggregate limit AAL. Also as noted in that section, Albrecher et al. (2017) refers AAL as a type of *global protections.*

Example 7.9. Separate Coverages by Peril. Sometimes risks are split by cause of loss, also known as *peril,* with different arrangements such as sublimits made for each peril. Sublimits by peril are common in property and casualty insurance. For example, the ANU property risk, described in Appendix Section 8.3.4, is subdivided into two parts, each having different sets of sublimits. One part is for material loss or damage that covers real and personal property in which ANU has an insurable interest. The second part is for consequential loss that covers other damages suffered by ANU resulting from the interruption or interference of business. See Frees and Butt (2022) for more information about ANU sublimits.

To illustrate in a simple setting, one might introduce a sublimit u_1 that limits how much of risk X_1 can be transferred. Then, if X_2 represents all other risks and $u_{overall}$ is an overall limit, we can use

$$Y_{firm} = (X_1 \wedge u_1 + X_2) \wedge u_{overall}$$

to express the amount retained by a firm.

7.5.3 Frequency Severity Approaches to Modeling Risks

For risks that occur frequently in a calendar period, it is convenient to define risk transfers on a *per event* or *per occurrence of a claim* basis. So, instead of working directly with X_j, the overall type j loss amount, we might decompose this into a number of claims in a year, say N_j, and examine each individual claim outcome, $X_{j1}, \ldots, X_{jN_j}$. This is sometimes known as the *collective risk* model, cf. Actuarial Community (2020), Chapter 5. With the collective model approach, we express aggregate type j losses as $X_j = \sum_{i=1}^{N_j} X_{ji}$.

As an example, it is common for a large firm, the risk owner, to maintain a fleet of motor vehicles where insured claims occur with a high frequency. In this case, it is common to define risk transfer mechanisms (such as deductibles and upper limits) for each claim. For example, a firm might retain claims up to u_j and so the total retained claims is

$$Y_{j,retained} = \sum_{i=1}^{N_j} X_{ji} \wedge u_j.$$

As another example, suppose that an insurer provides protection to a large firm that has many buildings. As described in Section 4.1.4, the amount transferred to the insurer may be of the form

$$Y_{j,transfer} = \left[\sum_{i \in \text{event}} (X_{ji} - d_j)_+\right] \wedge u_j.$$

For each building claim, $(X_{ji} - d_j)_+$ represents the amount in excess of a deductible d_j. The sum within the square braces is the total excess for all claims but the transferred amount is limited by u_j. Analysis of this type of insurance protection is especially difficult when the buildings are concentrated geographically (such as buildings owned by a university) and there can be many claims due to a common event such as an earthquake or hurricane.

For the collective risk model, analysis by peril is a bit more complicated because, for an individual claim, only one of several types of perils can occur. See Frees et al. (2010) for further discussion of this model.

7.5.4 Comparing Risk Retention Functions

As described in this section, there can be many candidate risk retention plans depending on the nature of the risks. In addition, the experience of the risk manager and industry norms can also play important roles in selecting a risk retention plan.

As summarized in this chapter, optimal parameter values of each risk retention plan can be determined. However, risk retention parameters are not directly comparable as they quantify different types of contributions to retained risk. What can be quantified and compared is the overall measure of retained risk uncertainty.

For example, Figure 7.4 shows expected shortfall versus maximal risk transfer costs for two risk retention plans, the quota share problem described in Example 7.4 and the excess of loss problem described in Section 7.4.1. Both use the same distributions based on fourteen risks from the ANU case. Figure 7.4 shows that the solid plotting symbols, corresponding to quota share, are consistently higher than the open symbols corresponding excess of loss. The ratio difference can be as large as 12 percent for the middle values of RTC_{max}. Although some risk managers may have difficulty in interpreting an expected shortfall, the interpretation of a 12 percent reduction in uncertainty (by moving from quota share to excess of loss) can be appealing.

7.5.5 Market Loading of Risk Transfer Costs

This section has focused on extensions of risk retention functions needed to determine optimal risk transfer solutions. To promote interpretability, I have focused the development on fair risk transfer costs. Naturally, to implement these procedures, adjustments to fair

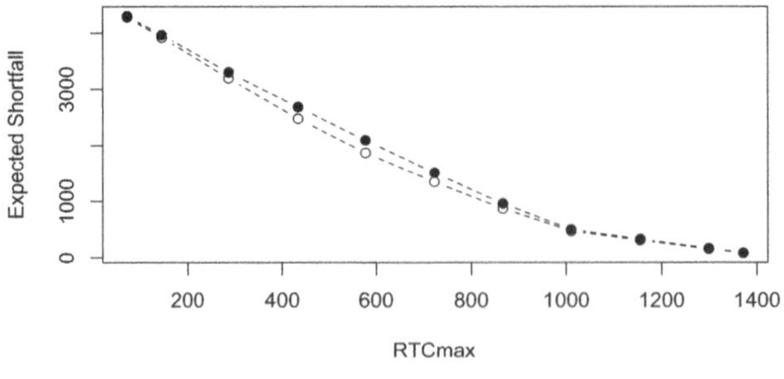

FIGURE 7.4
Comparison of Expected Shortfall Frontiers from Excess of Loss and Quota Share Programs.

costing will be needed in order to reflect administrative expenses and the market availability of coverages.

So, as a small extension of the fair risk transfer costs, one can consider *market loadings* (ML) of risk transfer costs. These are exogenous coefficients for which equation (7.11) gives two ways of introducing market loadings. In the first expression, we consider the portfolio as a whole and so this may be an adjustment for overall administrative expenses that are common to all risk types. To motivate the second approach, one might assume that the portfolio risk transfer costs can be subdivided into separate contracts as in Section 7.5.1. In this case, we can think about different market loadings for each risk (ML_j) and use

$$RTC_R(\boldsymbol{\theta}) = \begin{cases} \frac{ML}{R}\sum_{r=1}^{R}\left\{\left(\sum_{j=1}^{p} X_{rj}\right) - g(\mathbf{X}_r;\boldsymbol{\theta})\right\} \\ \frac{1}{R}\sum_{r=1}^{R}\left\{\sum_{j=1}^{p} ML_j[X_{rj} - g_j(\mathbf{X}_r;\boldsymbol{\theta}_j)]\right\} \end{cases}. \tag{7.11}$$

In some exercises that follow, such as Exercise 8.7 on varying a cyber risk premium, I implement these simple adjustments. Naturally, there are other methods of determining risks transfer costs, such as using quantiles in lieu of expectations for costing, that the reader is invited to explore.

7.6 Simulation Uncertainty

As seen in Chapter 5, with only two risks we are able to use precise deterministic methods to construct optimal insurable risk portfolios. When the number of risks is larger than two, simulation methods become preferred and are the focus of this chapter (and the remainder of the book). However, simulation methods introduce uncertainty and risk managers want some assurance that recommendations are not being clouded by unpredictable errors introduced by this calculation technique.

As analysts, we know that simulation methods typically converge to deterministic quantities due to the law of large numbers. However, these results do not help analysts decide how large a sample size, R, to use. There is, of course, a trade-off; larger values of R are more accurate but take longer to compute. For practicing analysts, one approach is to use a smaller number of replications during the design phase of a study when broad patterns and tendencies are of interest. Later, at the final stages of the study when precision is paramount, a greater number of replications can be used to enhance the reliability of the study.

To calibrate the size of simulation uncertainty, one option is to use central limit theorem results to get estimates of the size of simulation errors, cf. Actuarial Community (2020), Chapter 6. In addition, to reassure risk managers that simulation uncertainty has been addressed, this section demonstrates the use of resampling techniques to show effects of simulation uncertainty.

To assess simulation variability, I generated 20 independent samples of the ANU distribution, each of size 5,000. The sample size 5,000 is small relative to other simulation applications; with this relatively small number we can visualize the uncertainty induced by simulation. For each sample, I determined optimal values of the upper limits over a range of maximal risk transfer costs.

Figure 7.5 summarizes the results. The left-hand panel shows the risk transfer cost RTC versus the expected shortfall ES for each level of RTC_{max}. In the same way, the middle and right-hand panels show optimal values of the first two optimal retention parameters, u_1 and u_2. At each level, there are 20 points plotted, each corresponding to a different simulation scenario. For these runs, it turned out that the RTC equals the maximal value RTC_{max} so this constraint is binding.

For large values of RTC, it is difficult to distinguish among the different values of ES, u_1 and u_2, suggesting that using only $R = 5,000$ simulations is adequate. For small values of RTC, the figure provides an indication of the uncertainty imposed by the simulation. If one is dealing with a situation where most of the risk is being retained but a great deal of precision is needed, one should increase the number of simulations to increase the precision. The strength of the type of display in Figure 7.5 is that the simulation uncertainty is interpretable for a risk manager and can help to provide comfort when interpreting results from simulation optimization techniques.

A similar type of resampling technique that can show simulation uncertainty is a *bootstrap* method. For example, I started with a large number of replications, $R = 100,000$, and then resampled (with replacement) from this initial sample to generate $B = 10$ bootstrap replicates. Each bootstrap replicate is optimized and the variability among these optimizations can be used to portray simulation uncertainty, similar to Figure 7.5. This output is not provided because, as it turns out, there was practically no simulation uncertainty in such a graph due to the large number of initial replicates, $R = 100,000$. This type of exercise provided motivation for selecting $R = 100,000$ as the basis for the examples in this chapter so that readers need not worry about simulation error. An exercise asks readers to verify this finding.

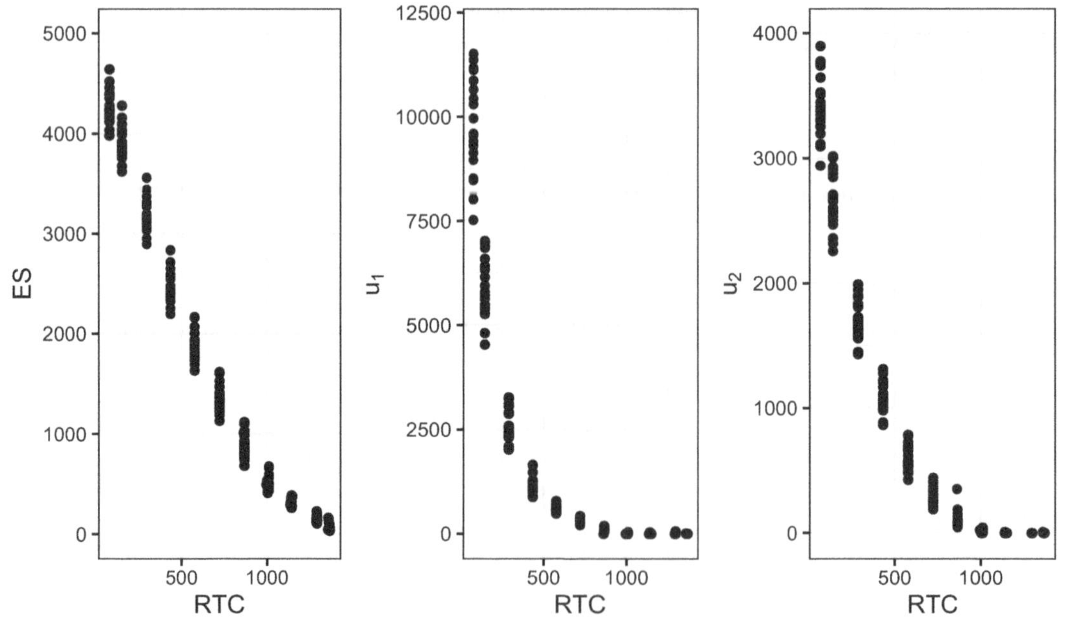

FIGURE 7.5
ANU Excess of Loss Optimization Results for Several Subsamples.

7.7 Supplemental Materials

7.7.1 Further Resources and Reading

As noted earlier, for readers wanting a refresher in simulation techniques, see the online book Actuarial Community (2020), Chapter 6. In addition, this book provides a short introduction to kernel estimation in Section 4.1.

The approach used in this book is based on minimizing a function that is approximated using simulation. In the operations research literature, this is known as *sample average approximation*. See, for example, Shapiro (2013) for a brief introduction or Kim et al. (2015) for a more detailed introductory guide. Within this literature, there are many variations of this basic approach that may be employed to improve the initial efforts presented here. In particular, I have adopted a smoothing approach based on kernel estimation in Section 7.3. Smoothing of non-continuous functions has been described more broadly, cf. Fu (2015b), with kernel estimation being a specific case. For example, Liu and Hong (2009) uses kernel estimation to derive quantile sensitivities. See Givens and Hoeting (2013) for a background on kernel estimation, including computational considerations.

The quota share model that was discussed in Section 7.2.3 had been put forth earlier in Sun et al. (2017). This paper worked, not using simulation, but in a more general context of nonparametric estimation, including the kernel techniques described in Section 7.3.

7.7.2 Exercises

See the *Online Supplement* for selected exercise solutions.

Section 7.1 Exercises

Exercise 7.1. Quantile Regression Approach. Many data analysts have some familiarity with *quantile regression*. As the name suggests, this is a technique for determining the value at risk, or quantile, in the presence of linear covariates. See, for example, Koenker (2005). The goal of this exercise is to establish the equivalence between this approach and the ES minimization in Display (7.4), at least for risk retention problems of interest to us.

a. Define a function $\phi_\alpha(y) = 0.5|y| + (\alpha - 0.5)\,y$ that is known as *loss function* in classical statistics. Consider a generic random variable Y having distribution function F that has a unique value, $F^{-1}(\alpha)$, for its α quantile. Check that $F^{-1}(\alpha)$ is the solution of the unconstrained minimization problem

$$\boxed{\text{minimize}_{z_0} \qquad \mathrm{E}\ \phi_\alpha(Y - z_0)}$$

b. Now, suppose that we have $i.i.d$ replications of $Y_r = g(\mathbf{X}_r;\boldsymbol{\theta})$, $r = 1, \ldots, R$. Show that one can summarize the quantile regression retention problem as:

$$\boxed{\begin{array}{ll} \text{minimize}_{z_0,\boldsymbol{\theta}} & \mathrm{E}_R\ \phi_\alpha[g(\mathbf{X};\boldsymbol{\theta}) - z_0] \\ \text{subject to} & RTC_R(\boldsymbol{\theta}) \le RTC_{max}. \end{array}} \qquad (7.12)$$

c. To see that the expected shortfall problem is essentially identical to the regression quantile risk problem, consider the following.

- In Display (7.12), we seek to minimize the objective function $f_{QR,0}(z_0,\boldsymbol{\theta}) = \mathrm{E}_R\ \phi_\alpha[g(\mathbf{X};\boldsymbol{\theta}) - z_0]$.
- In Display (7.4), we seek to minimize the objective function $f_{ES,0}(z_0,\boldsymbol{\theta}) = z_0 + \frac{1}{(1-\alpha)}\mathrm{E}_R[g(\mathbf{X};\boldsymbol{\theta}) - z_0]_+$.

Show that these are related through the expression

$$f_{ES,0}(z_0,\boldsymbol{\theta}) = \mathrm{E}_R\ g(\mathbf{X};\boldsymbol{\theta}) + \frac{1}{1-\alpha} f_{QR,0}(z_0,\boldsymbol{\theta}). \tag{7.13}$$

Equation (7.13) implies that the optimal value z_0^* is the same for both procedures. We defer the demonstration of the equivalence between these two methods for the other decision variables to Exercise 11.5 where it arises naturally as part of a discussion on risk transfer conditions.

Exercise 7.2. No Kernel Smoothing: Simulated Expected Shortfall Derivatives. Consider the simulation version of the expected shortfall objective function *without* kernel smoothing. Specifically, as in Display (7.4), use an objective function that depends on $\mathbf{z} = (z_0,\boldsymbol{\theta})'$ decision variables and can be expressed as

$$ES1_R(\mathbf{z}) = z_0 + \frac{1}{(1-\alpha)}\mathrm{E}_R\ \{[g(\mathbf{X};\boldsymbol{\theta}) - z_0]_+\}.$$

a. Show that the gradient is $\partial_{\mathbf{z}}\ ES1_R(\mathbf{z}) = (\partial_{z_0}\ ES1_R(z_0,\boldsymbol{\theta}), \partial_{\boldsymbol{\theta}}\ ES1_R(z_0,\boldsymbol{\theta}))$ where

$$\partial_{z_0}\ ES1_R(\mathbf{z}) = 1 - \frac{1}{1-\alpha}\mathrm{E}_R\ \{I\,[g(\mathbf{X};\boldsymbol{\theta}) > z_0]\}$$

and

$$\partial_{\boldsymbol{\theta}}\ ES1_R(\mathbf{z}) = \frac{1}{1-\alpha}\mathrm{E}_R\ \{I\,[g(\mathbf{X};\boldsymbol{\theta}) > z_0]\,\partial_{\boldsymbol{\theta}} g(\mathbf{X};\boldsymbol{\theta})\}.$$

b. Show that the Hessian is

$$\partial_{\mathbf{z}}\partial_{\mathbf{z}'}\ ES1_R(\mathbf{z}) = \begin{pmatrix} 0 & \mathbf{0}' \\ \mathbf{0} & \partial_{\boldsymbol{\theta}}\partial_{\boldsymbol{\theta}'}\ ES1_R(z_0,\boldsymbol{\theta}) \end{pmatrix},$$

where

$$\partial_{\boldsymbol{\theta}}\partial_{\boldsymbol{\theta}'}\ ES1_R(z_0,\boldsymbol{\theta}) = \frac{1}{1-\alpha}\mathrm{E}_R\ \{I\,[g(\mathbf{X};\boldsymbol{\theta}) > x]\,\partial_{\boldsymbol{\theta}} g(\mathbf{X};\boldsymbol{\theta})\partial_{\boldsymbol{\theta}'} g(\mathbf{X};\boldsymbol{\theta})\}.$$

Note, from this expression, that the Hessian is not invertible.

Section 7.2 Exercises

Exercise 7.3. Quota Sharing of ANU Risks. To see how this works, let us revisit the *ANU Risk Data*, as in Example 7.4. Replicate the results in Table 7.6 that summarizes optimization results which extends results from Example 7.4 over a range of costs. As in Example 7.4, the minimization is done over the set of allocations $c_1,\ldots,c_{14}$ although only the first three quota share coefficients are printed, for brevity. Optimization calculations should be done using the `R` convex optimization package CVXR.

TABLE 7.6
ANU Quota Share Optimization Results Using CVXR.

RTC_{max}	RTC	VaR	ES	Std Dev	c_1	c_2	c_3
1371.2	1371.2	78.7	81.4	5.1	0.000	0.000	0.000
1299.0	1299.0	157.5	162.7	10.3	0.000	0.000	0.000
1154.7	1154.7	319.2	332.8	24.2	0.000	0.000	0.000
1010.4	1010.4	487.7	510.7	42.5	0.000	0.000	0.000
866.0	866.0	787.9	1187.9	312.7	0.020	0.136	0.126
721.7	721.7	1169.3	2004.5	636.1	0.054	0.369	0.345
577.3	577.3	1574.3	2876.3	973.3	0.096	0.650	0.610
433.0	433.0	1992.9	3764.8	1314.8	0.138	0.939	0.885
288.7	288.7	2346.8	4765.8	1734.7	0.377	1.000	1.000
144.3	144.3	2583.1	5954.1	2263.0	0.688	1.000	1.000
72.2	72.2	2681.6	6576.2	2546.1	0.844	1.000	1.000

Exercise 7.4. Quota Sharing of ANU Risks Minimizing the Variance. This is a continuation of Exercise 7.3 but now minimizing the variance, not the expected shortfall. Provide `R` code to determine quota share allocations when minimizing the variance. The results are not displayed because of their similarity to Table 7.6.

Exercise 7.5. Comparing Quota Sharing of ANU Risks Using Different Objective Functions. In this exercise, you will compare results from Exercise 7.3 based on the (auxiliary) expected shortfall objective function to those of Exercise 7.4 based on the variance objective function. To show the similarity among the results, provide a graph similar to Figure 7.6 that compares values of VaR, ES, and two of the quota share parameters, c_1 and c_2.

We learned in Section 7.2.3 that results would be exactly the same for elliptically distributed risks yet have no theory for non-elliptical risks such as the ANU data. This exercise reinforces the observations made in Example 7.4, extending the analysis to a range of values of RTC_{max}.

Section 7.3 Exercises

Exercise 7.6. Kernel Smoothing of Expectations, with Derivatives. Consider the simulation version of the expectations, including expected shortfall, with kernel smoothing.

a. Verify the expression in equation (7.5).

b. Verify the expression in equation (7.6).

c. Show that the gradient is $\partial_{\mathbf{z}}\ ES1_{Rk}(\mathbf{z}) = (\partial_{z_0}\ ES1_{Rk}(z_0, \boldsymbol{\theta}), \partial_{\boldsymbol{\theta}}\ ES1_{Rk}(z_0, \boldsymbol{\theta}))$ where $\partial_{z_0}\ ES1_{Rk}(z_0, \boldsymbol{\theta})$ is given in equation (7.7) and $\partial_{\boldsymbol{\theta}}\ ES1_{Rk}(z_0, \boldsymbol{\theta})$ is given in equation (7.8).

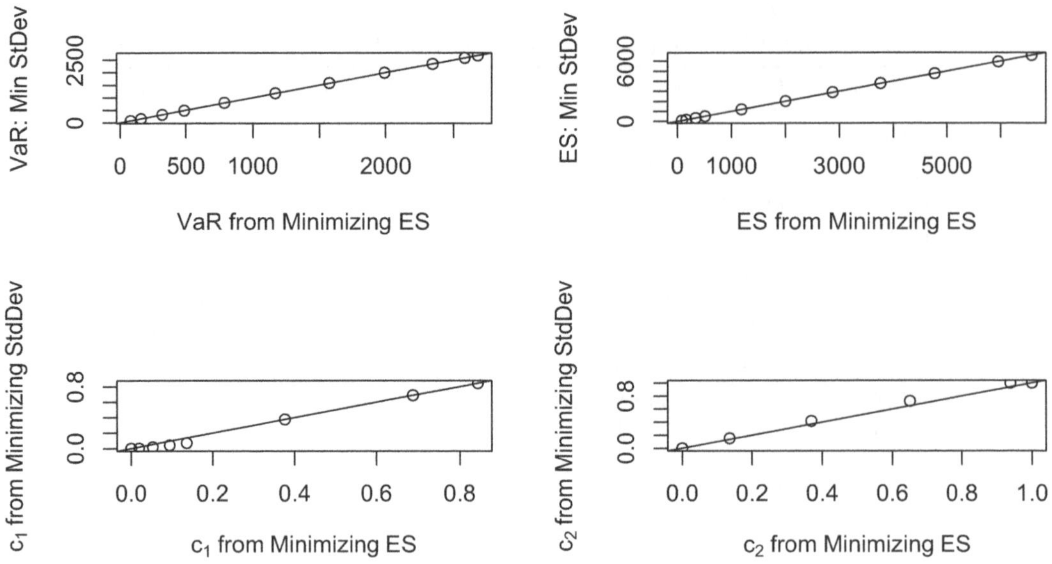

FIGURE 7.6
Comparison of Risk Measures and Allocation Parameters from Different Objective Functions.

Exercise 7.7. Kernel Smoothed Expected Shortfall Hessian. Provide an expression for the Hessian associated with kernel smoothed version of the expected shortfall. From the solution, you will see that this matrix is generally invertible, in contrast to the Hessian derived in Exercise 7.2.

Exercise 7.8. Value at Risk with Kernel Smoothing. Consider the simulation version of the value at risk with kernel smoothing. Define kernel smoothed value at risk VaR_{Rk} as the solution of the equation $F_{Rk}(y) = \alpha$. Further consider an excess of loss retention scheme with $\boldsymbol{\theta} = (u_1, \ldots, u_p)$. Show that the partial derivatives can be expressed as

$$\partial_{\mathbf{u}} VaR_{Rk} = \left\{ \sum_{r=1}^{R} k\left(\frac{y - S_r(\mathbf{u})}{b}\right) \begin{pmatrix} I(X_{1r} > u_1) \\ \vdots \\ I(X_{pr} > u_p) \end{pmatrix} \right\} \left[\sum_{r=1}^{R} k\left(\frac{y - S_r(\mathbf{u})}{b}\right) \right]^{-1} \Bigg|_{y=VaR_{Rk}},$$

that is, a weighted average.

Section 7.6 Exercise

Exercise 7.9. Bivariate Excess of Loss Using *ES* Optimization. Let us demonstrate the simulation uncertainty introduced in Section 7.6 but in a simpler context than the ANU data. Specifically, consider the distribution introduced in Example 7.3 that considers two risks that depend on one another through a Gaussian copula with dependence parameter $\rho = 0.5$. The first risk is X_1, with a gamma distribution having shape parameter 2 and scale parameter 2,000, and the second is X_2, with a Pareto distribution having shape parameter 3 and scale parameter 2,000. A level of confidence $\alpha = 0.95$ is used.

Replicate the work summarized in Figure 7.7. This was produced using 10 independent samples, each of size 10,000.

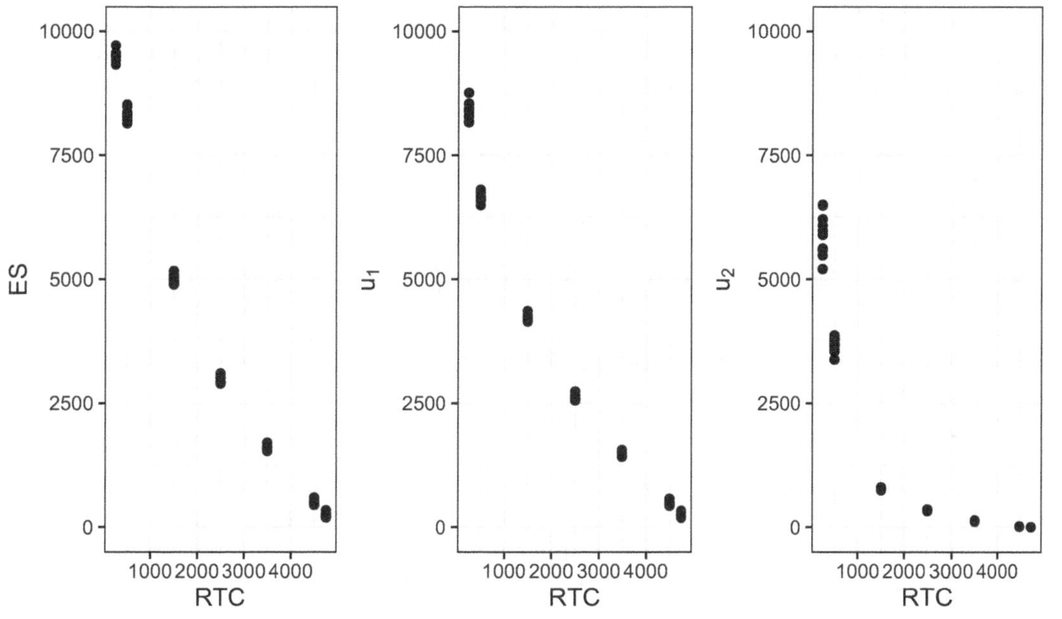

FIGURE 7.7
Gamma Pareto Excess of Loss Optimization Results for Several Subsamples.

7.7.3 Appendix. Starting Values for Multivariate Excess of Loss

We use numerical methods to solve constrained optimization problems as in Display (7.9). This problem is not in general convex and because of the large number of decision variables ($p = 14$ for the ANU case study), the optimization task is immensely simplified with good starting values. As suggested in Section 7.2.3, one approach to determine starting values is to utilize the *variance* as a risk measure. By making additional simplifying assumptions of independence among risks and assuming the constraint is active, we can rely on some classic results from the actuarial literature, as follows.

For numerical work, it helps to rescale risks. To this end, define scale factors sd_j and rescaled risks $X_j^s = X_j/sd_j$, $j = 1, \ldots, p$. You can think of sd as a standard deviation although it need not be.

With the rescaled risks and an active constraint, we seek to solve the problem

$$\boxed{\begin{array}{ll} \text{minimize}_{u_1^s,\ldots,u_p^s} & \sum_{j=1}^{p} \mathrm{Var}(X_j^s \wedge u_j^s) \\ \text{subject to} & RTC(u_1^s, \ldots, u_p^s) = RTC^s_{max}. \end{array}} \qquad (7.14)$$

Using the method of Lagrange multipliers, one can show that the solution is

$$\begin{aligned} LME^s &= 2(u_j^s - E[X_j^s \wedge u_j^s]) \qquad\qquad j = 1, \ldots, p \\ RTC^s_{max} &= \sum_{j=1}^{p} \mathrm{E}(X_j^s) - \mathrm{E}(X_j^s \wedge u_j^s)]. \end{aligned} \qquad (7.15)$$

The first result of this fashion that I am aware of appears in Bühlmann (1970), Section 5.2.2.

ⓘ *Under the Hood.* See the *Online Supplement* for the verification of Display (7.15).

Display (7.15) is a system of $p+1$ equations having $p+1$ unknowns, $\{LME^s, u_1^s, \ldots, u_p^s\}$, that can be solved via direct substitution. To this end, define the scaled distribution function $F_j^s(x) = \Pr(X_j^s \leq x) = F_j(x \times sd_j)$ and the corresponding integrated distribution function

$$H_j^s(x) = \int_0^x F_j^s(z)dz = x - \int_0^x [1 - F_j^s(z)]dz = x - E(X_j^s \wedge x).$$

With this notation, we may determine LME^s as the solution of the equation

$$\sum_{j=1}^{p} \mathrm{E}(X_j^s) - RTC_{max}^s = \sum_{j=1}^{p} \left\{ H_j^{-1s}\left(\frac{LME^s}{2}\right) - \frac{LME^s}{2} \right\}. \tag{7.16}$$

With a value of LME^s, we can then calculate the optimal scale retention limits using $u_j^s = H_j^{-1s}(\frac{LME^s}{2})$.

Under the Hood. See the *Online Supplement* for the verification of Equation (7.16).

After solving this problem, we revert to the original scale and define the rescaled upper limits to be $u_j^{rs} = sd_j \times u_j^s$. For the scaled problem, we know from the first line of Display (7.15) that the separation between the upper bound and its limited expected value, $u_j^s - \mathrm{E}(X_j^s \wedge u_j^s)$, is constant for all risks. However, reverting to the original scale

$$sd_j\left[u_j^s - \mathrm{E}(X_j^s \wedge u_j^s)\right] = u_j^{rs} - \mathrm{E}(X_j \wedge u_j^{rs}),$$

this gap is larger for risks with larger scale factors. This seems to be intuitively desirable.

To select a value of RTC_{max}^s, we propose $RTC_{max}^s = RTC_{max}/\overline{sd}$, where $\overline{sd} = \frac{1}{p}\sum_{j=1}^{p} sd_j$ is the average value of the scale factors.

Under the Hood. See the *Online Supplement* for the verification of the scaling factor to maintain the budget constraint.

The results are summarized in Table 7.7. We remark that when we used the classical result without scaling and including the property risk, the starting values produced were not helpful. Intuitively, this is because the gap for property risk is the same as for all the small risks, producing very large upper limits for those smaller risks. As a consequence, use of these starting values caused the constrained optimization algorithms to fail. The rescaling method did allow us to determine sensible starting values when including property (although we ultimately excluded property from the optimization as described above).

TABLE 7.7
Rescaling Method for Producing Starting Values of Upper Limits.

RTC_{max}	LME	SV_{u1}	SV_{u2}	SV_{u3}	SV_{u4}	SV_{u5}	SV_{u6}
1374	0	47	6	7	6	1	14
1302	0	79	11	12	10	1	16
1157	1	191	28	29	25	3	18
1013	2	387	59	62	54	6	20
868	5	700	112	118	101	12	22
723	9	1164	195	205	176	20	24
579	14	1825	318	334	287	33	27
434	23	2785	501	526	453	52	31
289	37	4300	792	832	716	83	36
145	64	7147	1337	1403	1207	140	47
72	93	10158	1905	2000	1720	199	57

8

Risk Retention Case Studies

Chapter Preview. This text features two case studies introduced in Section 1.2 and further elaborated upon in Sections 6.3.2 and 6.3.3. This chapter shows how to construct optimal insurable risk portfolios in each case. These cases illustrate the flexibility offered by the framework introduced in Chapter 7.

The Wisconsin Property Fund case, presented in Section 8.1, examines an insurance risk pool of 1,098 fund members, each managing up to $p = 6$ risk types. This case exhibits many features exhibited by property and casualty insurance companies. Typically, insurers possess extensive claim histories on their policyholders; similarly, for the Property Fund case, there are extensive fund loss histories available to calibrate the risk distributions necessary for constructing optimal portfolios. One potential application described in Section 8.1.5 is the use of portfolio construction methods when designing business owners policies, cf. Section 6.3.

The Australian National University (ANU) case, detailed in Section 8.2, involves a portfolio of $p = 15$ risks, a scenario commonly encountered by risk managers at medium and large corporations. Although empirical historical data are available for high-frequency risk types, losses associated with most risk types occur only infrequently, resulting in limited data availability for calibrating the risk distributions. Motivated by these data limitations, this chapter demonstrates how to use approximate risk distributions based in part on market conditions. In addition, it explores the impact of deviations from model assumptions on optimal risk retentions.

8.1 Wisconsin Property Fund

This section demonstrates portfolio construction techniques using the Wisconsin Local Government Property Insurance Fund case. As discussed in Section 6.3.3, we will demonstrate the creation of optimal portfolios from *three* perspectives: that of a typical member, the fund itself, and a reinsurer to the fund. Even with the same data and risk models, different perspectives yield different objective functions, resulting in different risk retention policies. Regardless of which perspective the analyst focuses on, it is important to understand alternative viewpoints.

8.1.1 Property Fund Data and Risk Models

The fund data consists of losses from a collection of pool members, each of which is observed over time so that the data underpinning the risk modeling are longitudinal. Pool members differ by type (e.g., school, county, and other types of governmental units), size (measured by exposure variables), and other features (e.g., presence of sprinkler systems) so that known characteristics are critical to the empirical modeling strategy. Additionally, as described in Section 6.3.3, six risk types are of interest when establishing risk retention programs.

DOI: 10.1201/9781003469568-8

Calibrating a model to predict future loss distributions is a complex task, yet it is routinely performed by insurers; after all, their entire business model is predicated based on their ability to forecast losses. Fortunately, we can utilize the work documented in Frees et al. (2016) to predict future loss distributions that can be used to construct optimal risk retention programs. Compared to the other case study that will be analyzed in Section 8.2, the estimated Property Fund model is firmly grounded in empirical observations.

Specifically, the Frees et al. (2016) study used 2006-2010 data to estimate parameters from a claim generating model. Several covariates, including coverage, no claim credit, entity type, and deductibles, are available for each pool member (see their Table 8) meaning that regression-type models were employed at the parameter estimation stage. Although pool members were assumed independent of one another, the dependence among the six risk types was calibrated using a Gaussian copula. Following parameter estimation, covariates from a 2011 data sample were used to simulate a forecast loss distribution.

With the 2011 forecast distribution, we have available information from $1,098$ pool members each of whom may have up to $p = 6$ risks. For notation, let X_{ij} represent the $j = 1, \ldots, p$ risk types for each of $i = 1, \ldots, n$ pool members. For each member and risk, a known exposure e_{ij} is available, with $e_{ij} = 0$ indicating that the ith pool member does not have a type j risk. Two approaches for estimation and forecasting are possible, both empirically calibrated in Frees et al. (2016). One is a frequency-severity approach, where the aggregate loss X_{ij} is considered the random sum of independent claims. The second involves a direct model of the distribution of X_{ij} using the Tweedie distribution (see Section 8.3.3). Here, we use the simpler Tweedie approach to demonstrate risk retention calculations. Tables A11 and 17 of Frees et al. (2016) provide parameter estimates of the Tweedie distribution and the copula, respectively. Implementation of the simulated forecast loss distribution is summarized in Appendix Section 8.3.4.

8.1.2 Risk Retention for a Specific Member

As a technical analyst, you may be employed to advise any of the 1,098 fund members. Or, you may be an advisor to the fund, and the fund aims to develop a process so that each pool member can devise its own risk retention program. To be concrete, in this example we will develop an optimal retention program for Member 2, Ashland County.

Coverage and Forecast Loss Distribution. To begin, let us examine the relative size of each of the six risk types. Table 8.1 displays Ashland County's distribution of coverages, in millions of USD, for each of the six risks. Readers may refer to the descriptions of these coverages provided in Table 1.3. Table 8.1 shows that the building and contents (BC) type is by far the largest among the six risk types.

Next, let us describe the range of potential losses by examining the forecast distribution. To gain insight into this forecast distribution, Table 8.2 presents the aggregate losses in thousands of USD, summarized over 10,000 simulations. The sum of means over the six risks is 119.06 with BC representing nearly half of this sum.

TABLE 8.1
Six Coverage Amounts for Member 2.

BC	IM	CN	CO	PN	PO
118.771	12.497	1.592	4.946	1.592	5.011

Note: Coverages are in millions of U.S. dollars.

TABLE 8.2
Forecast Loss Distribution for Member 2.

	BC	IM	CN	CO	PN	PO
Minimum	0.00	0.00	0.00	0.00	0.00	0.00
Median	0.00	0.00	5.47	4.48	3.25	0.00
Mean	55.75	10.83	14.66	23.90	8.14	5.77
95th Percentile	298.26	64.63	57.09	102.96	30.99	31.00
99th Percentile	581.21	118.66	91.55	168.22	48.79	56.13
Mean Percent of Total	46.83	9.10	12.31	20.08	6.84	4.85

Note: Losses are in thousands of U.S. dollars.

Multivariate Excess of Loss Risk Retention. To manage these risks, let us assume that Member 2 would like to enter into one or more contracts providing multivariate excess of loss protection. With this, the retained risk is

$$S_2(\mathbf{u}) = \sum_{j=1}^{6} \min(X_{2j},\ u_j).$$

Our objective is to identify the best set of upper limit parameters u_j, $j = 1, \ldots, p = 6$ that minimizes the uncertainty of $S_2(\mathbf{u})$ while adhering to a constraint on the risk transfer cost. Consistent with earlier work, we utilize expected shortfall to summarize the uncertainty of retained risks and a fair risk transfer cost to represent the cost of offloading risks in excess of the upper limits. Based on Table 8.2, a value of $RTC_{max} = 119.06$ indicates full transfer, resulting in all upper limits being zero. This provides an upper end for a range of maximal risk transfer costs, RTC_{max}.

Constructing an Optimal Frontier. With this structure, the optimization problem is a special case of Display (7.4) that we have already learned how to solve. So, we can now proceed to interpretation of findings, beginning with Table 8.3 that summarizes the optimization results for each level of RTC_{max}.

To interpret Table 8.3, let us first examine the building and contents (BC) risk type. When a high risk transfer cost is expended, say $RTC_{max} = 113$, the value of the upper

TABLE 8.3
Fund Member 2 Excess of Loss Frontier.

RTC_{max}	RTC	VaR	ES	Std Dev	u_{BC}	u_{IM}	u_{CN}	u_{CO}	u_{PN}	u_{PO}
113	113	9.6	10.8	3.7	1.6	0.5	4.1	2.3	2.6	0.2
107	107	19.6	22.4	7.7	4.1	1.8	7.2	5.7	4.8	0.7
95	95	40.0	47.6	16.4	12.3	5.3	12.7	12.7	8.7	2.8
83	83	60.8	75.3	26.3	23.2	12.1	18.5	19.1	12.0	6.4
71	71	82.7	105.3	37.3	37.1	20.7	26.3	26.3	15.5	9.7
60	60	104.4	137.8	49.6	54.2	33.9	35.6	34.9	18.5	14.3
48	48	127.7	173.3	63.5	76.5	51.2	44.9	46.6	23.2	20.6
36	36	150.4	213.1	79.4	105.7	68.0	60.7	67.2	31.7	28.2
24	24	175.7	258.1	98.3	151.7	112.3	80.9	97.0	48.1	45.4
12	12	193.3	313.7	121.6	245.8	174.8	152.0	171.8	82.3	95.9
6	6	193.3	343.4	136.4	361.7	193.8	167.5	197.8	90.5	105.6

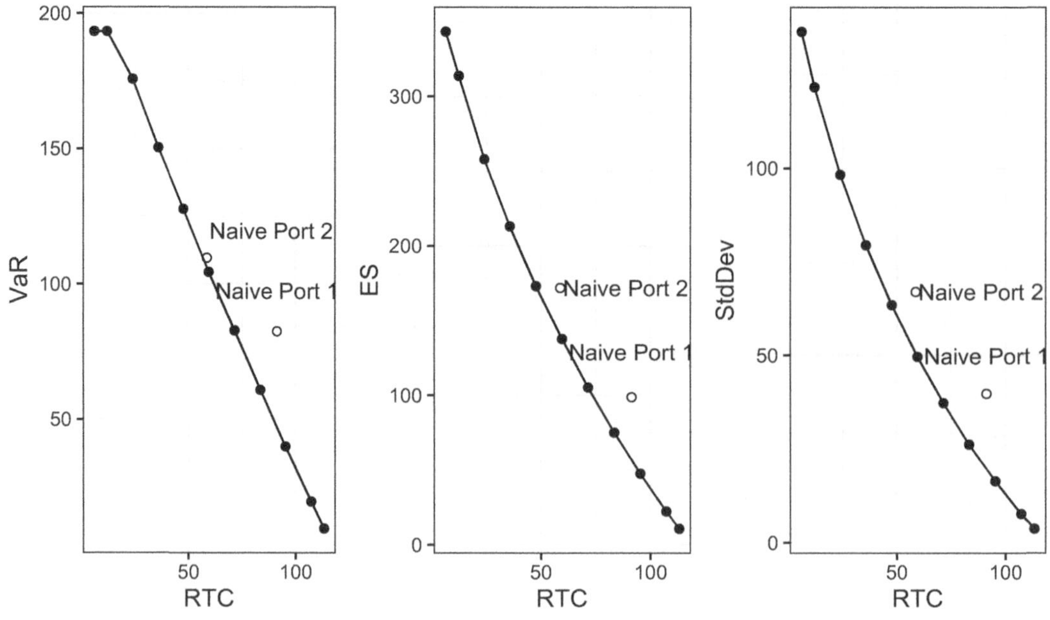

FIGURE 8.1
Fund Member 2 Efficient Frontier. Plots of retained risk measures versus risk transfer costs. Also shown are two naive portfolios that do not lie on the efficient frontier.

limit is small, $u_{BC} = 1.6$, in thousands of USD. In contrast, when a low risk transfer cost is expended, $RTC_{max} = 12$, the value of the upper limit is high, $u_{BC} = 245.8$, in thousands of USD. To provide context, referring to Table 8.2, this is close to the 95th (298.26) percentile of the forecast distribution.

For comparison, we also can look to the collision old (CO) coverage. With a high $RTC_{max} = 113$, the value of the upper limit is a low $u_{CO} = 2.3$, in thousands of USD. With a low $RTC_{max} = 12$, the value of the upper limit is high, $u_{CO} = 171.8$, in thousands of USD. To interpret this, from Table 8.2, the 99th percentile of the simulated distribution is 168.22, in thousands of USD. Any upper limit that exceeds this value is effectively infinite, indicating that the Property Fund retains all of the risk.

To discern patterns across different levels of RTC_{max}, Figure 8.1 illustrates the development of retained risk measures. As the budget increases and more is spent on risk transfer, the uncertainty of retained risks decreases for each risk measure. The curves connecting different RTC_{max} levels represent an efficient frontier.

Additionally, the figure includes two naive portfolios: one with an open plotting symbol indicating retention up to a limit of 100 for BC risk ($u_{BC} = 100$) and full transfer for the other risks, and another with a solid plotting symbol indicating full transfer for BC risk ($u_{BC} = 0$) and a limit of 100 for other risks. Both naive portfolios lie above and to the right of the efficient frontier, indicating excessive uncertainty of retained risks for the given level of risk transfer costs.

Figure 8.2 provides an overview of the progression of optimal upper limits across different RTC_{max} levels. Similar to the risk measures, as more is spent on risk transfer, each optimal upper limit decreases.

Constructing Risk Portfolios Summary. This section has demonstrated a process that analysts can use in their own construction of optimal insurable risk portfolios.

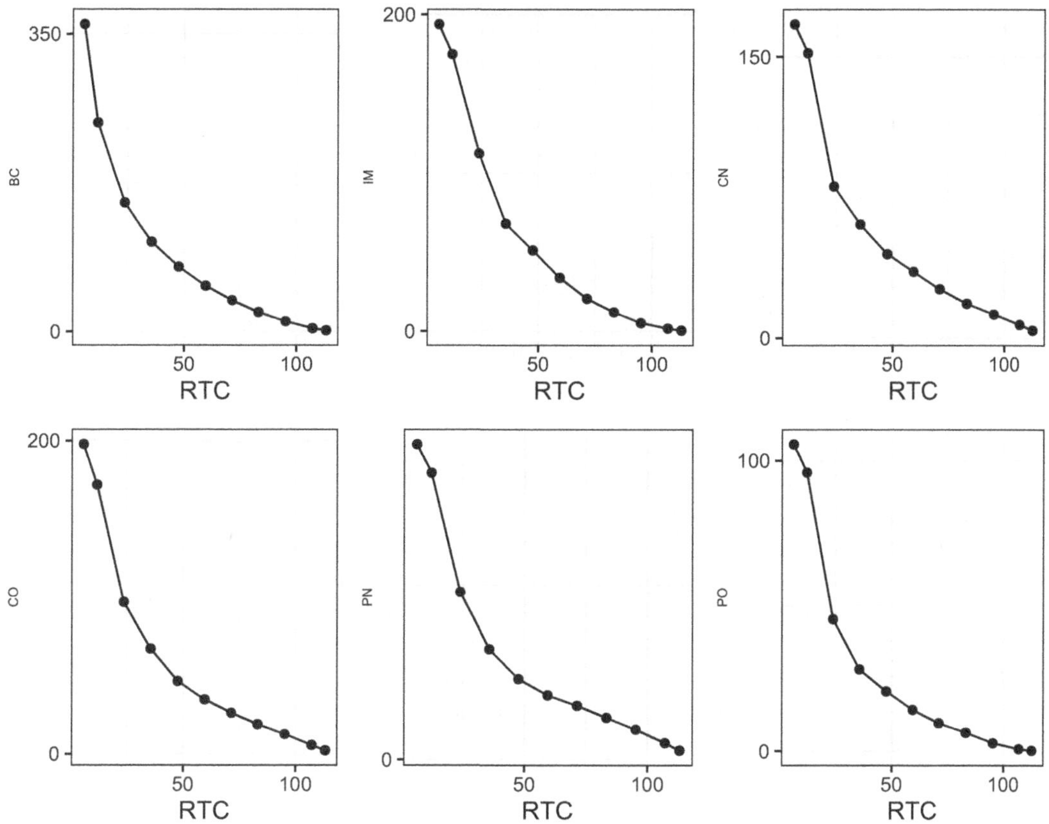

FIGURE 8.2
Fund Member 2 Optimal Upper Limits versus Risk Transfer Costs.

- Start by understanding the multivariate loss distribution, as in Tables 8.1 and 8.2 that summarize the coverage and forecast loss distributions.
- With a contract structure identified, the risk owner's retained loss variable can be determined.
- This, a risk measure to summarize the uncertainty, and a principle to determine risk transfer costs, is sufficient for constructing an optimal frontier by solving the problem in Display (7.4).
- Results over a frontier can be summarized as in Table 8.3, with visualization tools as in Figures 8.1 and 8.2.

Let us see how this process works in subsequent sections.

8.1.3 Fund Risk Retention

The prior section took the perspective of an individual fund member that sought to construct an optimal portfolio of risks with the aid of the resources of the general fund, in particular using data from other fund members to create forecasts of distributions.

In this section, the pool itself is viewed as the risk owner that seeks protection from the portfolio of potential losses. For simplicity, all losses are assumed to be the responsibility of the pool. It is straightforward to modify the analysis by assuming that the pool only takes a subset of losses, e.g., losses in excess of an agreed upon deductible from each pool member.

Coverage and Forecast Loss Distribution. As with the analysis for an individual pool member, we can get a sense of the size of the fund by examining coverage amounts and the distribution of forecast losses. To begin, Table 8.4 displays the distribution over members of coverages, in millions of USD, for each of the six risks. The table shows that the building and contents coverage (BC) is by far the largest risk type in terms of coverage.

To get a sense of the distribution of forecast losses, the aggregate (sum over the pool) losses are presented in Table 8.5, in thousands of USD. This summarizes the distribution over 10,000 simulations of all pool 1,098 members. Building and contents represents about 80% of mean losses.

The sum of means over the six risks is 23,648.49.

Excess of Loss Proportional Limits Risk Retention. To manage these risks, we propose that the portfolio owner seeks multivariate excess of loss protection. If one thinks about all fund members pooling their risks and the fund seeking protection for this pool of risks, this set-up is similar to the program developed for Fund Member 2 in the prior section. Nonetheless, this is a potentially important application and so readers are asked to verify their understanding of this situation in Exercise 8.1.

To demonstrate the flexibility of the risk retention set-up, I propose a variation that is inspired by the recent risk-sharing rules of decentralized insurance described in Section 6.5.2. As before, the protection is parameterized by u_j, $j = 1, \ldots, p$, that provides the upper

TABLE 8.4
Distribution of Six Coverages.

	BC	IM	CN	CO	PN	PO
Minimum	0.00	0.00	0.00	0.00	0.00	0.00
Median	12.77	0.19	0.00	0.00	0.00	0.00
Mean	42.35	0.97	0.09	0.38	0.17	0.69
95th Percentile	162.91	5.09	0.70	2.31	0.86	3.75
Maximum	2394.10	56.58	4.22	17.84	27.65	96.27
Number of Observations	1095	904	268	375	287	394

Note: Coverages are in millions of U.S. dollars.

TABLE 8.5
Simulated Distribution of 2011 Aggregate Losses.

	BC	IM	CN	CO	PN	PO
Minimum	12131.53	303.64	415.28	751.05	272.29	203.85
Median	18832.49	1046.98	861.66	1531.66	566.30	573.19
Mean	19024.33	1059.46	867.72	1545.60	569.61	581.78
95th Percentile	23371.96	1481.90	1108.67	2026.12	722.61	805.36
Maximum	30056.41	2154.13	1458.01	3023.36	954.07	1163.31
Mean Percent of Total	80.45	4.48	3.67	6.54	2.41	2.46

Note: Losses are in thousands of U.S. dollars.

limit amount but is now *proportional* to an exposure. Specifically, with this, the risk owner retains $\min(X_{ij}, u_j e_{ij})$ and the excess is transferred to another party such as insurer(s), reinsurer(s), or other type, such as a pool. For retention parameters $\mathbf{u} = (u_1, \ldots, u_p)$, the amount retained by the portfolio owner for the ith policyholder is

$$S_i(\mathbf{u}) = \sum_{j=1}^{p} \min(X_{ij},\ u_j e_{ij}).$$

With this, the total risk retained is $S(\mathbf{u}) = \sum_{i=1}^{n} S_i(\mathbf{u})$ and the risk transferred is $S(\infty, \ldots, \infty) - S(\mathbf{u})$. Note that the choice of $u_j = \infty$ represents full risk retention (no insurance) and $u_j = 0$ represents full transfer.

Constructing an Optimal Frontier. As in the prior Section 8.1.2, we now have defined a retained risk (as well as the risk measure and its risk transfer cost) so that constructing an optimal frontier can be viewed as a special case of Display (7.4). Because we have already learned how to solve this problem, we can proceed to interpretation of findings. The results are summarized in Table 8.6 and Figure 8.3, based on 10,000 simulations. For consistency, let us interpret these results using the same approach as in Table 8.3.

Starting with the building and contents (BC) coverage, we see that:

- From Table 8.4, the median value of the BC coverage is 12.77 and the maximum value is 2,394.10, both in millions of USD.
- From Table 8.6, when RTC_{max} = 22,466, the value of the upper limit is $u_{BC} =$ 0.02. This corresponds to an upper limit of 307 at the median value of the BC coverage and to an upper limit of 57,458 at the maximum value of the BC coverage, both in thousands of USD.
- In the same way, when RTC_{max} = 1,182, the value of the upper limit is $u_{BC} =$ 11.692. This corresponds to an upper limit of 149,365 at the median value of the BC coverage.
- Based on the information in Table 8.5, the maximum value of the simulated distribution turns out to be 30,056. Any upper limit that exceeds this value is effectively infinite, meaning that the Property Fund retains all of the risk.

TABLE 8.6
Property Fund Excess of Loss Frontier.

RTC_{max}	RTC	VaR	ES	Std Dev	u_{BC}	u_{IM}	u_{CN}	u_{CO}	u_{PN}	u_{PO}
22466	22466	1265	1323	99	0.02	0.97	1.00	1.00	1.00	0.98
20101	20101	3826	3993	321	0.17	0.54	1.13	1.14	1.07	0.72
17736	17736	6448	6783	614	0.34	0.20	1.31	1.42	1.20	0.57
15372	15372	9059	9602	921	0.56	0.13	1.60	1.99	1.42	0.70
13007	13007	11669	12419	1218	0.81	0.72	2.11	3.04	1.84	1.13
10642	10642	14276	15265	1527	1.15	1.37	2.62	3.99	2.29	1.49
8277	8277	16850	18098	1832	1.64	2.32	3.28	5.20	2.90	2.06
5912	5912	19470	20923	2143	2.59	2.64	3.64	5.80	3.24	2.32
3547	3547	22002	23601	2364	4.13	6.22	5.41	9.15	5.04	4.59
1182	1182	24534	26229	2547	11.69	14.72	13.77	20.14	12.32	10.06

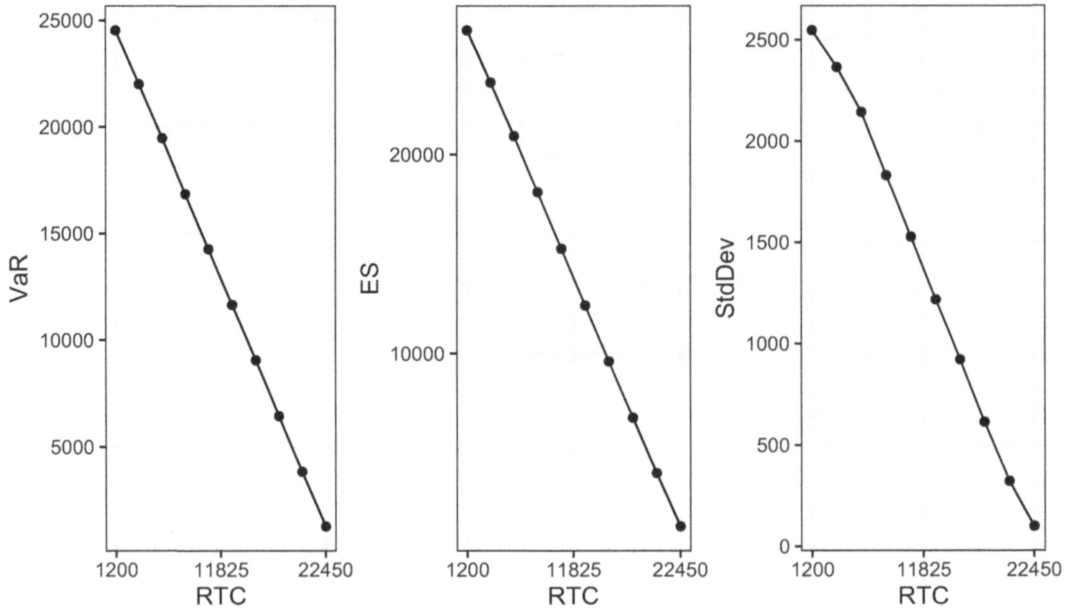

FIGURE 8.3
Property Fund Efficient Frontier. Plots of retained risk measures versus risk transfer costs.

For comparison, we also can look to the collision old (CO) coverage:

- From Table 8.6, when RTC_{max} = 22,466, the value of the upper limit is u_{CO} = 1.00. This corresponds to an upper limit of 374 at the mean value of the CO coverage and to an upper limit of 17,787 at the maximum value of the CO coverage.
- In the same way, when RTC_{max} = 1,182, the value of the upper limit is u_{CO} = 20.14. This corresponds to an upper limit of 7,558 at the mean value of the CO coverage.
- From Table 8.5, the maximum value of the simulated distribution is 3,023. Any upper limit that exceeds this value is essentially infinity, meaning that the Property Fund retains all of the risk.

R *Code.* The *Online Supplement* provides illustrative statistical code. The code employed for this section can be used either with excess of loss or quota share. It also utilizes kernel smoothing and gradients to speed convergence.

8.1.4 Reinsurer Risk Retention

We now adopt a third perspective, one where a reinsurer is the risk owner. Specifically, suppose that the pool and a reinsurer enter into a multivariate excess of loss contract parameterized by u_j, $j = 1, \dots, p$. As in the prior section, suppose that the pool is responsible for

$$S(\mathbf{u}) = \sum_{i=1}^{1098} \sum_{j=1}^{p} \min(X_{ij},\ u_j e_{ij}).$$

Denoting total losses as $S = S(\infty, \ldots, \infty)$, we have that the excess losses, $S - S(\mathbf{u})$, is the responsibility of the reinsurer.

As the risk owner, the reinsurer seeks to minimize the uncertainty of its risks subject to earning an income from taking on the risk. With fair costs, the reinsurer will receive

$$RTC = \mathrm{E}[S - S(\mathbf{u})] \geq RTC_{min},$$

where RTC_{min} is the minimum amount that the reinsurer requires. Using notation similar to equation (7.4), the problem can be written as

$$\begin{array}{ll} \text{minimize}_{z_0, \mathbf{u}} & z_0 + \mathrm{E}_{Rk}[S - S(\mathbf{u}) - z_0]_+ \\ \text{subject to} & RTC_R(\mathbf{u}) \geq RTC_{min} \\ & u_1 \geq 0, \ldots, u_p \geq 0. \end{array}$$

Constructing an Optimal Frontier. This can be solved using the same set of tools as before. As in the prior section, the number of simulations is 10,000. The results are summarized in Table 8.7 and Figure 8.4. Note from Figure 8.4 the positive relationship between RTC and each of the retained risk measures. As the reinsurance takes on more risks, it receives more income through the risk transfer cost but also has greater uncertainty of retained risks.

From Table 8.7, we can also observe the optimal retention limits. Note that these limits differ from those in Table 8.6. This is not surprising; after all, in each case the objective function differs and so values of upper limits that minimize each problem will naturally differ.

Code. See the *Online Supplement* for illustrative statistical code.

8.1.5 Designing Business Owners Policies

As described in Section 6.3.1, a business owners policy is a contract offered by a commercial insurer to small and moderate-sized businesses, covering a standard set of risks. Many such contracts offer a standard set of risk retention options for businesses to choose from. This section proposes a method to create a set of options that are desirable from an insurer's viewpoint.

To this end, as in prior sections we consider p risks and, at a broad design stage, assume that each risk may be subject to a coinsurance c and an upper limit u as in equation (2.6).

TABLE 8.7
Property Fund Excess of Loss Frontier—Reinsurer Perspective.

RTC_{min}	RTC	VaR	ES	Std Dev	u_{BC}	u_{IM}	u_{CN}	u_{CO}	u_{PN}	u_{PO}
22466	22466	26980	28333	2552	0.03	0.54	0.00	0.00	3.16	1.57
20101	20101	24354	25665	2404	0.19	0.79	0.00	0.00	0.00	0.42
17736	17736	21586	22892	2209	0.38	0.65	0.00	0.14	0.27	0.00
15372	15372	18834	20097	2010	0.56	2.79	0.03	0.07	0.01	0.97
13007	13007	15967	17081	1722	0.94	0.16	0.00	0.61	0.28	0.04
10642	10642	13074	13985	1419	1.43	0.21	0.00	0.00	0.08	0.14
8277	8277	10175	10807	1095	2.26	0.00	0.94	0.00	0.15	0.05
5912	5912	7226	7631	759	4.08	0.00	0.17	0.01	0.00	1.38
3547	3547	4375	4602	479	11.10	3.66	0.00	1.67	1.19	3.51
1182	1182	1488	1576	179	33.11	23.42	0.00	73.31	9.32	18.65

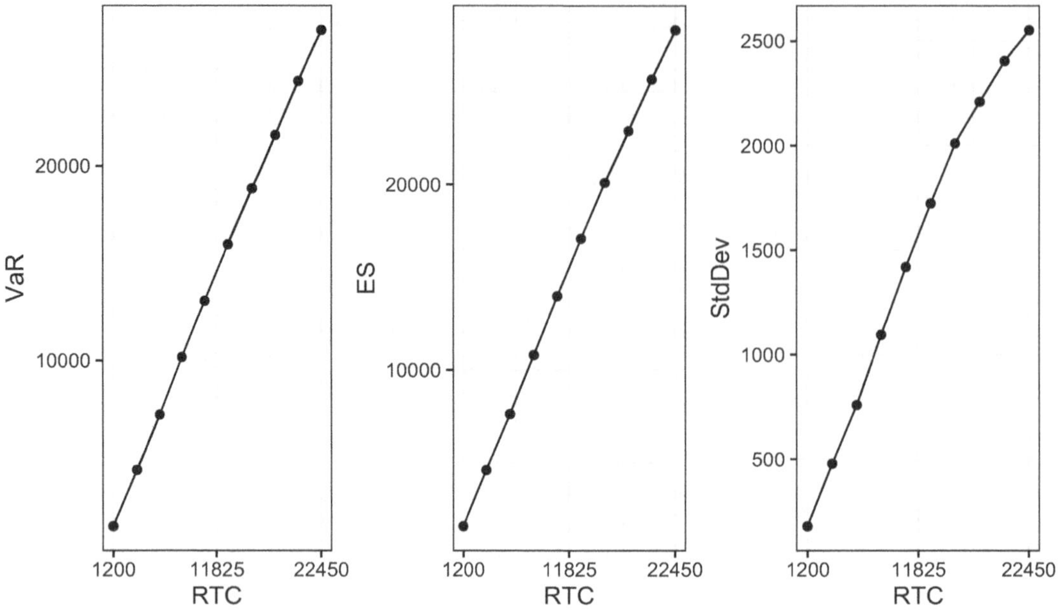

FIGURE 8.4
Reinsurer's Efficient Frontier. Plots of retained risk measures versus risk transfer costs.

To enhance flexibility at the design stage, we allow these risk parameters to vary by risk and by the size of the customer firm. To account for firm size adjustment, we again consider an exposure variable e so that upper limits are proportional to the exposure.

With n potential customers, the insurer's retained risk can be expressed as

$$S(\boldsymbol{\theta}) = \sum_{i=1}^{n} \sum_{j=1}^{p} \{X_{ij} - c_j[X_{ij} \wedge u_j e_{ij}]\},$$

where we recall the wedge operator $\wedge$ for a minimum. This expression for retained risks bears similarity to a reinsurer's risk retention problem discussed in Section 8.1.4 and could be optimized using similar methods.

Coinsurance parameters are linear and so are readily straightforward to account for in risk transfer costing and understanding the uncertainty of retained risks. That leaves us with upper limit parameters that were already accounted for in Section 8.1.4. Sometimes, the optimization process leads to very small upper limits which could be interpreted as a "deductible." Deductibles are commonly used by insurers, not to minimize potential claims liabilities, but to mitigate small "nuisance" claims that can be administratively expensive relative to the size of the claim. Therefore, one may adjust the design to include small, yet still positive, limits even when the algorithm suggests values near zero.

Building upon this base model, several variations, such as offering combined limits for several risks, could also be included as noted in Section 7.5. Additionally, once an upper limit for each risk proportional to exposure is established, it is likely to be presented to potential client firms as levels varying by category of size (essentially categorizing the exposure variable). With different options that vary by a firm's willingness to spend (RTC_{max}), our tools for creating optimal risk portfolios provide a rigorous approach for designing business owners policies.

8.2 Stress Testing the ANU Risk Portfolio

In Section 7.4.1 we explored how to optimize ANU's portfolio of risks. Although we learned how to solve the problem in Display (7.1), many of the assumptions underpinning this mathematical representation may not hold true. Managers often employ *stress testing* to examine how changes to underlying assumptions impact risk modeling results. One such assumption concerns the budget parameter RTC_{max}, which dictates the amount to be spent on transferring risk. As we have already discussed, risk managers may be interested in understanding how allocations change with variations in RTC_{max}.

There are other assumptions that may be of concern to risk managers. For instance:

- Industry norms regarding the most appropriate confidence level α for a risk retention policy have yet to be established. Thus, it is important to consider how the overall risk retention level and allocations change with variations in α.
- In the same way, there is no consensus on the type of risk measure to use. Therefore, it is worth exploring how allocations differ between Value at Risk (VaR) and Expected Shortfall (ES) risk measures.

Subsections 8.2.2 and 8.2.3 delve into these issues, providing insights into these considerations. Following this, an alternative to the Tweedie model based on frequency and severity assumptions is presented. Specifically, this section concludes with a discussion on introducing event-level retention parameters based on the discourse in Section 7.5.3. As a baseline model from which these alternatives will be examined, Section 8.2.1 demonstrates how to incorporate property risk when constructing optimal insurable risk portfolios.

8.2.1 Including the Property Risk

This section begins by demonstrating how to incorporate property risk, a significant source of uncertainty in ANU's portfolio. As mentioned in Section 7.4.1, the property risk dominates the portfolio of liability risks both in terms of premiums and uncertainty of outcomes. Although technically possible to include property risk in the portfolio alongside other risks, practical and numerical considerations suggest that it is more appropriate to treat this risk separately. Specifically, we assume that ANU has negotiated a separate property excess of loss policy.

For illustrative purposes, we assume that ANU transfers all property risks in excess of 5000, in thousands of AUD. Using notation, ANU's portfolio of retained risks is represented as:

$$S_{Prop}(u_2, \ldots, u_{15}) = \min(X_1, 5000) + \min(X_2, u_2) + \cdots + \min(X_{15}, u_{15}).$$

The objective is to determine the optimal values of the upper limits $\{u_2, \ldots, u_{15}\}$.

Interpreting the results in Table 8.8 is similar to the discussion in Section 7.4.1. Note that the risk transfer costs do not include the cost of transferring the property risk so these values are comparable to those in Table 7.4. In contrast, the retained risk portion does include property risk, resulting in significantly larger summary measures (VaR, ES, and standard deviation).

R *Code.* See the *Online Supplement* for illustrative statistical code.

8.2.2 Varying the Level of Confidence

Specifying a quantile-based risk measure, such as a value at risk VaR or the expected shortfall ES, may pose a challenge for a risk manager. If the risk owner is a firm, then firm attributes may play into the risk manager's choice of the confidence level. For example, confidence levels may vary by whether the firm is public (for profit or non-profit) or private. In addition, when seeking to transfer costs, the risk owners may be concerned with the credit quality of the party taking on the risk. For example, counter-party risk of reinsurers is a major concern for insurers who seek to offload some of their risks. As mentioned previously, unlike solvency valuations, no industry norms have been established for either the choice of the measure or the associated confidence level.

For general commercial firms, the amount of retained risk summarized by a risk measure does not appear in financial statements. In contrast, the budget expenditure RTC represents the amount allocated for transferring risks, an expense, and expenses do appear in financial statements. Additionally, retention parameters are specified in risk transfer contracts. To summarize:

- Budget expenditures and retention parameters are contractually documented and therefore likely to be scrutinized carefully.
- Risk measures, which summarize the degree of uncertainty, receive less scrutiny and may vary substantially depending on opinions of different decision-makers involved in the risk management process.

Although determining a desirable amount of uncertainty of retained risks can be ambiguous, we know that it can be changed by varying the confidence level α. Therefore, it is of interest to examine how varying α affects not only the overall uncertainty but also the retention parameters.

This is achieved by utilizing the optimization process outlined in Table 8.8 as a baseline while varying α. To keep the number of comparisons manageable, only results for $RTC_{max} =$ 722 are reported, which corresponds to a typical budget constraint (actually, the middle value in Table 8.8).

Results are presented in Table 8.9. The first row corresponds to an optimization performed using $\alpha = 0.50$. From this optimization, values of upper limits have been determined (e.g., $u_2 = 274$, $u_3 = 266$, and so on) and the value at risk is $VaR = 744$. Recall that the VaR is simply a quantile and so, with $\alpha = 0.50$, this means that the median level of retained risk is 744 (thousands of AUD). The expected shortfall $ES = 3{,}280$ can be interpreted to be the expected amount of retained risks given that it exceeds the median.

TABLE 8.8
ANU Excess of Loss *ES* Optimization Frontier, Including the Property Risk.

RTC_{max}	RTC	VaR	ES	Std Dev	u_2	u_3	u_4
1371	1366	5071	5078	2031	1	0	0
1299	1294	5144	5151	2031	2	1	0
1155	1087	5330	5371	2038	9	9	1
1010	953	5391	5570	2071	42	40	30
866	865	5390	5765	2129	219	407	31
722	709	5401	6110	2251	504	600	333
577	565	5405	6434	2406	1106	959	827
433	411	5410	6786	2639	2528	1627	1635
289	185	5456	7600	3252	10499	3941	4111
144	144	5481	7473	3140	5254	5746	5798
72	72	5481	7834	3385	9246	6330	6393

TABLE 8.9
ANU Excess of Loss *ES* Optimization by Level of Confidence (α).

α	*RTC*	*VaR*	*ES*	Std Dev	u_2	u_3	u_4
0.50	722	744	3280	2225	274	266	270
0.55	722	804	3558	2225	278	268	268
0.60	722	897	3897	2225	294	268	274
0.65	722	997	4318	2225	281	277	258
0.70	722	1163	4860	2225	291	295	255
0.75	722	1993	5545	2228	200	475	181
0.80	722	5400	6070	2227	268	548	221
0.85	722	5584	6264	2227	209	506	215
0.90	722	5777	6554	2225	275	307	250
0.95	722	6184	7174	2226	208	352	214
0.98	722	7500	7529	2229	600	291	292

For many purposes, a confidence level $\alpha = 0.50$ may be insufficient. Thus, Table 8.9 provides optimization results over a range of α beginning at 0.50 up to 0.98. Thus, for example, with $\alpha = 0.95$, we see that with a fund of 6,184 there is a 95% probability that this is sufficient to cover retained risks.

Figure 8.5 summarizes the relationship between the confidence level α and the uncertainty of retained risks as summarized by the ES measure. Not surprisingly, as α increases, so does ES. Surprising however is the relatively constant level of the standard deviation. This suggests that using the standard deviation as an optimization criterion can produce very different results from the quantile-based risk measures used here.

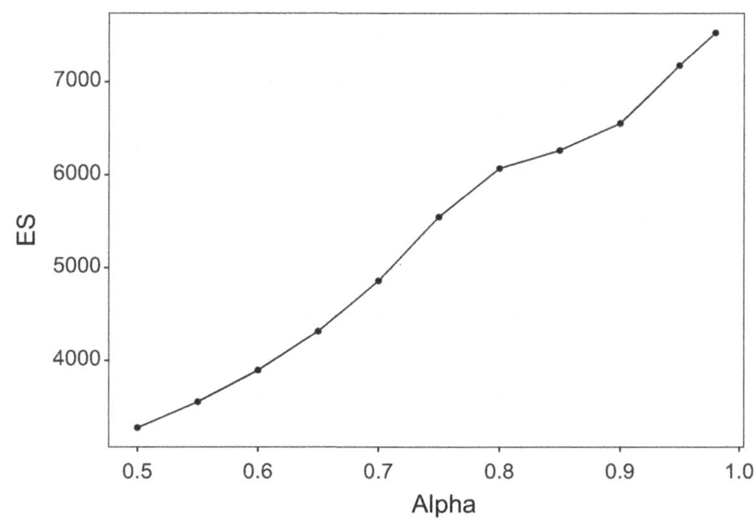

FIGURE 8.5
Retained Risk (Measured by *ES*) versus Level of Confidence (Measured by Alpha α). Based on the ANU Excess of Loss Case Study where upper limits are determined by minimizing the ES criterion.

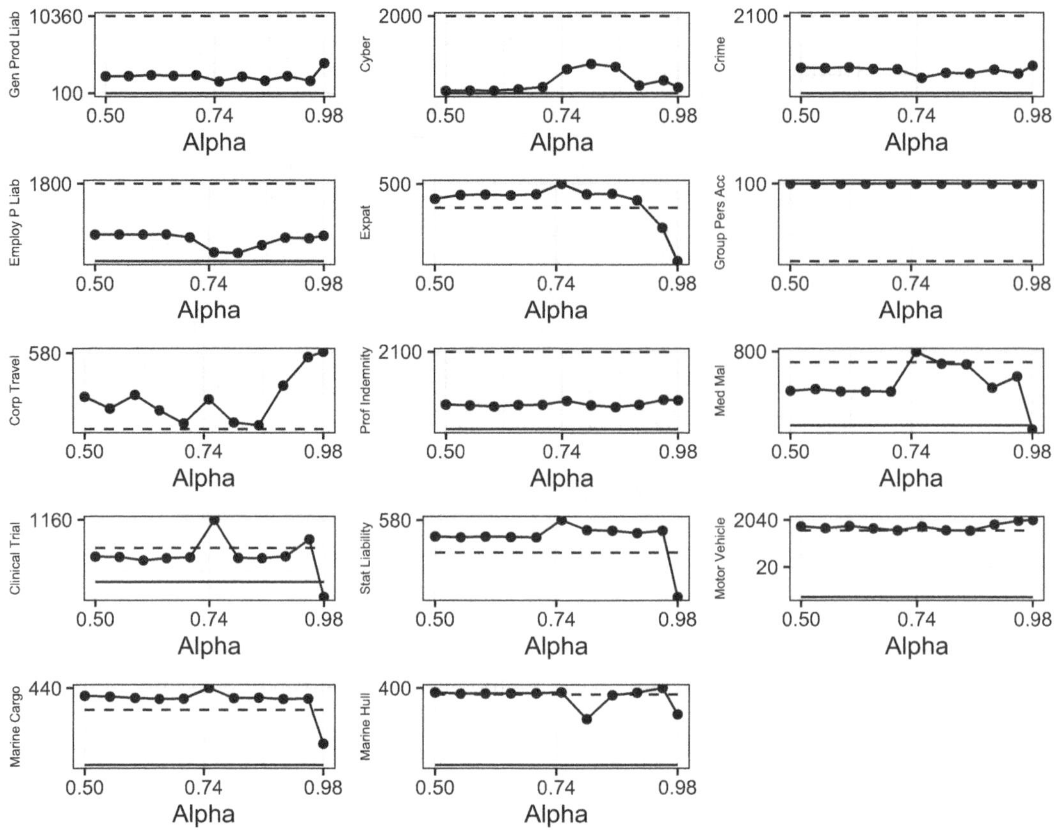

FIGURE 8.6
Upper Limits versus Alpha α. Based on the ANU Excess of Loss Case Study where upper limits are determined by minimizing the ES criterion.

Table 8.9 also shows the progression of the first three upper limits u_2, u_3, u_4 as α increases. This is supplemented by Figure 8.6 which depicts the progressions of all upper limits. As previously noted, the optimal retention policy adjusts the mixture of risks as α changes; the optimal values of the upper limits seem to be relatively stable as α changes. Only when α is very large, say at 0.98, do we see big changes. This is encouraging news for risk managers - the choice of the confidence level does not materially affect optimal risk retention results.

Code. See the *Online Supplement* for illustrative statistical code.

8.2.3 Varying the Measure of Retained Risk

Some risk managers may be confident in selecting a confidence level α such as $\alpha = 0.95$ or 0.98, which are commonly used in certain fields. However, even with a fixed confidence level, there can be substantial variation in measures of uncertainty within the set of quantile-based risk measures. In particular, we focus on the two most well-known choices: the value at risk VaR and the expected shortfall ES. To explore trade-offs between these two measures, this subsection investigates another tool for balancing risk measures, the range value at risk $RVaR$. Using equation (2.4), we can express this measure as

TABLE 8.10
ANU Excess of Loss Optimization—RVaR Criterion—by Levels of Beta.

β	*RTC*	*VaR*	*ES*	Std Dev	*RVaR*	u_2	u_3	u_4
0.00	722	5788	6571	2234	5788	338	349	264
0.01	722	5781	6569	2235	5800	293	273	287
0.02	722	5781	6569	2235	5824	288	273	288
0.03	722	5781	6569	2235	5855	285	279	289
0.04	722	5781	6569	2235	5891	293	274	290
0.05	722	5781	6569	2235	5932	283	279	279
0.06	722	5782	6569	2235	5992	285	288	275
0.07	722	5805	6597	2242	6079	8	385	380
0.08	722	5723	6623	2267	6159	56	265	35
0.09	722	5708	6615	2266	6312	116	166	118
0.10	722	5708	6615	2266	6615	140	146	116

$$RVaR_{(\alpha,\beta)}[X] = F^{-1}_{\alpha+\beta} + \frac{1}{\beta}\left\{(1-\alpha)(F^{-1}_{\alpha} - F^{-1}_{\alpha+\beta}) + [\mathrm{E}(X \wedge F^{-1}_{\alpha+\beta}) - \mathrm{E}(X \wedge F^{-1}_{\alpha})]\right\}.$$

To clarify, this measure provides an intermediate alternative between VaR and ES. Specifically, with $\beta = 0$, we have $RVaR = VaR$ and with $\beta = 1-\alpha$, we have $RVaR = ES$. Thus, β is the parameter that adjusts the $RVaR$ between VaR and ES.

Table 8.10 summarizes the optimization results for several different choices of β. Unlike other portions of this book, in this analysis we minimized the $RVaR$ criterion, not the ES. To facilitate comparisons, optimization results are presented in the same manner as before. Therefore, the row with $\beta = 0$ corresponds to $VaR_{0.90}$ minimization and the row with $\beta = 1 - 0.90 = 0.10$ corresponds to $ES_{0.90}$ minimization. Figure 8.7 illustrates that although the optimal value of $RVaR$ changes substantially with different choices of β, the corresponding values of *StdDev* (standard deviation), VaR, and ES are remarkably stable.

Similarly, Figure 8.8 depicts the progression of optimal upper limits. Some of these values do change with a change in β such as those for medical malpractice and clinical trials. Nonetheless, overall there is relatively little change in the limits. This supports the intuition that the risk retention problem is relatively robust to the choice between the VaR and ES risk measures.

Code. See the *Online Supplement* for illustrative statistical code.

Minimization of the $RVaR$ criterion is a bit more complex than the ES minimization. Readers who examine the code will observe several differences that made the calculations feasible. First, the code minimizes the risk measure directly, not the auxiliary function introduced in the ES calculation. Second, no implementation of kernel smoothing was attempted. Third (as a consequence of the second), gradient functions were not utilized in the optimization. Happily, despite all of these simplifications, reasonable results could be achieved as reported in this section.

8.2.4 Claim Level Retention

As demonstrated in Appendix Section 8.3.5, the three-parameter Tweedie distribution is flexible and can be used to represent a wide variety of aggregate losses. Moreover, when information about individual claims is available, an even more flexible model can be used

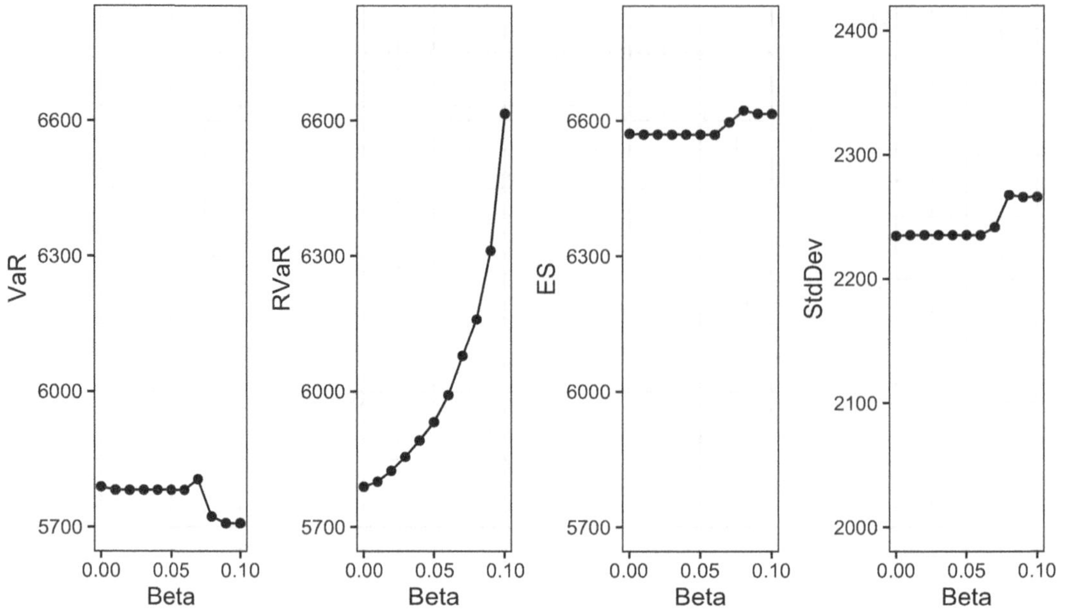

FIGURE 8.7
Retained Risk Measures versus Beta β. Based on minimizing the $RVaR$ criterion. Here, $\beta = 0$ corresponds to VaR and $\beta = 1 - \alpha$ corresponds to ES.

to model the number, or frequency, of claims and, for each claim, the severity, or amount, of a claim. As described in Section 7.5.3, by modeling claims at this more granular level we are able to introduce risk retention parameters at the **claim level** instead of the aggregate level.

For the ANU data, let us now consider an upper limit for each occurrence of a claim for three risks: group personal accident, corporate travel, and motor vehicle. Other upper limits remain at the aggregate level. The number of retention parameters stays the same but now risk retention is operating at a more granular level, at the occurrence of a claim, as opposed to the overall aggregate. This mirrors the way actual contracts are written for ANU and is common for coverages that have relatively high frequency.

Table 8.11 shows the results from this optimization, again varying different levels of RTC_{max}. As before, note that the upper limit parameters increase as RTC_{max} decreases and the risk owner takes on more risk but pays less in transfer costs. Figure 8.9 depicts how the summary measures of retained risk uncertainty decrease as risk transfer costs increase.

Figure 8.10 illustrates optimal upper limits as risk transfer costs vary. The progression of the optimal upper limits for the three claim-level risks, group personal accident, corporate travel, and motor vehicle, appears less volatile than when at the aggregate level as shown in Figure 7.1.

Code. See the *Online Supplement* for illustrative statistical code.

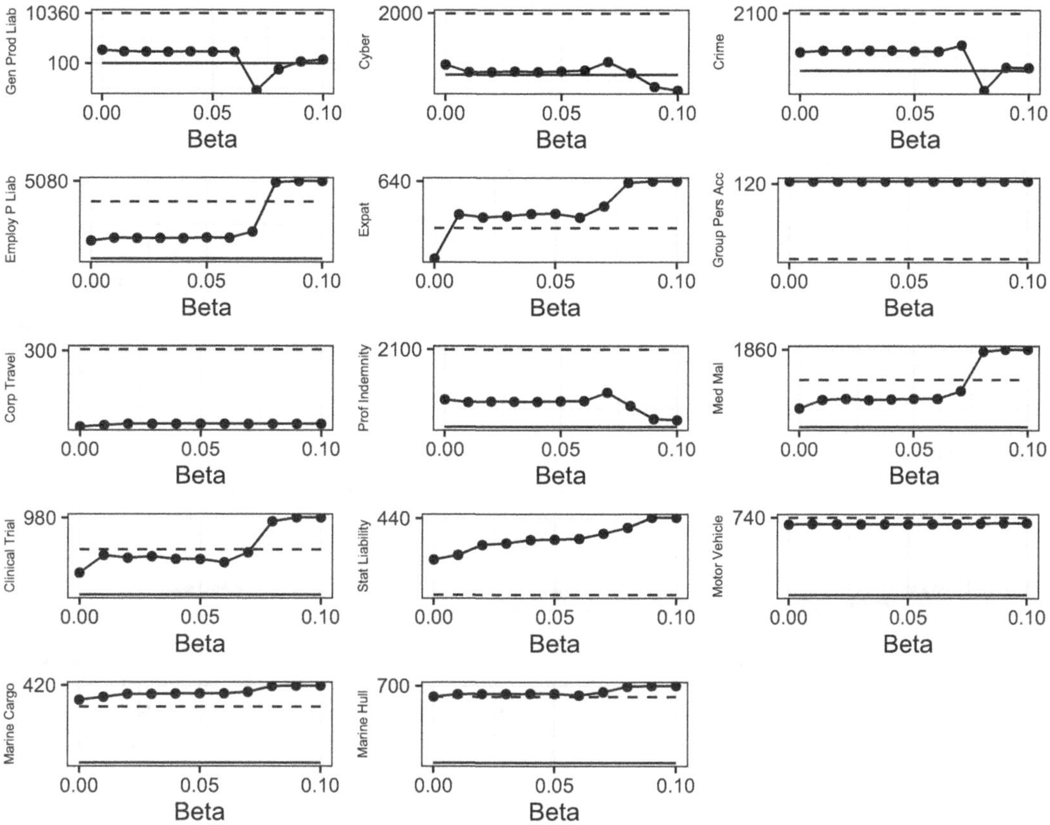

FIGURE 8.8
Upper Limits versus Beta. Based on minimizing the $RVaR$ criterion. Here, $\beta = 0$ corresponds to VaR and $\beta = 1 - \alpha$ corresponds to ES.

TABLE 8.11
ANU Excess of Loss Optimization—Includes Claim Level Upper Limits.

RTC_{max}	RTC	VaR	ES	Std Dev	u_{GPA}	u_{Trav}	u_{MV}
1374	1374	5056	5150	2039	0	0	0
1302	1302	5135	5205	2036	0	0	0
1157	1157	5295	5327	2032	1	0	3
1013	1013	5445	5482	2034	8	5	8
868	868	5615	5976	2115	16	22	62
723	723	5776	6547	2224	12	34	65
579	579	5985	7156	2362	21	14	70
434	434	6016	7783	2563	26	44	146
289	289	6085	8439	2775	37	89	232
145	145	6166	9159	3086	630	490	917
72	72	6353	9801	3312	1210	903	1629

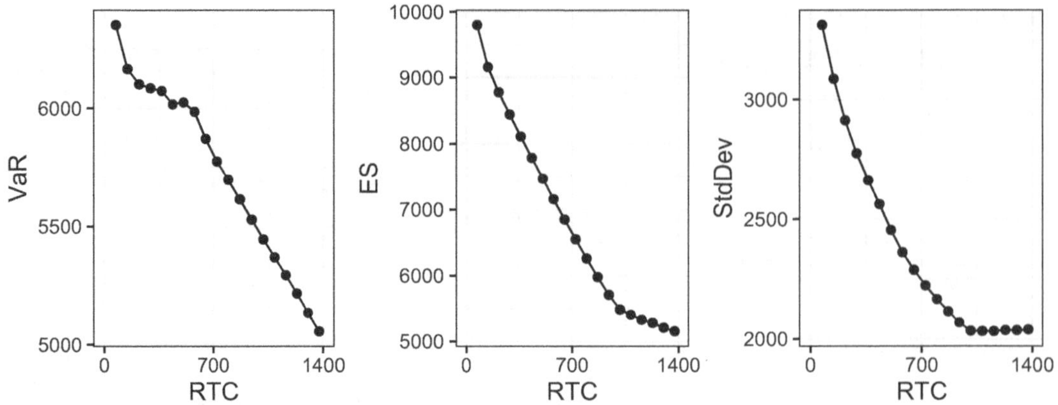

FIGURE 8.9
Retained Risk Measures—Includes Claim Level Upper Limits.

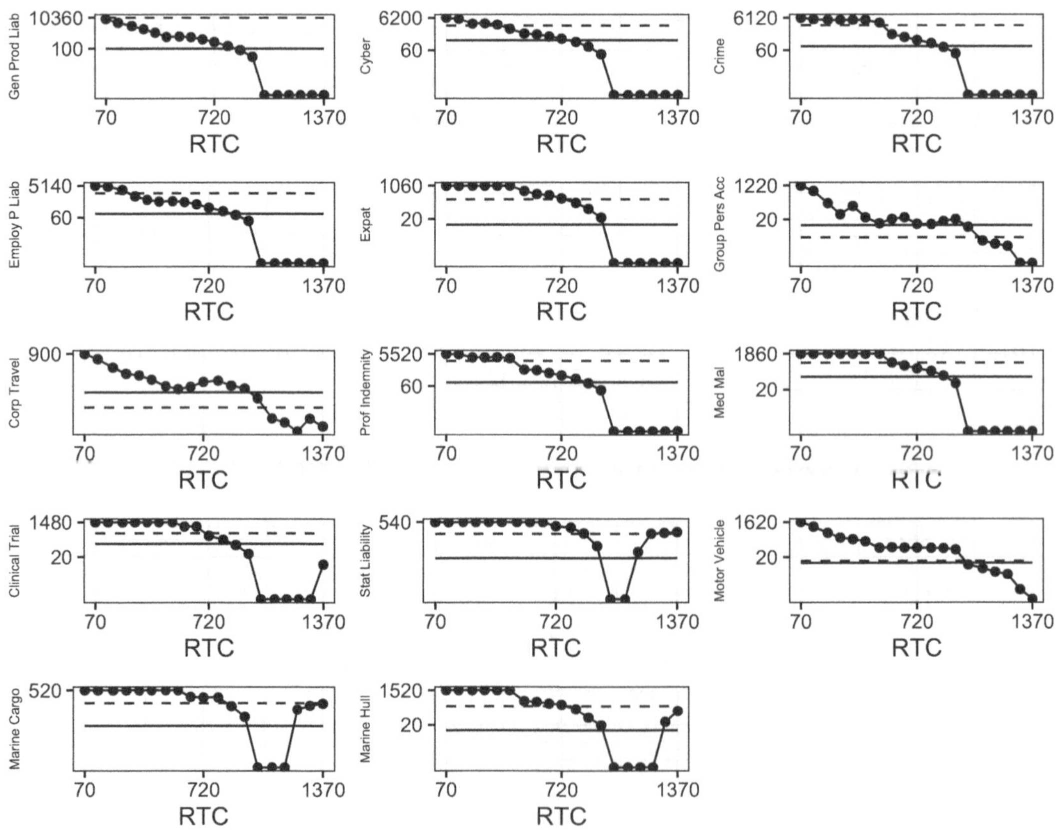

FIGURE 8.10
Upper Limits—Claim Level Upper Limits.

8.3 Supplemental Materials

8.3.1 Further Resources and Reading

As seen in Section 4.1, reinsurance provides a natural setting for introducing multivariate risk retention. Although the methods introduced in this book provide a foundation for interpreting the best choices of risk retention, practical reinsurance applications require additional care. For example, the models presented here do not handle considerations over time and so can be thought of as "one-period" models. To illustrate, it is common practice for reinsurers to raise premiums after claim occurrence in order to adjust (or "reinstate") retention limits to their original levels, e.g. Albrecher et al. (2017), Section 2.3.2. This variation is not addressed directly in the basic one-period model presented here because it does not account for realized claims.

As another example, in reinsurance, there may be an interplay between the price of risk transfer cost and a risk owner's level of confidence. That is, prices of reinsurers are affected by the amount of capital that they have which in turn affects their ability to pay (large) claims. This potential credit risk can complicate matters, particularly when there are many reinsurers involved in taking responsibility for different layers of coverage, as introduced in Section 4.1.3.

In addition to the stress testing topics introduced in Section 8.2, risk managers are also likely to be concerned with the effects of dependence on risk retention parameters. That is, risk managers have a sense that risks are positively dependent but often are not able to quantify the extent of this dependence. How do the optimal retention limits vary with different levels of dependence? Chapter 12 will take up this question.

Section 8.1 explored optimization challenges from three distinct perspectives: that of an individual fund member, the fund itself, and a reinsurer. As a consultant, you may find yourself advising any of these entities. Since each perspective naturally leads to different objective functions, it is expected that the perspective significantly influences the construction of an optimal portfolio. Indeed, there is a robust academic discourse dedicated to reconciling these varied viewpoints. For further insights into harmonizing these different perspectives, you may refer to works such as Boonen and Ghossoub (2023), Cai et al. (2017), Bernard et al. (2020), Jiang et al. (2021), and Assa (2015).

8.3.2 Exercises

For these exercises, you will need to generate simulated data using resources found in:

- Appendix Section 8.3.4 shows how one obtains estimated risk model parameters and provides code to simulate the Wisconsin property fund data.
- Appendix Section 8.3.5 describes development of the parameters for the ANU risk distributions.
- Appendix Section 8.3.6 provides code to simulate the ANU data.

See the *Online Supplement* for selected exercise solutions.

Section 8.1 Exercises

Exercise 8.1. Wisconsin Property Fund Seeks Reinsurance Protection. In this exercise, the Wisconsin Property Fund is the risk owner and seeks excess of loss protection.

TABLE 8.12
Exercise 8.1. Fund Excess of Loss Optimization.

RTC_{max}	RTC	VaR	ES	Std Dev	u_{BC}	u_{IM}	u_{CN}	u_{CO}	u_{PN}	u_{PO}
22466	22466	1182	1182	0	197	197	197	197	197	197
20101	20101	3547	3547	0	2268	245	220	527	164	124
17736	17736	5912	5912	1	4205	0	447	772	292	197
15372	15212	8441	8441	21	6884	31	17	1074	435	0
13007	13007	10642	10642	1	8942	446	265	580	204	204
10642	10642	13007	13007	0	11527	304	416	556	0	205
8277	8277	15372	15372	22	13302	471	480	862	0	257
5912	5912	17758	17758	127	14697	637	613	1049	406	356
3547	3547	20319	20319	514	16586	814	722	1264	479	454
1182	1182	23591	23798	1386	18936	1208	978	1637	629	657

In contrast to the work in Section 8.1.3, the total risk retained is $S(\mathbf{u}) = \sum_{j=1}^{6}\{S_j \wedge u_j\}$ where S_j is the sum of claims from the jth risk. Provide a frontier of optimal retention limits using expected shortfall and $\alpha = 0.80$, comparable to Table 8.12. (*Hint*: Use the illustrative code from Example 7.6.)

Exercise 8.2. Excess of Loss Frontier Interpretation. One can explain the pattern of the excess of loss frontier that appears in Table 8.12 with a simple example. To motivate this exercise, note from Table 8.5 that the minimum of each risk distribution of each risk is positive.

For simplicity, use the expected shortfall objective function that appears in Display (7.4) that omits kernel smoothing. In addition, with fair costing of risk transfer, suppose that the difference between full insurance and the maximal risk transfer cost is $DELTA_{RTC} = \sum_{j=1}^{p} \mathrm{E}(X_j) - RTC_{max} > 0$.

Motivated by Table 8.5, let us suppose that each random variable can be expressed as $X_j = a_j + \tilde{X}_j$, where $\tilde{X}_j$ is a nonnegative random variable and a_j is a positive constant, for $j = 1, \ldots, p$.

a. Consider upper limits u_j such that $0 \leq u_j \leq a_j$. Show that *any* set of limits subject to these bounds and that sum to $\sum_{j=1}^{p} u_j = DELTA_{RTC}$ minimizes expected shortfall and that this minimal value is $DELTA_{RTC}$.

b. Use the results from part (a) to interpret the patterns in optimal upper limits in Table 8.12 as RTC_{max} varies.

Exercise 8.3. An Attractive Naive Portfolio. Consider the risk retention problem faced by Ashland County in Section 8.1.2. To diversify the portfolio, one solution is to set the upper limit for each risk at the quantile for that risk. In this way, each risk is well represented in the portfolio.

Determine an α quantile for each risk and use this to set the level of that upper limit. This set of upper limits can be used to form a portfolio. Determine the RTC and the ES (with a 95% level of confidence) for this portfolio. Determine ten such portfolios by varying α starting at 0.5, 0.55, …, 0.95, 0.98. Superimpose the resulting set of portfolios on the

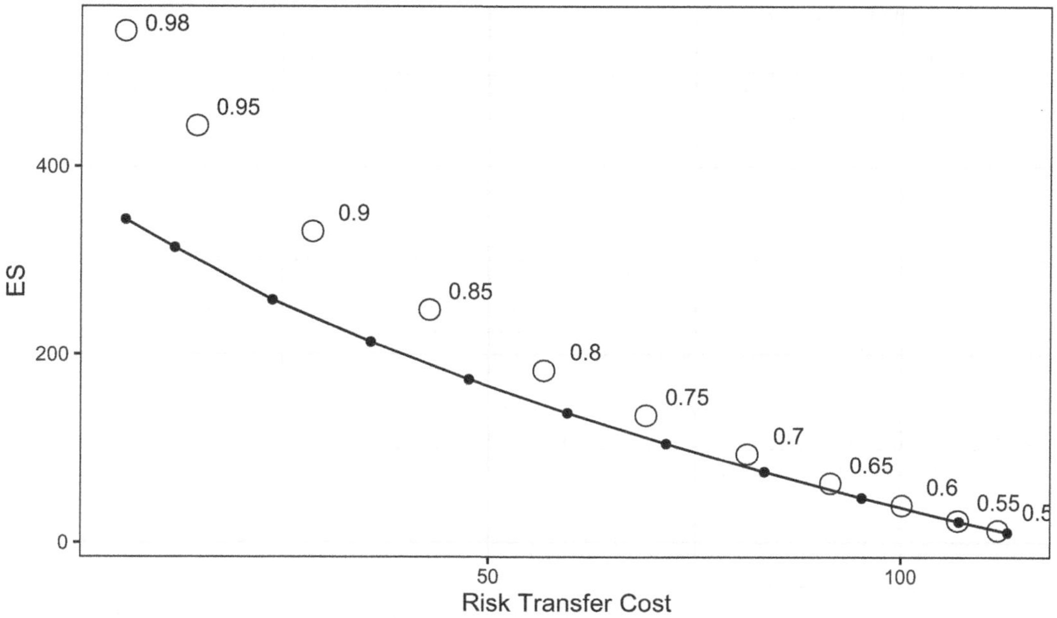

FIGURE 8.11

Exercise 8.3. Efficient Frontier of Retained Risk Measures versus Risk Transfer Costs, with Attractive Naive Portfolios Superimposed. The numbers on the graph provide the fractions used to create the quantiles of each risk in the portfolio.

efficient frontier from the middle panel of Figure 8.1. Your work should result in something comparable to Figure 8.11.

You will learn that this set of portfolios is quite competitive with the efficient frontier set. This new set of portfolios is "naive" in the sense that it does not require the dependencies nor the use of constrained optimization for its calculations. For theory underpinning this result, see Example 11.3.

Exercise 8.4. Fund Member 2 Optimal Portfolio with a Deductible. Modify the analysis in Table 8.3 by assuming that pool member 2 cedes property losses in excess of 1,000.

Exercise 8.5. Fund Risk Retention with Member Deductibles. Modify the analysis in Table 8.6 by assuming each pool member cedes only losses in excess of 20,000 to the pool.

Section 8.2 Exercises

Exercise 8.6. ***GlueVaR* and the ANU Case.** Instead of expected shortfall, think about how you might modify the ANU analysis in minimize the *GlueVaR* criterion. (Exercise 9.9 will provide additional guidance.)

TABLE 8.13
ANU Excess of Loss Optimization by Level of Cyber Load.

Cyber Load	*RTC*	*VaR*	*ES*	Std Dev	u_2	$u_{3:Cyber}$	u_4
0.5	723	6093	6859	2205	175	0	288
1.0	723	6226	7205	2248	186	432	252
1.5	723	6357	7322	2264	198	1506	144
2.0	723	6361	7314	2271	204	5633	145
2.5	723	6364	7315	2272	204	5848	147
3.0	723	6364	7315	2272	204	5848	147

Exercise 8.7. Varying the Cyber Risk Premium. For the cyber risk, the third in our notation, we now use risk transfer cost

$$RTC_{Cyber}(u) = CyberLoad \times \mathrm{E}[X_3 - \min(X_3, u)].$$

Here, we introduce a *CyberLoad* coefficient that can be used to account for administrative loadings or market availability factors. Table 8.13 shows results from the excess of loss optimization over several choices of the loading factor. Provide code to replicate this table and interpret the results.

8.3.3 Appendix. Tweedie Distribution

The Tweedie distribution is defined as a Poisson sum of gamma random variables. Specifically, suppose that N has a Poisson distribution with mean λ, representing the number of claims. Let $\{X_j\}$ be an i.i.d. sequence, independent of N, with each X_j having a gamma distribution with parameters α and γ, representing the amount of an individual claim. Then, $S_N = X_1 + \cdots + X_N$ is a Poisson sum of gammas.

To understand the mixture aspect of the Tweedie distribution, first note that it is straightforward to compute the probability of zero claims as $\Pr(S_N = 0) = \Pr(N = 0) = e^{-\lambda}$. The distribution function can be computed using conditional expectations,

$$\Pr(S_N \le y) = e^{-\lambda} + \sum_{n=1}^{\infty} \Pr(N = n) \Pr(S_n \le y), \qquad y \ge 0.$$

We calculate the moments using iterated expectations as $\mathrm{E}(S_N) = \lambda \frac{\alpha}{\gamma}$ and $\mathrm{Var}(S_N) = \frac{\lambda\alpha}{\gamma^2}(1+\alpha)$. Now, define three parameters μ, ϕ, p through the relations

$$\lambda = \frac{\mu^{2-p}}{\phi(2-p)}, \qquad \alpha = \frac{2-p}{p-1} \qquad \text{and} \qquad \frac{1}{\gamma} = \phi(p-1)\mu^{p-1}.$$

The Tweedie distribution is a member of the linear exponential family; thus, it can readily be used in regression type applications and so has found favor in the insurance industry, cf. Frees (2009). Further, easy calculations show that $\mathrm{E}(S_N) = \mu$ and $\mathrm{Var}(S_N) = \phi\mu^p$, where $1 < p < 2$.

8.3.4 Appendix. Generate Property Fund Data

As described in Section 8.1.1, the theory underpinning the risk modeling of the Wisconsin Property Fund is in Frees et al. (2016). The corresponding code and data to support development of the modeling are available from Frees et al. (2024). The associated code implements this development.

Code. See the *Online Supplement* for illustrative statistical code.

8.3.5 Appendix. ANU Risk Distribution

As described in Section 6.3.2, the set of insurable risks faced by ANU is complex. An earlier study, Frees and Butt (2022), documented the experience available that can be used to capture the distributions of each risk. Specifically, there was ample recent experience to fit frequency and severity distributions for three risks: Group Personal Accident, Corporate Travel, and Motor Vehicle. For each risk, a Poisson distribution was used for the number of annual losses and a lognormal distribution was used for the loss severity. The summary statistics for these distributions appear in Table 8.14.

Not surprisingly, the prior study indicated that the experience for the other 12 risks was sparse. Rather than split losses into frequency and severity components, we used a single aggregate distribution, the Tweedie, that is used extensively in many studies of insurance losses, see, for example, Frees (2014). The prior Appendix Section 8.3.3 describes this distribution.

To determine parameters of these 12 marginal distributions, market information through the premiums and deductibles was utilized. The first three columns of Table 8.14 provide the deductible, premium, and expected number, that are based on ANU contracts and data (see also Table 6.1 for the deductible and premium). For each risk, we used these inputs to match the parameters of a Tweedie distribution. That is, we used the expression

$$premium = \mathrm{E}[(X - deductible)_+],$$

where "+" denotes the positive part and X is a random variable with a Tweedie distribution. For most of the risks, we used a power parameter equal to 1.67 as is common in many actuarial studies. However, for both the Property and the General and Products Liability lines, we used a 1.9 power parameter to account for their longer tails.

To summarize each of the marginal risk distributions, we generated several summary measures including the mean, standard deviation, as well as the 95th and 99th percentiles. These appear in Table 8.14.

8.3.6 Appendix. Generate ANU Simulated Distributions

In this stage, we use the marginal risk parameters from the prior section to generate multivariate claims. The copula is specified as a mixture between independence and a so-called *comonotonic* copula that will be described in Section 9.1. With a copula, we can randomly generate (dependent) uniform numbers, the marginal risk parameters, and quantile functions from the appropriate distributions to generate dependent claims. See the general description in the first two steps of the Section 7.1 strategy and the appended code for more details.

Code. See the *Online Supplement* for illustrative statistical code.

TABLE 8.14
ANU Risk Summary Statistics.

	Deductible	Premium	Expected Number	Mean	Standard Dev	95th Percentile	99th Percentile
Property	5000	23565	0.67	24753	95631	147455	481482
General and Products Liability	100	448	0.50	467	2090	2585	10354
Cyber	250	76	0.25	112	389	748	1992
Crime	100	100	0.25	117	408	785	2090
Employment Practices Liability	100	84	0.25	101	351	675	1799
Expatriate	0	12	0.25	12	41	78	208
Group Personal Accident	0	105	20.00	23	7	35	40
Corporate Travel	0	75	220.00	140	70	255	303
Professional Indemnity	100	100	0.25	117	408	785	2090
Medical Malpractice	100	20	0.25	33	116	223	594
Clinical Trial	100	10	0.25	21	75	143	382
Statutory Liability	0	8	0.25	8	29	56	149
Motor Vehicle	1	85	80.00	276	198	601	736
Marine Cargo	5	6	0.25	7	24	47	125
Marine Hull	0	12	0.25	12	40	78	207

9

Stress Testing, Sensitivity, and Robustness

Chapter Preview. In previous chapters, we explored how to utilize constrained optimization to construct desirable insurable risk portfolios. To recap, building on Display (7.1), many risk retention problems can be formulated as:

$$\boxed{\begin{array}{ll} \text{minimize}_{z_0,\boldsymbol{\theta}} & f_0(z_0,\boldsymbol{\theta}) \\ \text{subject to} & RTC(\boldsymbol{\theta}) \leq RTC_{max} \\ & \mathbf{P}\boldsymbol{\theta} \leq \mathbf{p}_0 \quad . \end{array}} \tag{9.1}$$

Here, the objective function $f_0(z_0, \boldsymbol{\theta})$ represents either a risk measure summarizing the uncertainty of retained risks or a function that, when optimized, yields a risk measure (as developed in Section 7.2). The set of decision variables is denoted as $\mathbf{z} = (z_0, \boldsymbol{\theta})$, consisting of an auxiliary variable and the risk retention coefficients. The constraints include a budget constraint, $RTC(\boldsymbol{\theta}) \leq RTC_{max}$, as well as linear constraints on the risk retention parameters of the form $\mathbf{P}\boldsymbol{\theta} \leq \mathbf{p}_0$. Both the objective function f_0 and the risk transfer cost RTC can be determined analytically or via simulation.

Analysts find great satisfaction in producing an optimal set of results, knowing they have achieved the best possible outcome within a defined structure. However, like any mathematical representation, constrained optimization results are reliant on model assumptions. Therefore, when interpreting findings, it is natural to question the reliability of the results, something that risk managers are keen to understand. Particularly, one would like to know how the results respond to changes in assumptions.

Stress Testing. How can an analyst quantify these effects? The most direct approach, often termed *stress testing*, involves allowing a quantity of interest to vary over a range of potential values and observing how the optimization performs over this range. This approach is readily interpretable and requires no new theory. We already have witnessed stress testing in action for the ANU case in Section 8.2. In that section, we examined the reliability of results due to changes in the budget constraint, different choices of the confidence level, and by varying the risk measure. Section 9.1 will further explore this theme by evaluating alternative levels of dependence.

Sensitivity Analysis. Although stress testing is intuitively appealing, it can be time-consuming to rerun optimization procedures across many different plausible scenarios. Hence, Section 9.2 introduces a tool known in economics as the *Envelope Theorem*, used to calculate *local* sensitivities. By "local," we emphasize that the Envelope theorem captures marginal changes assuming small changes in an assumption, i.e., a differential approach, which we refer to as a *sensitivity*.

Section 9.3 demonstrates how the Envelope theorem can be applied to risk retention problems. As shown in this section, this tool is particularly valuable for understanding the impact on assumptions that solely affect the objective function, such as the choice of the risk measure. The next chapter presents an extension of the Envelope theorem that will allow us to handle more complex alternatives.

DOI: 10.1201/9781003469568-9

Robustness. Sensitivity analysis is limited to measuring the impact of small, or local, changes in assumptions. In the operations research literature, alternative techniques designed to study the impact of larger changes have been developed. Section 9.4 concludes the chapter with a review of different approaches to computing potential variations in optimization results to large changes in assumptions, known as *robustness.*

9.1 Dependence Stress Testing

As with financial risks, it is widely understood that the dependence among risks is crucial when making decisions to manage them. For instance, if a risk owner is responsible for properties within the same earthquake fault region, a single event like a major earthquake could cause serious losses across many risks. These risks are dependent, highlighting the importance of considering dependency when constructing portfolios. Further applications where dependence may impact a collection of risks will be discussed in Chapter 12. For this section, we assume that dependence is a key aspect of modeling risk portfolios.

Selecting an approach to measure dependence is another important aspect of dependence modeling. While historically prominent methods like multivariate normal distributions and Pearson correlations are familiar to many analysts, they are better suited for risks with thin tails and linear relationships. However, for insurable risks, distributions with heavier tails and non-linear relationships are more common. Copulas, introduced in Section 4.2, are often used to model dependence among multivariate outcomes in risk and insurance. We adopt this approach in this book.

How accurately can one measure dependence? We have already seen the Wisconsin Property Fund example in Section 8.1, where very detailed data allowed for reasonably precise estimation of the dependence model. In contrast, for the ANU case in Section 8.2, many of the risk occurred infrequently. This makes it challenging to calibrate the marginal risk distribution and impossible to accurately estimate dependence. To address these challenges, this section demonstrates how stress testing can be used to calibrate the impact of dependence.

To establish a range of potential dependency models, we begin with a worst-case scenario known as the *comonotonic* case. You can think of this as a model for a set of properties within the same earthquake fault region. Mathematically, in copula modeling, a widely used restriction is the Fréchet-Höeffding upper bound, given as

$$C(v_1, \ldots, v_p) \leq \min(v_1, \ldots, v_p) = C^{upper}(v_1, \ldots, v_p).$$

To see this equation, note that for any copula,

$$C(v_1, \ldots, v_p) = \Pr(V_1 \leq v_1, \ldots, V_p \leq v_p) \leq \Pr(V_j \leq v_j) = v_j,$$

for $j = 1, \ldots, p$. Because the copula is bounded by any argument v_j, it must be bounded by the minimum of all such arguments, which is the Fréchet-Höeffding upper bound. Further, note that this bound is attainable when $V_1 = \cdots = V_p$. This is the comonotonic case; you can think of these risks as being perfectly positively dependent upon one another. See Denuit et al. (2006), Chapter 2, for a general introduction to comonotonicity.

If there are only two risks, then it is possible to define perfectly negatively dependent risks such as (V_1, V_2) where $V_2 = 1 - V_1$. However, with more than two risks, there is less agreement in the literature on what constitutes perfect negative dependence, cf. Puccetti and Wang (2015). As an alternative, we utilize the case of *independence* as the lower bound for our range for stress testing. Denote $C^{ind}(v_1, \ldots, v_p) = v_1 \cdots v_p$ to be the independence copula.

TABLE 9.1
ANU Excess of Loss Optimization—Varying Dependency.

w	RTC	VaR	ES	Std Dev	u_2	u_3	u_4
1.0	722	5557	6947	2653	440	504	299
0.9	722	5510	6851	2610	392	498	296
0.8	722	5465	6713	2549	320	538	326
0.7	722	5445	6558	2479	256	550	335
0.6	722	5437	6428	2419	313	570	300
0.5	722	5431	6353	2380	349	593	298
0.4	722	5423	6214	2313	355	551	294
0.3	722	5422	6095	2252	337	550	312
0.2	722	5419	5981	2194	350	541	318
0.1	722	5416	5865	2130	286	542	367
0.0	722	5409	5727	2062	377	494	399

For stress testing, we consider the mixture distribution, defined as

$$C^w(v_1, \ldots, v_p) = w\ C^{upper}(v_1, \ldots, v_p) + (1-w)\ C^{ind}(v_1, \ldots, v_p),$$

where w is a constant, $0 \leq w \leq 1$. This distribution is easy to quantify using simulation. To illustrate, see Section 8.3.6 for a description of the ANU data generation process that is based on mixture copulas. By allowing the mixing weight w to vary, we are able to assess the effects of dependence.

Example 9.1. Effects of Dependence on Optimal an ANU Risk Portfolio. To exhibit the effects of varying dependence, we return to stress testing of the ANU case described in Section 8.2. As before, To keep the number of comparisons manageable, only results for $RTC_{max} = 722$ and $\alpha = 0.80$ are reported.

Results are summarized in Table 9.1 and Figures 9.1 and 9.2. In Table 9.1 and Figure 9.1, we see that each of the retained risk measures, VaR, ES, and standard deviation, increases as the dependency increases (that is, as w increases.)

With Table 9.1 and Figure 9.2, we can get a sense as to how the dependency affects the optimal upper limits. These displays suggest that the optimal limits appear to be relatively stable as one moves from the comonotonic case ($w = 1$) to the case of independence ($w = 0$). Such a result would be pleasing to the risk manager, suggesting that the optimal insurable risk portfolios are relatively robust to the dependence assumption.

9.2 Envelope Theorem

This section introduces the Envelope theorem as a tool for calibrating the sensitivity of the optimal results to changes in exogenous (non-decision) variables. To this end, now introduce another variable, "a" for *auxiliary*, into the optimization problem, that represents the assumption that we wish to vary. So that you can appreciate this tool across various fields, this section provides a general introduction without a restriction to risk retention applications. Therefore, instead of focusing solely on risk retention problems such as in

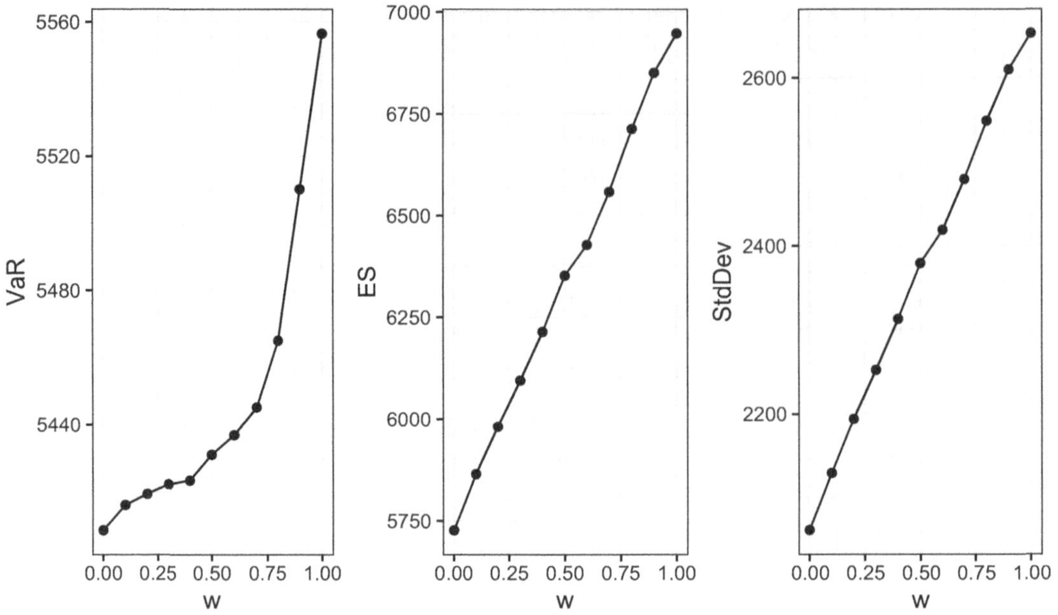

FIGURE 9.1
Trade-off of Retained Risk versus Dependency Weight. Based on the ANU Excess of Loss Case Study where upper limits are determined by minimizing the ES criterion.

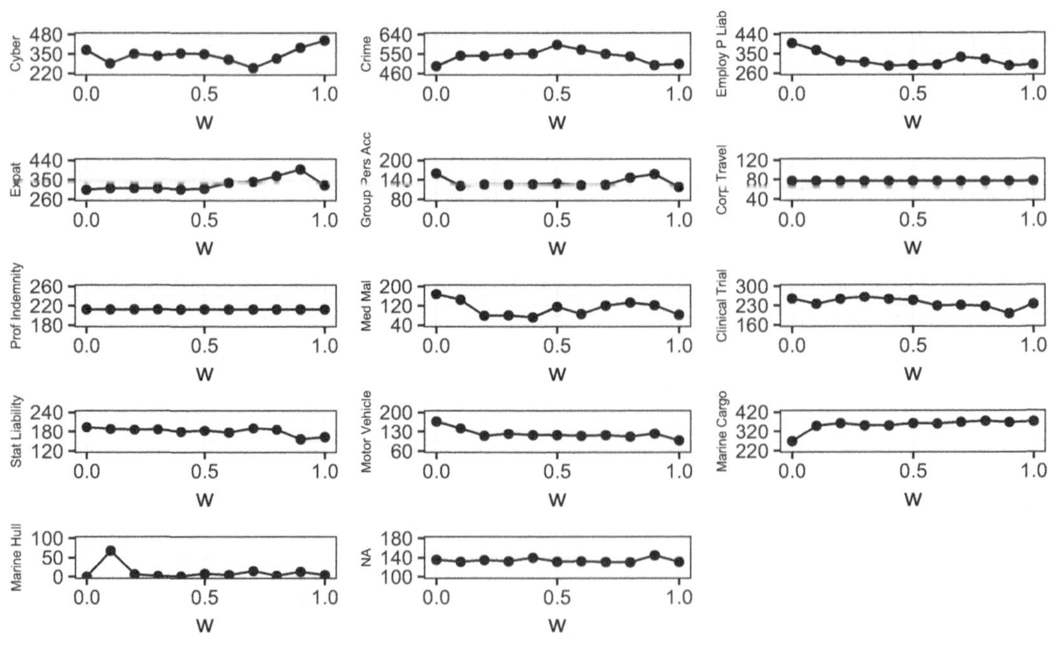

FIGURE 9.2
Upper Limits versus Mixing Weight. Based on minimizing the ES criterion.

Display (9.1), we work with a modified version of a general problem from Display (3.9) given as:

$$\boxed{\begin{array}{lll} \text{minimize}_{\mathbf{z}} & f_0(\mathbf{z}, a), & \\ \text{subject to} & f_{con,j}(\mathbf{z}, a) \leq 0 & j \in CON_{in} \\ & f_{con,j}(\mathbf{z}, a) = 0 & j \in CON_{eq}. \end{array}} \tag{9.2}$$

In Display (9.2), both the objective function f_0 and any of the constraint functions $f_{con,j}$ may depend on the auxiliary variable a. To emphasize, a is not one of the variables that is used in the optimization; the decision or choice variables are still represented by the vector $\mathbf{z}$. The interest lies in calibrating the sensitivity of the optimization results to changes in a.

Active Sets of Constraints. Assuming that the problem in Display (9.2) is resolved for a fixed a, we now suppose that $\mathbf{z}^*(a)$ is a stationary point on the feasible set and that there exists a set of optimal Lagrange multipliers. At the solution, some of the inequalities are binding $f_{con,j}[\mathbf{z}^*(a), a] = 0$, while others are not. For those not binding, a small movement in a may produce a small change in the constraint value but, because it is not binding, it will not affect the associated Lagrange multiplier. Thus, as we verify in the proofs, we can essentially ignore non-binding (at the solution) constraints in the sensitivity analysis. To ease notation, let m be the number of equality plus binding inequality constraints and use $f_{con,j}$ for $j = 1, \ldots, m$ for these constraints, and similarly for the Lagrange multipliers, LME_j.

Define the differentials $f_{0,a}(\mathbf{z}, a) = \partial_a f_0(\mathbf{z}, a)$ for $j = 1, \ldots, m$ and $i = 1, \ldots, p_z$,

$$f_{con,j,a}(\mathbf{z}, a) = \partial_a f_{con,j}(\mathbf{z}, a) \qquad \text{and} \qquad f_{con,j,i}(\mathbf{z}, a) = \partial_{z_i} f_{con,j}(\mathbf{z}, a).$$

With this notation, we have a version of the

Envelope Theorem. Assuming the objective and constraint functions are sufficiently smooth, we have

$$\partial_a f_0[\mathbf{z}^*(a), a] = f_{0,a}[\mathbf{z}^*(a), a] + \sum_{j=1}^{m} LME_j^*(a)\; f_{con,j,a}[\mathbf{z}^*(a), a], \tag{9.3}$$

where

$$f_{con,j,a}[\mathbf{z}^*(a), a] = -\sum_{i=1}^{p_z} f_{con,j,i}[\mathbf{z}^*(a), a]\; \partial_a z_i^*(a), \quad j = 1, \ldots, m. \tag{9.4}$$

A sketch of the proof of the Envelope theorem is in Appendix Section 9.5.3. Many economic applications focus on the first equation (9.3), providing an expression for the rate of change in the optimal value of the function being optimized, $f_0[\mathbf{z}^*(a), a]$. Note that the m equalities in equation (9.4) provide information about the trade-offs among the decision variable changes ($\partial_a z_i^*(a)$) at the optimum.

The Envelope theorem may seem complex, but many problems do not require this full generality. In some problems, the constraint functions are not affected by the auxiliary variable. In such instances, from equation (9.3), we observe that $\partial_a f_0[\mathbf{z}^*(a), a] = f_{0,a}[\mathbf{z}^*(a), a]$, indicating that small changes in the optimal value of the objective function can be computed as a marginal change in the objective function evaluated at the optimum. Furthermore, we have $f_{con,j,a}[\mathbf{z}^*(a), a] = 0$ for the left-hand side of equation (9.4), making it easy to discern the trade-offs among the decision variable changes ($\partial_a z_i^*(a)$) at the optimum.

In other cases, the objective function is not affected by the auxiliary variable, and thus $f_{0,a}[\mathbf{z}^*(a), a] = 0$. This is illustrated in the following example.

Example 9.2. Maximal Risk Transfer Cost as an Auxiliary Variable. Consider the risk retention problem in Display (9.1), letting a generic risk measure RM of retained risks serve as the objective function so that $f_0(z_0, \boldsymbol{\theta}) = RM[g(\mathbf{X};\boldsymbol{\theta})]$. In addition, let the maximal risk transfer cost be the auxiliary variable, $a = RTC_{max}$. Intuitively, if the budget constraint is not binding, then a small change in RTC_{max} will not have an effect on the optimization result. So, we assume that the budget constraint is binding. Further, for illustration purposes, let us assume that one of the thetas is at the boundary and, for simplicity, take it to be the first one so that $\theta_1^*(a) = 0$. With this, we have $m = 2$ and with some mild smoothness assumptions can evaluate the differentials

$$
\begin{aligned}
f_{0,a}(\boldsymbol{\theta}, a) &= \partial_{RTC_{max}} RM[g(\mathbf{X};\boldsymbol{\theta})] = 0,\\
f_{con,1,a}(\boldsymbol{\theta}, a) &= \partial_{RTC_{max}} [RTC(\boldsymbol{\theta}) - RTC_{max}] = -1\\
f_{con,2,a}(\boldsymbol{\theta}, a) &= \partial_{RTC_{max}} [-\theta_1] = 0.
\end{aligned}
$$

Using equation (9.3), we have

$$
\partial_{RTC_{max}} RM[g(\mathbf{X};\boldsymbol{\theta}^*)] = -LME_1^* \ .
$$

Thus, the first Lagrange multiplier captures the marginal change in the minimal value of the objective function, per unit change in the maximal risk transfer cost. Note, too, that the sign reinforces our intuition. Thus, small increases in the maximal risk transfer costs, RTC_{max}, lead to small decreases in the uncertainty retained risk (as summarized by the risk measure); this is what we would expect.

Example 9.3. Interpreting ANU Lagrange Multipliers. Let us continue with Example 9.2 but now specialize this to the ANU situation developed in Example 7.5. The risk measure is ES. Figure 9.3 summarizes some of the earlier optimization results but now augments these with the Lagrange multipliers. This table and plot show the negative relationship between RTC_{max} and the optimal value of ES. This rate of change is captured by the Lagrange multiplier LME_1^*.

Because the Lagrange multipliers are also calculated using simulation methods, we can see unevenness in the early and later values of RTC_{max} where the calculations are less precise. For many audiences, analysts will find it valuable to demonstrate the relationship between RTC_{max} and the optimal value of ES by giving a range over different values of RTC_{max}, as in this example. However, it is also useful to be able to interpret the Lagrange multiplier as a partial derivative. In this way, an analyst can provide information about the rate of change with a single optimization. For example, consider the table that is part of Figure 9.3. If one only ran the optimization at $RTC_{max} = 866$, then we could interpret the corresponding $LME_1 = 3.1$ to mean a decrease of 3.1 in the optimal ES for each unit increase in RTC_{max}.

RTCmax	RTC	ES	LME1
1371	1370	79	0
1299	1293	154	0
1155	1145	310	0
1010	1010	472	2.3
866	866	875	3.1
722	722	1350	3.5
577	577	1870	3.8
433	433	2477	4.8
289	289	3197	5
144	144	3919	5
72	72	4280	5

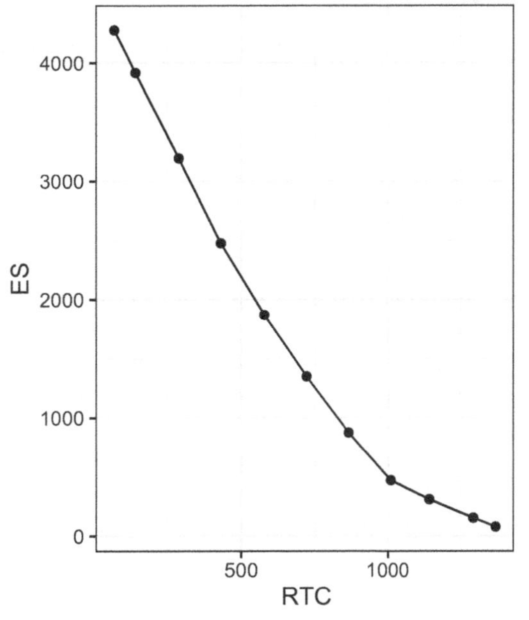

FIGURE 9.3
ANU Excess of Loss Optimization Results—ES Criterion.

Extending the idea of interpreting Lagrange multipliers as partial derivatives, consider the following

Special Case. Perturbed Problem. By letting one of the equality constraints equal the auxiliary variable, we can investigate small changes in the effect of that constraint on the problem solution, as follows. Consider the problem:

$$\boxed{\begin{array}{lll} \text{minimize}_{\mathbf{z}} & f_0(\mathbf{z}), & \\ \text{subject to} & f_{con,j}(\mathbf{z}) \leq 0 & j \in CON_{in} \\ & f_{con,k}(\mathbf{z}) = a & \\ & f_{con,j}(\mathbf{z}) = 0 & j \in CON_{eq}, j \neq k. \end{array}} \tag{9.5}$$

From the Envelope Theorem, we have $LME_k^*(a) = -\partial_a f_0[\mathbf{z}^*(a)]$. That is, the multipliers measure the sensitivity of the optimal value of the objective function to changes in the limiting values of the constraints.

In economics, one can interpret Lagrange multipliers as providing a natural measure of value of scarce resources. For example, one of the classical economic situations is the profit maximization problem where each variable represents an input (e.g., Simon and Blume (1994)). Then, the jth multiplier can be interpreted as the change in the maximum profit per unit change in the jth input. In this sense, it is sometimes known as the *shadow price*. See also Boyd and Vandenberghe (2004), Section 5.6.

9.3 Sensitivity Analysis for Risk Retention Problems

Now consider the modified version of the risk retention problem from Display (9.1)

$$\boxed{\begin{array}{ll} \text{minimize}_{\mathbf{z}=(z_0,\boldsymbol{\theta})} & f_0(\mathbf{z},a) \\ \text{subject to} & f_{con,1}(\mathbf{z},a) = RTC(\boldsymbol{\theta},a) - RTC_{max} \leq 0 \\ & \mathbf{P}\boldsymbol{\theta} \leq \mathbf{p}_0 \quad . \end{array}} \tag{9.6}$$

Here, the objective function f_0 and the first constraint $f_{con,1}$ are potentially affected by the auxiliary variable a.

The other $m-1$ constraints $\mathbf{P}\boldsymbol{\theta} \leq \mathbf{p}_0$ trace out *edges* of the feasible region. They can ensure non-negativity of retention parameters (e.g., $\theta_j \geq 0$), provide an upper bound on the parameters (e.g., $\theta_j \leq 1$), or enforce a balance among the parameters (e.g., $\theta_1+\cdots+\theta_p = 1$). Recall that for sensitivity purposes, we only consider those constraints that are binding at the optimum. To simplify the presentation, I assume that the constraints $\mathbf{P}\boldsymbol{\theta} \leq \mathbf{p}_0$ are not affected by a. As these constraints do not depend on the auxiliary variable,

$$f_{con,j,a}(\mathbf{z},a) = \partial_a f_{con,j}(\mathbf{z},a) = 0,$$

a partial derivative with respect to a is zero and they can be ignored for sensitivity analysis.

With only $m=1$ binding constraint that is affected by the auxiliary variable a, from the Envelope Theorem we have

$$\partial_a f_0[\mathbf{z}^*(a),a] = f_{0,a}[\mathbf{z}^*(a),a] + LME_1^*(a)\ f_{con,1,a}[\mathbf{z}^*(a),a], \tag{9.7}$$

where

$$f_{con,1,a}[\mathbf{z}^*(a),a] = -\sum_{i=1}^{p_z} \partial_{z_i}\ RTC(\mathbf{z},a)|_{\mathbf{z}=\mathbf{z}^*(a)}\ \partial_a z_i^*(a). \tag{9.8}$$

For interpretation, now consider the case where the budget constraint $f_{con,1}$ does not depend on a. This means that

$$\partial_a f_0[\mathbf{z}^*(a),a] = f_{0,a}[\mathbf{z}^*(a),a]. \tag{9.9}$$

That is, to determine the partial derivative of the objective function at the optimal value, we need only determine the partial derivative of the objective function and then evaluate it at the optimal value. In addition, the left-hand side of equation (9.8) is 0. To illustrate the trade-off among sensitivities, assume that $p_z = 2$ which yields

$$\partial_{z_1}\ RTC(\mathbf{z},a)|_{\mathbf{z}=\mathbf{z}^*(a)} \times \partial_a z_1^*(a) = -\partial_{z_2}\ RTC(\mathbf{z},a)|_{\mathbf{z}=\mathbf{z}^*(a)} \times \partial_a z_2^*(a).$$

Typically, as one optimal parameter changes, the other changes in a proportional (and opposite in sign) way, subject to the signs of the partial derivatives of the constraint function.

Example 9.4. Level of Confidence. For the risk retention problem in Display (9.6), take the confidence level as the auxiliary variable, $a = \alpha$. Further assume that the confidence level does not affect the budget constraint, so $f_{con,1,a}(\mathbf{z},a) = \partial_a f_{con,1}(\mathbf{z},a) = 0$.

The goal is to minimize the ES as the measure of uncertainty but one can use $ES1_F(x)$ for the objective function as described in Section 7.2.2. In this case, with $\alpha = a$, we have

$$\begin{aligned} f_{0\alpha}(\mathbf{z},\alpha) &= \partial_\alpha ES1_{F(\mathbf{z})}(z_0) = \partial_\alpha \left\{ z_0 + \tfrac{1}{1-\alpha} \{ \mathrm{E}[Y(\mathbf{z}) - z_0]_+ \} \right\} \\ &= \left(\tfrac{1}{1-\alpha} \right)^2 \{ \mathrm{E}[Y(\mathbf{z}) - z_0]_+ \}. \end{aligned}$$

At the optimum (where $z_0^* = VaR^*$ is the α quantile), we can express this as

$$\left(\tfrac{1}{1-\alpha} \right)^2 \{ \mathrm{E}[Y(\mathbf{z}^*) - z_0^*]_+ \} \ = \tfrac{1}{1-\alpha} \{ ES(z_0^*) - VaR^* \}.$$

From this expression, we see that the derivative is non-negative, meaning that as the confidence level increases so does the optimal value of ES.

To get information about allocation trade-offs, from equation (9.8) we have

$$0 = \sum_{i=1}^{p_z} \partial_{z_i} RTC(\mathbf{z})\big|_{\mathbf{z}=\mathbf{z}^*} \ \partial_\alpha z_i^*(\alpha).$$

This reveals an interesting balance among the retention parameters that is driven by the risk transfer cost function.

Example 9.5. Level of Confidence and the ANU Case. Let us continue with Example 9.4 but now specialize this to the ANU situation developed in Example 7.5. The decision variables are $\mathbf{z} = (z_0, u_1, \ldots, u_p)$.

Figure 9.4 summarizes some of the earlier optimization results from Table 8.9 and Figure 8.6; it now augments these with the partial derivative of the optimal ES. The fourth column of the table, labelled `Deriv ES` is given in changes in the percentages of α. These partial derivatives are consistent with the right-hand side plot of Figure 9.4.

Assuming fair risk transfer costs, we have $\partial_{u_j} RTC(u_1, \ldots, u_p) = \partial_{u_j} \mathrm{E}(X_j \wedge u_j) = 1 - F_j(u_j)$. Thus,

$$0 = \sum_{j=1}^{p} \left[1 - F_j(u_j^*) \right] \times \partial_\alpha u_j^*(\alpha).$$

This expression is interesting. In Figure 8.6 we saw the progressions of all upper limits. Further, the optimal retention policy changes the mixture of risks as α changes; some of the u_j values increase as α increases whereas others decrease. This expression provides greater insight into this trade-off of upper limit changes.

Example 9.6. Range Value at Risk and the ANU Case. For the risk retention problem in Display (9.6), now use the range value at risk $RVaR$ for the objective function and let $a = \beta$ be the auxiliary variable. Recall that β is the parameter that moves the $RVaR$ between the VaR and the ES.

From equation (2.4), if $\beta > 0$, we have

$$\begin{aligned} \partial_\beta \ RVaR_{(\alpha,\beta)}(X) &= \partial_\beta \left(\frac{1}{\beta} \int_\alpha^{\alpha+\beta} VaR_z(X)\, dz \right) \\ &= \frac{-1}{\beta^2} \int_\alpha^{\alpha+\beta} VaR_z(X)\, dz + \frac{1}{\beta} VaR_{\alpha+\beta}(X) \\ &= \frac{1}{\beta} \left\{ VaR_{\alpha+\beta}(X) - RVaR_{(\alpha,\beta)}(X) \right\}. \end{aligned}$$

Alpha	VaR	ES	Deriv ES
0.5	751	3299	50.97
0.55	807	3580	61.62
0.6	898	3921	75.56
0.65	1012	4344	95.23
0.7	1178	4889	123.68
0.75	1971	5579	144.31
0.8	5408	6080	33.59
0.85	5580	6276	46.41
0.9	5782	6569	78.72
0.95	6176	7204	205.63
0.98	7585	7602	8.72

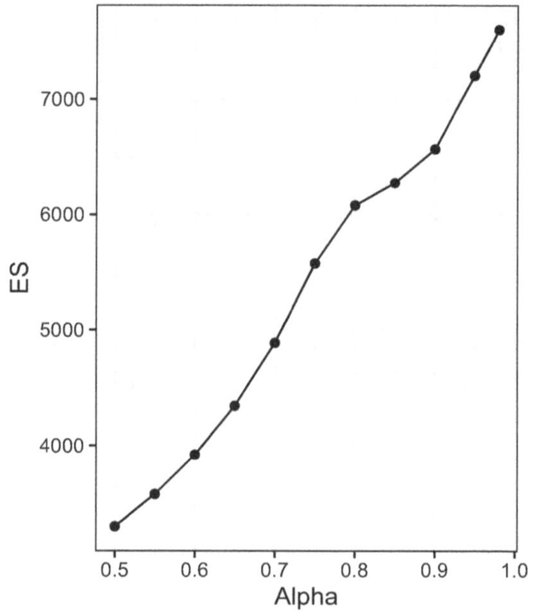

FIGURE 9.4
Retained Risk (Measured by *ES*) versus Level of Confidence (Measured by Alpha α). Based on the ANU Excess of Loss Case Study where upper limits are determined by minimizing the *ES* criterion.

As in Example 9.5, β does not affect the constraints and so equation (9.9) holds:

$$\partial_\beta \; RVaR_{(\alpha,\beta)}[g(\mathbf{X};\mathbf{z}^*)] \quad = \tfrac{1}{\beta}\left\{VaR_{\alpha+\beta}[g(\mathbf{X};\mathbf{z}^*)] - RVaR_{(\alpha,\beta)}[g(\mathbf{X};\mathbf{z}^*)]\right\},$$

where $\mathbf{z}^* = \mathbf{z}^*(\beta)$.

Figure 9.5 illustrates the steady increase in the optimal range value at risk as a function of the range parameter β. The corresponding table also provides numerical values for the sensitivities that are small for small values of β and increase as β increases.

Example 9.7. Value at Risk Sensitivity. As we saw in Section 7.2.1, sometimes we can treat a risk measure as the solution of an optimization problem. In these cases, we can utilize the Envelope Theorem to quantify small changes in them and calibrate their sensitivity to auxiliary/exogenous changes.

Specifically, consider the problem in Display (9.1) where $f_0(z_0) = z_0$ with $m = 1$ constraint $f_{con,1}(z_0,\mathbf{z}) = -[F_{g(X;\mathbf{z})}(z_0) - \alpha]$. As we saw, the best value of z_0 is $z_0^* = F^{-1}_{g(X;\mathbf{z})}(\alpha)$ with corresponding Lagrange multiplier $LMI_1^* = 1/f_{g(X;\mathbf{z})}(z_0^*)$ (from the proof). For this example, we are only working with minimization over z_0 and so may treat risk retention variables z_j as auxiliary variables.

With this, one can use the Envelope theorem to establish for any $j = 1,\ldots,p$,

$$\partial_{\theta_j} VaR_\alpha[g(X;\mathbf{z})] = -\frac{1}{f_{g(X;\mathbf{z})}[F^{-1}_{g(X;\mathbf{z})}(\alpha)]} \; \partial_{\theta_j} F_{g(X;\mathbf{z})}(z_0)\Big|_{z_0=F^{-1}_{g(X;\mathbf{z})}(\alpha)}. \tag{9.10}$$

ⓘ *Under the Hood.* See the *Online Supplement* for confirmation of this result.

Beta	VaR	RVaR	Deriv RVaR
0	5788.44	5788.44	0
0.01	5781.41	5799.87	0.19
0.02	5781.27	5823.93	0.44
0.03	5781.28	5854.97	0.76
0.04	5781.38	5891.48	1.15
0.05	5780.99	5932.19	1.59
0.06	5781.5	5991.8	2.24
0.07	5805.13	6078.65	2.94
0.08	5722.76	6159.39	4.75
0.09	5708.04	6311.83	6.64
0.1	5708	6615.07	10.08

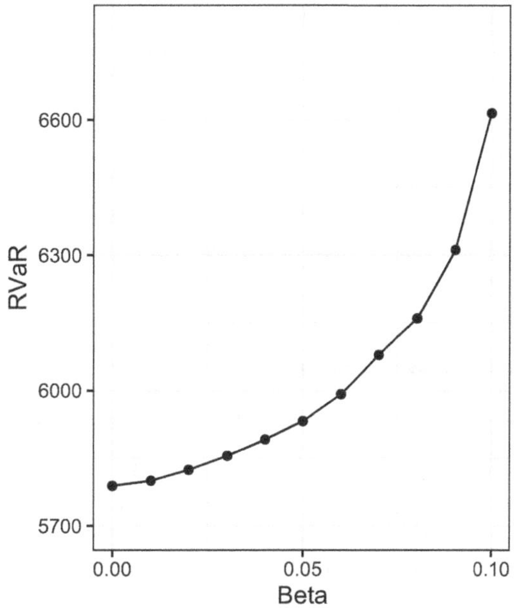

FIGURE 9.5
Retained Risk Measures versus Beta β. Based on minimizing the $RVaR$ criterion. Here, $\beta = 0$ corresponds to VaR and $\beta = 1 - \alpha$ corresponds to ES.

This corroborates the result established earlier in Section 5.3.1.

The right-hand side of equation (9.8) is a sum over the number of decision variables p_z, not the number of random variables. For a multivariate example where the number of random variables exceeds the number of decision variables, consider the following:

Example 9.8. A Contract with Mixed Retention Functions. Consider 3 risk types, where the first risk, X_1, is limited by an excess of loss policy with parameter u_1 so the retained risk is $X_1 \wedge u_1$. For the second risk, X_2, a quota share is agreement is in place with parameter c_2, so the retained risk is $c_2 X_2$. For the third risk type, a per occurrence excess of loss agreement is in place with parameter u_3, so the retained risk can be written as $S(u_3) = \sum_{j=1}^{N} \{X_{2+j} \wedge u_3\}$, where N represents a random number of claims independent of the *iid* individual claims $\{X_3, X_4, \ldots, \}$. With fair pricing, the cost of this insurance is

$$\begin{aligned} RTC(u_1, c_2, u_3) &= \mathrm{E}[X_1 + X_2 + S(\infty)] - \mathrm{E}\{X_1 \wedge u_1 + c_2 X_2 + S(u_3)\} \\ &= \mathrm{E}(X_1) + \mathrm{E}(X_2) + \mathrm{E}(N)\mathrm{E}(X_3) \\ &\quad - \{\mathrm{E}(X_1 \wedge u_1) + c_2 \mathrm{E}(X_2) + \mathrm{E}(N) \times \mathrm{E}(X_3 \wedge u_3)\}. \end{aligned}$$

The budget constraint is $RTC(u_1, c_2, u_3) \leq RTC_{max}$. If this constraint is binding at an optimal choice of retention parameters $\{u_1^*, c_2^*, u_3^*\}$, then we can use equation (9.8) to investigate their relationship. Straightforward calculations yield

$$\begin{aligned} \partial_{u_1} RTC(u_1, c_2, u_3) &= -[1 - F_1(u_1)] \\ \partial_{c_2} RTC(u_1, c_2, u_3) &= -\mathrm{E}(X_2) \\ \partial_{u_3} RTC(u_1, c_2, u_3) &= -\mathrm{E}(N)[1 - F_3(u_3)]. \end{aligned}$$

With equation (9.8), we have

$$\begin{aligned}0 &= [1 - F_1(u_{1,opt})]\ \partial_a\ u_{1,opt}(a) + \mathrm{E}(X_2)\ \partial_a\ c_{2,opt}(a)\\ &\quad + \mathrm{E}(N)[1 - F_3(u_{3,opt})]\ \partial_a\ u_{3,opt}(a).\end{aligned}$$

In this expression, a is a generic exogenous variable (not part of the budget constraint). For our applications, it could be a confidence level in the value at risk or expected shortfall or a parameter summarizing the dependence among risks. This expression provides an interesting balance among the decision variable sensitivities.

9.4 Asset Allocation Sensitivity and Robustness

When examining the sensitivity to parameter assumptions, it is particularly intriguing to consider the asset allocation problem introduced in Section 1.3. Although this framework has been well received intellectually, its practical implementations have been limited, partly due to parameter uncertainty.

9.4.1 Asset Allocation Sensitivity

The asset allocation optimization model's inputs include means, variances, and covariances/correlations, typically estimated from historical financial return data, as illustrated in Example 1.1. It is known that the optimal asset weights are highly sensitive to estimates of expected returns. Many investment managers find the optimal portfolio weights extreme and counter-intuitive (e.g., He and Litterman (2002)). Furthermore, the instability of these weights can result in poor out-of-sample performance for the optimal portfolio as demonstrated by De Prado (2016). Additionally, uncertainty in means can lead to highly concentrated portfolios. For instance, Best and Grauer (1991) demonstrate how slight changes in expected returns can result in heavy investments in one return balanced by a strong "negative" investment (shorting an asset) when assets are correlated.

Diversification of a portfolio of assets is one of the primary objectives of asset allocation, with variances and covariances at the core of this theory. Unfortunately, these inputs are also subject to estimation error which can lead to unstable portfolio weights. Interestingly, Chopra and Ziemba (1993) find that uncertainty in expected returns is roughly ten times as impactful as uncertainty in the covariance matrix. Additionally, the variance–covariance matrix can become unstable as asset correlations increase, which occurs precisely when diversification is crucial and yet more difficult to achieve (De Prado (2016)).

As summarized in Beketov et al. (2018), in practical applications, all optimization model inputs are adjusted, substituted, and/or complemented by various methods, including robo-advising methods outlined in Section 6.5. One objective of this section is to put forth a method for quantifying the sensitivity of these inputs using methods developed earlier in the chapter.

To this end, building on the notation from Section 3.4, we now let $\boldsymbol{\mu}$ be the vector of mean returns and, as before, let $\boldsymbol{\Sigma}$ be the variance-covariance matrix. We wish to minimize the objective function as $f_0(\mathbf{c}, \boldsymbol{\Sigma}) = -(\mathbf{c}'\boldsymbol{\mu} - \lambda_{Port}\ \mathbf{c}'\boldsymbol{\Sigma}\mathbf{c})$ subject to a single ($m = 1$) budget constraint $f_{con,1}(\mathbf{c}) = \mathbf{c}'\mathbf{1} - 1 = 0$. In this minimal setting with no requirement that the allocation parameters be nonnegative, closed-form expressions for the sensitivities can be obtained, providing intuition for the results.

To begin, consider the uncertainty arising from mean parameters and choose the auxiliary variable to be $a = \mu$, where μ is an element of $\boldsymbol{\mu}$ and use $\partial_\mu \boldsymbol{\mu}$ for the change in the vector with respect to that element. So, for example, if μ is the first mean, then $\partial_\mu \boldsymbol{\mu} = \mathbf{1}_1$ which represents a vector with 1 in the first entry and 0 for the other entries. Then, one can use first principles to establish

$$\partial_\mu \mathbf{c}^*(\mu) = \frac{1}{2\lambda_{Port}} \boldsymbol{\Sigma}^{-1} \left\{ [\partial_\mu \boldsymbol{\mu}] - \mathbf{1}\ \partial_\mu LMI_1^*(\mu) \right\} \tag{9.11}$$

and

$$\partial_\mu LMI_1^*(\mu) = \frac{\mathbf{1}'\boldsymbol{\Sigma}^{-1}[\partial_\mu \boldsymbol{\mu}]}{\mathbf{1}'\boldsymbol{\Sigma}^{-1}\mathbf{1}}. \tag{9.12}$$

Under the Hood. See the *Online Supplement* for confirmation of this result.

With this, the portfolio sensitivity for the mean, variance, and standard deviation are

$$\begin{aligned}
\partial_\mu \left[\boldsymbol{\mu}'\mathbf{c}^*\right] &= \boldsymbol{\mu}'[\partial_\mu \mathbf{c}^*] + [\partial_\mu\ \boldsymbol{\mu}]'\mathbf{c}^* \\
\partial_\mu \left[\mathbf{c}^{*\prime}\boldsymbol{\Sigma}\mathbf{c}^*\right] &= [\partial_\mu \mathbf{c}^*]'\boldsymbol{\Sigma}\mathbf{c}^* + \mathbf{c}^{*\prime}\boldsymbol{\Sigma}[\partial_\mu \mathbf{c}^*] \\
\partial_\mu \sqrt{\mathbf{c}^{*\prime}\boldsymbol{\Sigma}\mathbf{c}^*} &= \frac{1}{2\sqrt{\mathbf{c}^{*\prime}\boldsymbol{\Sigma}\mathbf{c}^*}} \partial_\mu[\mathbf{c}^{*\prime}\boldsymbol{\Sigma}\mathbf{c}^*] = \partial_\mu\ SD[c^{*'}X],
\end{aligned}$$

that can be evaluated using equations (9.11) and (9.12).

Example 9.9. Allocation Sensitivities for a Portfolio of Insurance Stock Returns. This is a continuation of Example 4.10 which is based on Examples 1.1 and 3.3. However, unlike these earlier examples, we now no longer require nonnegative allocation parameters.

Table 9.2 summarizes results. The first two rows provide estimates of the mean and standard deviation based on five years of the training sample. These are annualized estimates based on daily data.

The third row provides the optimal allocation parameters $\mathbf{c}^*$ using a penalty parameter $\lambda_{Port} = 5$ as in Example 3.3. Note that the estimate for the French insurer AXA is negative, suggesting a negative investment or "shorting" this asset. For benchmarking, the optimal portfolio mean is 0.197 and the corresponding standard deviation is 0.126.

The fourth row provides allocation coefficient sensitivities for that risk. For example, if the mean for United Health increases by 1 unit (holding other means fixed), then the allocation for United Health increases by 1.262. The allocation parameters for the other seven stocks also change as displayed in Table 9.3.

TABLE 9.2
Mean Sensitivities to Asset Allocations.

	United Health	AIA	Ping An	Chubb	Allianz	Aon	AXA	Tokio Marine
μ	0.249	0.166	0.208	0.097	0.159	0.181	0.139	0.148
σ	0.213	0.239	0.264	0.160	0.193	0.169	0.241	0.267
c^*	0.269	0.125	0.080	0.005	0.270	0.329	-0.164	0.086
$\partial_\mu\ c^*(\mu)$	1.262	1.222	1.075	3.021	3.802	2.780	2.493	0.781
$\partial_\mu\ [c^{*'}\mu]$	0.405	0.107	0.122	-0.271	0.330	0.465	-0.229	0.071
$\partial_\mu\ SD[c^{*'}X]$	0.054	-0.007	0.017	-0.110	0.024	0.054	-0.026	-0.006

TABLE 9.3
Mean Sensitivities to Allocation Coefficients—$\partial_\mu\ c^*(\mu)$.

	United Health	AIA	Ping An	Chubb	Allianz	Aon	AXA	Tokio Marine
United Health	1.262	-0.057	-0.055	-0.517	-0.165	-0.375	-0.035	-0.058
AIA	-0.057	1.222	-0.591	-0.102	-0.206	-0.055	-0.032	-0.178
Ping An	-0.055	-0.591	1.075	0.013	-0.122	-0.134	-0.051	-0.136
Chubb	-0.517	-0.102	0.013	3.021	-0.636	-1.799	0.125	-0.105
Allianz	-0.165	-0.206	-0.122	-0.636	3.802	-0.259	-2.294	-0.120
Aon	-0.375	-0.055	-0.134	-1.799	-0.259	2.780	-0.090	-0.068
AXA	-0.035	-0.032	-0.051	0.125	-2.294	-0.090	2.493	-0.116
Tokio Marine	-0.058	-0.178	-0.136	-0.105	-0.120	-0.068	-0.116	0.781

The fifth and sixth rows of Table 9.2 display the marginal changes of the portfolio mean and standard deviation for changes in each mean parameter. It is interesting to see that the changes in the standard deviations tend to be smaller in magnitude than mean changes.

Table 9.3 provides more information about the allocation coefficient sensitivities. For example, as in the fourth row of Table 9.2, we see that if the mean for United Health increases by 1 unit (holding other means fixed), then the allocation for United Health increases by 1.262. Furthermore, the first row of Table 9.3 provides changes in allocations for the other risks, all due to the 1 unit increase in the mean of United Health. In particular, note that there is a significant decrease in the Chubb allocation, -0.517. For each row, the sum of these changes is zero due to the budget constraint $\mathbf{c}'\mathbf{1} = 1$. Thus, the entire portfolio of asset allocations is highly sensitive to the uncertainty of even a single input mean, suggesting the need for more robust approaches as will be taken up in Section 9.4.2.

Exercise 9.12 shows how to construct sensitivities due to variance and covariance parameters. Readers will find it interesting to compare these sensitivities to the mean sensitivities described above to re-affirm the industry perception that mean changes have a greater effect than variance and covariance changes.

9.4.2 Asset Allocation Robustness

Stress testing, such as illustrated in Section 8.2, and sensitivity analysis (via the Envelope Theorem) in the previous sections of this chapter, aimed to understand the impact of changes made to underlying assumptions on risk modeling results. In areas such as financial asset allocation problems, it is known that extensive estimation error can occur in parameters like means, variances, and covariances of stocks. This parameter uncertainty has prompted the development of alternative optimization techniques that guard against, or are robust to, this underlying parameter uncertainty.

To illustrate this robustness concept, let us consider the asset allocation problem outlined in Section 3.4. As we saw in that section, in asset allocations it is common to maximize expected returns subject to a penalty for the uncertainty, expressed as

$$\text{maximize}_{\mathbf{c}} \left\{ \mathbf{c}'\boldsymbol{\mu} - \frac{LMI_1}{2} \times \mathbf{c}'\boldsymbol{\Sigma}\mathbf{c} \right\}.$$

A robust version is

$$\text{maximize}_{\mathbf{c}} \left\{ \text{minimize}_{\boldsymbol{\mu}\in \mathbf{U}} \left(\mathbf{c}'\boldsymbol{\mu}\right) - \frac{LMI_1}{2} \times \mathbf{c}'\boldsymbol{\Sigma}\mathbf{c} \right\}. \tag{9.13}$$

Here, $\mathbf{U}$ represents the *uncertainty set*. Intuitively, the mean parameters are unknown and vary within the uncertainty set. The minimization step yields the worst portfolio mean within $\mathbf{U}$. Practitioners perceive this as a conservative and hence prudent form of portfolio construction, with parameter uncertainty incorporated directly into the portfolio optimization process. This approach aims to maximize penalized worst-case portfolio returns.

In typical applications, the uncertainty set is centered about $\bar{\boldsymbol{\mu}} = (\bar{\mu}_1, \ldots, \bar{\mu}_p)$ and takes on one of two forms:

$$\underset{\textit{box uncertainty set}}{\mathbf{U} = \{|\mu_j - \bar{\mu}_j| \le \xi_j, j = 1, \ldots, p\}} \qquad \underset{\textit{quadratic uncertainty set}}{\mathbf{U} = \{(\boldsymbol{\mu} - \bar{\boldsymbol{\mu}})'\boldsymbol{\Omega}^{-1}(\boldsymbol{\mu} - \bar{\boldsymbol{\mu}}) \le \kappa^2\}.}$$

For the box uncertainty set, it can be shown (c.f., Yin et al. (2021)) that equation (9.13) reduces to

$$\text{maximize}_{\mathbf{c}} \left\{ \mathbf{c}'\bar{\boldsymbol{\mu}} - \kappa \ \times \max_j(|c_j|) - \frac{LMI_1}{2} \times \mathbf{c}'\boldsymbol{\Sigma}\mathbf{c} \right\}.$$

In the same way, for the quadratic uncertainty set, equation (9.13) reduces to

$$\text{maximize}_{\mathbf{c}} \left\{ \mathbf{c}'\bar{\boldsymbol{\mu}} - \kappa\sqrt{\mathbf{c}'\boldsymbol{\Omega}\mathbf{c}} - \frac{LMI_1}{2} \times \mathbf{c}'\boldsymbol{\Sigma}\mathbf{c} \right\}.$$

Both problems are convex, specifically, second-order cone programming problems and so can be readily solved with modern tools, c.f., Boyd and Vandenberghe (2004) (Section 4.4.2). In asset allocation applications, it is common to also introduce uncertainty sets for the variance-covariance parameters, see, for example, Yin et al. (2021). Moreover, because this is a convex optimization problem, it is straight-forward to add the convex constraints on allocation parameters described in Section 3.4.2.

Example 9.10. Robust Allocationss for a Portfolio of Insurance Stock Returns. This is a continuation of Example 9.9. We illustrate the variability based on the quadratic uncertainty set. As before, we use sample training data to estimate the baseline mean vector $\bar{\boldsymbol{\mu}}$ and variance-covariance matrix $\boldsymbol{\Sigma}$. The uncertainty associated with the means is assumed to be proportional to $\boldsymbol{\Sigma}$, so $\boldsymbol{\Omega} \propto \boldsymbol{\Sigma}$. This proportionality constant can be absorbed into the parameter κ, so the constrained optimization problem of interest is

$$\boxed{\begin{array}{ll} \text{maximize}_{\mathbf{c}} & \mathbf{c}'\bar{\boldsymbol{\mu}} - \kappa\sqrt{\mathbf{c}'\boldsymbol{\Sigma}\mathbf{c}} - \frac{LMI_1}{2} \times \mathbf{c}'\boldsymbol{\Sigma}\mathbf{c} \\ \text{subject to} & \mathbf{c}'\mathbf{1} = 1 \\ & c_j \ge 0, \quad j = 1, \ldots, p. \end{array}} \tag{9.14}$$

We are now ready to compute the portfolio specified Display (9.14) with the `CVXR` package.

Figure 9.6 summarizes results over four optimizations. The first is the Markowitz portfolio optimum, with $\kappa = 0$, that serves as a benchmark. The second is a base robust optimization using Display (9.14). For the third and fourth, we perturbed values of $\bar{\boldsymbol{\mu}}$ by adding small Gaussian noise and used these perturbed values in the Display (9.14) optimization problem. Figure 9.6 shows that the three robust allocations are similar when compared to the Markowitz allocations.

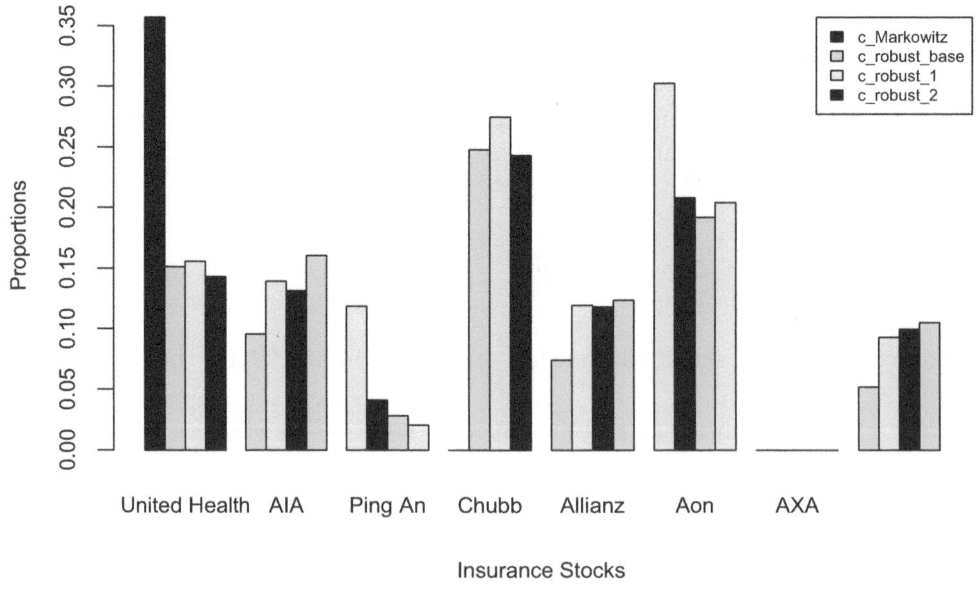

FIGURE 9.6
Four Sets of Robust Portfolio Allocations, by Insurer.

9.5 Supplemental Materials

9.5.1 Further Resources and Reading

Descriptions of the Envelope theorem may be found in standard mathematical economic textbooks such as Simon and Blume (1994) and in a handy online resource from Osborne (2021).

Descriptions of the investment portfolio problem have already been provided in Sections 1.3, 3.4, and 4.4.1. In addition to the references on this problem that appear in Section 9.4, another resource is Chalabi and Würtz (2015) for an introduction to portfolio allocation in R from an actuarial perspective. In the same spirit, Blanchet et al. (2019) provide an investigation of robustness in the context of insurance risk analysis.

9.5.2 Exercises

See the *Online Supplement* for selected exercise solutions.

Section 9.2 Exercises

Exercise 9.1. Minimizing Risk Transfer Costs with an Auxiliary Level of Confidence and a VaR Constraint. This exercise switches the role of objective and (budget) constraint functions. The objective is to minimize risk transfer costs, $f_0(\mathbf{z}) = RTC(\mathbf{z})$, subject to a constraint on the maximal value at risk, $f_{con,1}[\mathbf{z}(\alpha), \alpha] = VaR_{\alpha}[\mathbf{z}(\alpha)] - VaR_{max}$ where VaR_{max} as a maximal allowed value at risk. Further, take the auxiliary variable to be the confidence level of the risk measure so $a = \alpha$.

Thus, the objective function does not depend on the auxiliary variable. Use the Envelope Theorem to check that

$$\partial_\alpha RTC[\mathbf{z}^*(\alpha)] = LME_1^*(\alpha) \; \frac{1}{f_{g(X;\mathbf{z}^*)}[VaR_\alpha(\mathbf{z}^*(\alpha), \alpha)]},$$

where $f_{g(X;\mathbf{z}^*)}(\cdot)$ is the probability density function of the random variable $g(X;\mathbf{z})$ evaluated at the optimum $\mathbf{z}^*$. In problems we work with, typically $LME_1^* > 0$, indicating that $\partial_\alpha RTC[\mathbf{z}^*(\alpha)]$ is also positive.

This is interesting because if f_0 represents risk transfer costs and a represents a confidence level, then typically one would not think of the confidence level as affecting risk transfer costs. However, this special case shows that optimal cost does change through the changes in the optimal risk retention parameters $\mathbf{z}^*$.

Exercise 9.2. Minimizing Risk Transfer Costs with an Auxiliary Level of Confidence and an *ES* Constraint. Use the same set-up as in Exercise 9.1 but now use a constraint function in terms of the expected shortfall. Use the Envelope Theorem to show that

$$\partial_\alpha RTC[\mathbf{z}^*(\alpha)] = \frac{LME_1^*(\alpha)}{1-\alpha} \left\{ES_\alpha(\mathbf{z}^*(\alpha)) - VaR_\alpha(\mathbf{z}^*(\alpha))\right\}.$$

For interpretation, note that if $LME_1^* > 0$, then $\partial_\alpha RTC(\mathbf{z}^*(\alpha)) > 0$. So, in many situations, an increase in the confidence level α means an increase in the optimal risk transfer cost.

Exercise 9.3. Minimizing Risk Transfer Costs with an Auxiliary β and a *RVaR* Constraint. Use a similar set-up as in Exercises 9.1 and 9.2 but now assume the constraint function in terms of the range value at risk and take the auxiliary variable to be $a = \beta$. Recall that β is the parameter that moves the $RVaR$ between the VaR and the ES. Use the Envelope Theorem to show that

$$\partial_\beta RTC[\mathbf{z}^*(\beta)] = \frac{LME_1^*(\beta)}{\beta} \left\{VaR_{\alpha+\beta}[\mathbf{z}^*(\alpha)] - RVaR_{(\alpha,\beta)}[\mathbf{z}^*(\alpha)]\right\}.$$

As in Exercises 9.1 and 9.2, in many situations (that is, $LME_1^* > 0$), an increase in the parameter β means an increase in the smallest risk transfer cost. As VaR and ES represent opposite ends of the $RVaR$ spectrum, we anticipate the smallest risk transfer costs for the expected shortfall measure to be larger than the small risk transfer costs for the value at risk measure.

Exercise 9.4. Minimizing the *VaR* Using Distribution Functions. Consider the VaR minimization problem as in Display (7.3). Let the variable a represent an (as yet) unspecified auxiliary variable. Use the Envelope Theorem to show that

$$\partial_a VaR[g(X;\mathbf{z}^*), a] = \sum_{j=1}^{2} LME_j^*(a) \; f_{con,j,a}(z_0^*, \mathbf{z}^*, a) \quad , \tag{9.15}$$

where $f_{con,1}(z_0, \mathbf{z}, a) = -[F_{g(X;\mathbf{z})}(z_0) - \alpha]$ and $f_{con,2}(z_0, \mathbf{z}, a) = RTC(\mathbf{z}) - RTC_{max}$.

Exercise 9.5. Interpreting Multipliers. Continue with the Exercise 9.4 framework.

a. Now suppose in addition that the auxiliary variable is the confidence level for the value at risk, that is, $a = \alpha$. Show that $\partial_\alpha VaR^*(\alpha) = LME_1^*$. Interpret this result.

b. Now suppose in addition that the auxiliary variable is the maximal risk transfer cost, that is, $a = RTC_{max}$. Show that $\partial_{RTC_{max}} VaR^*(a) = -LME_2^*$. Interpret this result.

Section 9.3 Exercises

Exercise 9.6. Compare Changes in the *ES* and *GlueVaR*. Now we use a special case of *GlueVaR* for the objective function. Our first exposure to this flexible risk measure, due to Belles-Sampera et al. (2014), was in Exercise 2.4 where we saw that it exhibits many of the same qualities as the *RVaR*. We now consider the special case

$$GlueVaR_\omega(X) = (1-\omega)VaR_\alpha(X) + \omega ES_\alpha(X), \quad 0 \leq \omega \leq 1.$$

Show that

$$\partial_\alpha ES_\alpha(X) = \frac{1}{1-\alpha}\{ES_\alpha(X) - VaR_\alpha(X)\} = \frac{1}{1-\alpha}\,\partial_\omega\, GlueVaR_\omega(X).$$

Exercise 9.7. *GlueVaR* Risk Retention. Use the Envelope Theorem to show that

$$\partial_\omega\, GlueVaR_\omega[g(\mathbf{X};\mathbf{z}^*)] = ES_\alpha[g(\mathbf{X};\mathbf{z}^*)] - VaR_\alpha[g(\mathbf{X};\mathbf{z}^*)].$$

Exercise 9.8. *GlueVaR* and the ANU Case. In Example 9.5, we used the *RVaR* measure to move smoothly between the two most well-known choices, the value at risk *VaR* and the expected shortfall *ES*. In this way, we were able to calibrate the effects of using one measure versus the other. A simpler way of accomplishing this task is via the *GlueVaR* measure; in particular, the special case considered in Exercises 9.6 and 9.7 is simply a linear combination of the two measures.

Figure 9.7 shows although the optimal value of *GlueVaR* changes substantially with different choices of ω, the corresponding values of *VaR* and *ES* are relatively stable.

Create a table of values underpinning this figure, comparable to the work done in Example 9.6 for the *RVaR* measure. The table should summarize the optimization results over several different choices of the weight ω. Use $\alpha = 0.90$ for this demonstration. So, the row with $\omega = 0$ will corresponds to $VaR_{0.90}$ minimization and the row with $\omega = 1$ corresponds to $ES_{0.90}$ minimization. Your results should be similar to Table 9.4.

Exercise 9.9. *GlueVaR* and the ANU Case - Upper. As a follow-up to Exercise 9.8, develop a set of graphs that show the progression of upper limits as weights change. Your results should be similar to Figure 9.8.

This figure shows the progression of optimal upper limits as one moves from *VaR* minimization ($\omega = 0$) to *ES* minimization ($\omega = 1$). There is relatively little change in the limits. This provides a comforting result to risk managers; it suggests that the choice of the risk measure is not as critical. Put another way, the upper limit values are robust to this decision.

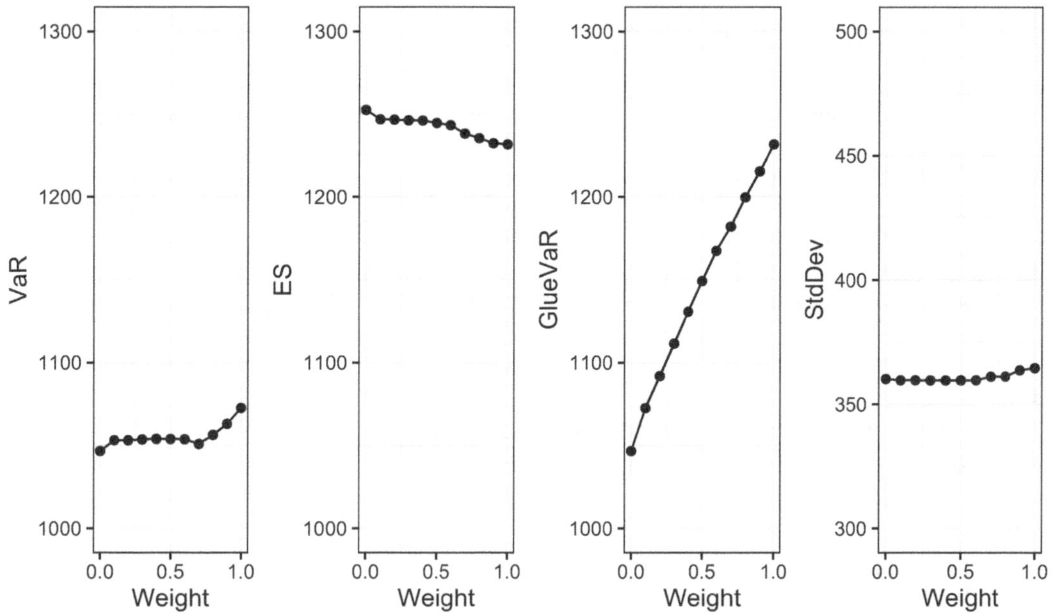

FIGURE 9.7
Retained Risk Measures versus Weight. Based on minimizing the $GlueVaR$ criterion. Here, $Weight = 0$ corresponds to VaR and $Weight = 1$ corresponds to ES.

TABLE 9.4
ANU Excess of Loss Optimization—GlueVar Criterion—Varying Weight.

Weight	RTC	VaR	ES	Std Dev	$GlueVaR$	LME	u_1	u_2	u_3
0.0	576	1047	1253	360	1047	2.61	596	178	130
0.1	576	1053	1247	360	1073	2.56	606	192	123
0.2	576	1053	1247	360	1092	3.24	607	189	126
0.3	576	1054	1246	360	1111	2.94	607	190	126
0.4	576	1054	1246	360	1131	3.64	607	192	124
0.5	576	1054	1245	360	1149	3.59	612	184	131
0.6	576	1054	1243	360	1168	3.61	614	199	125
0.7	576	1051	1238	361	1182	3.74	644	170	143
0.8	576	1056	1236	361	1200	3.66	656	176	141
0.9	576	1063	1233	364	1216	3.71	696	167	141
1.0	576	1073	1232	364	1232	3.72	708	157	139

Section 9.4 Exercises

Exercise 9.10. Quota Share. We now revisit the quota share agreement described in Section 4.3.2. As in that section, let $\boldsymbol{\Sigma}$ denote the variance-covariance of risks and now let $\boldsymbol{\mu} = \mathrm{E}(\mathbf{X})$ denote their mean. For quota share with parameter $\mathbf{c} = \mathbf{z}$, the amount retained is $\mathbf{c}'\mathbf{X}$ with uncertainty $f_0(\mathbf{c}, \boldsymbol{\Sigma}) = \mathrm{Var}(\mathbf{c}'\mathbf{X}) = \mathbf{c}'\boldsymbol{\Sigma}\mathbf{c}$. There is a single ($m = 1$) budget constraint based on $f_{con,1}(\mathbf{c}) = \boldsymbol{\mu}'(\mathbf{1} - \mathbf{c}) - RTC_{max}$. The Lagrangian is $LA = \mathbf{c}'\boldsymbol{\Sigma}\mathbf{c} + LME_1[\boldsymbol{\mu}'(\mathbf{1} - \mathbf{c}) - RTC_{max}]$. The auxiliary variable is $\mu = a$, where μ is an element of $\boldsymbol{\mu}$. Further, let c^* be the optimal retention parameter corresponding to the choice of the μ.

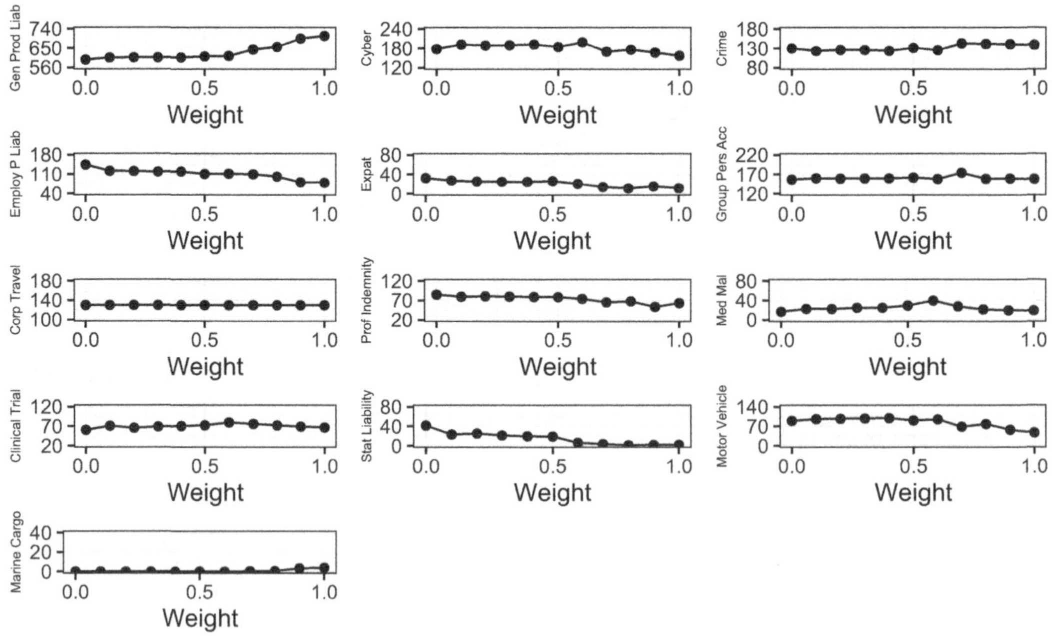

FIGURE 9.8
Upper Limits versus Weight. Based on minimizing the *GlueVaR* criterion. Here, $Weight = 0$ corresponds to VaR and $Weight = 1$ corresponds to ES.

Use first principles to establish

$$\partial_\mu \mathbf{c}^* = (\partial_\mu LME_1^*)\frac{1}{2}\ \boldsymbol{\Sigma}^{-1}\ \boldsymbol{\mu} + \frac{LME_1^*}{2}\ \boldsymbol{\Sigma}^{-1}\ \partial_\mu \boldsymbol{\mu} \tag{9.16}$$

and

$$\partial_\mu\ LME_1^* = \frac{2}{\boldsymbol{\mu}'\boldsymbol{\Sigma}^{-1}\boldsymbol{\mu}}\left[1 - LME_1^*\ (\partial_\mu \boldsymbol{\mu})'\ \boldsymbol{\Sigma}^{-1}\boldsymbol{\mu}\right]. \tag{9.17}$$

Equation (9.16) provides a closed-form expression for quota share sensitivities using the Lagrange multiplier sensitivity in equation (9.17).

Exercise 9.11. Quota Share . Use the same set-up as in Exercise 9.10 except now take the auxiliary variable to be $\sigma = a$, where σ is an element of $\boldsymbol{\Sigma}$.

Use first principles to establish

$$\partial_\sigma \mathbf{c}^* = \frac{1}{2}\boldsymbol{\Sigma}^{-1}\boldsymbol{\mu}\ \partial_\sigma LME_1^* - \boldsymbol{\Sigma}^{-1}(\partial_\sigma \boldsymbol{\Sigma})\mathbf{c}^* \tag{9.18}$$

and

$$\partial_\sigma LME_1^* = \frac{2}{\boldsymbol{\mu}'\ \boldsymbol{\Sigma}^{-1}\boldsymbol{\mu}}\boldsymbol{\mu}'\boldsymbol{\Sigma}^{-1}(\partial_\sigma \boldsymbol{\Sigma})\mathbf{c}^*. \tag{9.19}$$

Exercise 9.12. Asset Allocation. As a follow-up to Section 9.4.1, now the interest is in sensitivity due to the uncertainty parameters σ. As background, Table 9.5 shows strong relationships among daily asset returns. This motivates the need to examine the effects of covariances as well as variances.

TABLE 9.5
Pearson Correlations among Daily Asset Returns.

	United Health	AIA	Ping An	Chubb	Allianz	Aon	AXA	Tokio Marine
United Health	1.00	0.07	0.11	0.36	0.24	0.34	0.23	0.08
AIA	0.07	1.00	0.55	0.06	0.26	0.08	0.27	0.28
Ping An	0.11	0.55	1.00	0.09	0.28	0.14	0.28	0.28
Chubb	0.36	0.06	0.09	1.00	0.33	0.61	0.29	0.09
Allianz	0.24	0.26	0.28	0.33	1.00	0.32	0.80	0.25
Aon	0.34	0.08	0.14	0.61	0.32	1.00	0.30	0.10
AXA	0.23	0.27	0.28	0.29	0.80	0.30	1.00	0.28
Tokio Marine	0.08	0.28	0.28	0.09	0.25	0.10	0.28	1.00

a. Now choose the auxiliary variable to be $a = \sigma$, where σ is an element of $\mathbf{\Sigma}$. In the same way, prove that

$$\partial_\sigma \mathbf{c}^*(\sigma) = \frac{-1}{2\lambda_{Port}} \mathbf{\Sigma}^{-1}\mathbf{1}\ \partial_\sigma LME_1^*(\sigma) - \ \mathbf{\Sigma}^{-1}(\partial_\sigma \mathbf{\Sigma})\mathbf{c}^*(\sigma) \tag{9.20}$$

and

$$\partial_\sigma LME_1^*(\sigma) = 2\lambda_{Port} \frac{\mathbf{1}'\mathbf{\Sigma}^{-1}(\partial_\sigma \mathbf{\Sigma})\mathbf{c}^*(\sigma)}{\mathbf{1}'\ \mathbf{\Sigma}^{-1}\mathbf{1}}. \tag{9.21}$$

---

b. For the variance parameters, with $a = \sigma$, the portfolio sensitivity for the mean, variance, and standard deviation are

$$\partial_\sigma\,[\boldsymbol{\mu}'\mathbf{c}^*] = \boldsymbol{\mu}'[\partial_\sigma \mathbf{c}^*]$$
$$\partial_\sigma\,[\mathbf{c}^{*\prime}\mathbf{\Sigma}\mathbf{c}^*] = [\partial_\sigma \mathbf{c}^*]'\mathbf{\Sigma}\mathbf{c}^* + \mathbf{c}^{*\prime}[\partial_\sigma \mathbf{\Sigma}]\mathbf{c}^* + \mathbf{c}^{*\prime}\mathbf{\Sigma}[\partial_\sigma \mathbf{c}^*]$$
$$\partial_\sigma \sqrt{\mathbf{c}^{*\prime}\mathbf{\Sigma}\mathbf{c}^*} = \frac{1}{2\sqrt{\mathbf{c}^{*\prime}\mathbf{\Sigma}\mathbf{c}^*}}\ \partial_\sigma[\mathbf{c}^{*\prime}\mathbf{\Sigma}\mathbf{c}^*] = \partial_\sigma\ SD[c^{*'}X],$$

that can be evaluated using equations (9.20) and (9.21). Use these relationships to confirm the sensitivities for these data as found in Tables 9.6 and 9.7.

TABLE 9.6
Portfolio Mean Variance Sensitivities.

	United Health	AIA	Ping An	Chubb	Allianz	Aon	AXA	Tokio Marine
United Health	-1.11	-0.83	-0.67	0.66	-2.05	-2.65	1.32	-0.57
AIA	-0.83	-0.15	-0.25	0.31	-0.75	-0.98	0.49	-0.20
Ping An	-0.67	-0.25	-0.10	0.19	-0.61	-0.79	0.39	-0.17
Chubb	0.66	0.31	0.19	0.01	0.66	0.80	-0.40	0.21
Allianz	-2.05	-0.75	-0.61	0.66	-0.93	-2.43	1.20	-0.51
Aon	-2.65	-0.98	-0.79	0.80	-2.43	-1.57	1.56	-0.67
AXA	1.32	0.49	0.39	-0.40	1.20	1.56	-0.39	0.33
Tokio Marine	-0.57	-0.20	-0.17	0.21	-0.51	-0.67	0.33	-0.07

TABLE 9.7
Portfolio Standard Deviation Variance Sensitivities.

	United Health	AIA	Ping An	Chubb	Allianz	Aon	AXA	Tokio Marine
United Health	0.29	0.27	0.17	0.01	0.58	0.70	-0.35	0.18
AIA	0.27	0.06	0.08	0.01	0.27	0.33	-0.16	0.08
Ping An	0.17	0.08	0.03	0.00	0.17	0.21	-0.10	0.05
Chubb	0.01	0.01	0.00	0.00	0.01	0.01	-0.01	0.00
Allianz	0.58	0.27	0.17	0.01	0.29	0.70	-0.35	0.18
Aon	0.70	0.33	0.21	0.01	0.70	0.43	-0.43	0.22
AXA	-0.35	-0.16	-0.10	-0.01	-0.35	-0.43	0.11	-0.11
Tokio Marine	0.18	0.08	0.05	0.00	0.18	0.22	-0.11	0.03

Table 9.6 shows sensitivities for the mean portfolio $\mathbf{c}^{*\prime}\boldsymbol{\mu}$ for each of the 8 variance parameters and the 28 covariance parameters. (Note the symmetry in the table due to the covariance and that the diagonal elements are variances, not standard deviations.) Table 9.7 gives corresponding results for the standard deviation of the portfolio, $\sqrt{\mathbf{c}^{*\prime}\boldsymbol{\Sigma}\mathbf{c}^{*}}$. As with the mean sensitivities in Table 9.2, it is interesting to see that the changes in the standard deviations tend to be smaller in magnitude than changes in the means.

9.5.3 Appendix. Establishing the Envelope Theorem

To establish the Envelope theorem, we use the Karush-Kuhn-Tucker (KKT) conditions that will be introduced in Section 11.1.

Starting with Display (9.2), we first re-label those inequalities that are binding at $\mathbf{z}^*(a)$ as equalities. With this adjustment in the notation, the associated Lagrangian

$$\begin{aligned} &LA\left(\mathbf{z}, a, \mathbf{LME}, \mathbf{LMI}\right) \\ &= f_0(\mathbf{z}, a) + \sum_{j \in CON_{eq}} LME_j \; f_{con,j}(\mathbf{z}, a) + \sum_{j \in CON_{in}} LMI_j \; f_{con,j}(\mathbf{z}, a) \\ &= SLA\left(\mathbf{z}, a, \mathbf{LME}\right) + \sum_{j \in CON_{in}} LMI_j \; f_{con,j}(\mathbf{z}, a), \end{aligned} \tag{9.22}$$

where SLA is the shortened version, omitting the (assumed to be non-binding) inequality constraints.

For each a, suppose that $\mathbf{z}^*(a)$ is a local minimizer on the feasible set and that there exists Lagrange multipliers $\mathbf{LME}^*(a) = \left(LME_j^*(a);\ j \in CON_{eq}\right)'$ and $\mathbf{LMI}^*(a) = (LMI_1^*(a);\ j \in CON_{in})'$ that satisfy the KKT conditions in Display (11.1),

$$\boxed{\begin{array}{cc} LA_i\left[\mathbf{z}^*(a), a, \mathbf{LME}^*(a), \mathbf{LMI}^*(a)\right] = 0 & i = 1, \ldots, p_z \\ f_{con,j}(\mathbf{z}^*(a), a) = 0 & j \in CON_{eq} \\ LMI_j^*(a) = 0 & j \in CON_{in} \ . \end{array}}$$

We see that $LMI_j^*(a) = 0$, $j \in CON_{in}$, because these inequalities are assumed to be not binding at the optimum. So, for the purpose of developing the Envelope theorem, we can restrict consideration to the shortened version of the Lagrangian. Also note that this uses the notation $LA_i\left[\mathbf{z}, a, \mathbf{LME}, \mathbf{LMI}\right] = \partial_{z_i} LA\left(\mathbf{z}, a, \mathbf{LME}, \mathbf{LMI}\right)$.

We now establish equations (9.3) and (9.4).

Using the *KKT* conditions and taking a derivative of the equality constraint functions yields

$$\begin{aligned} 0 &= \partial_a\ f_{con,j}[\mathbf{z}^*(a), a] \\ &= \sum_{i=1}^{p_z} \left\{ \partial_{z_i} f_{con,j}(\mathbf{z}, a)\big|_{\mathbf{z}=\mathbf{z}^*(a)} \times \partial_a z_i^*(a) \right\} + \partial_a f_{con,j}(\mathbf{z}, a)\big|_{\mathbf{z}=\mathbf{z}^*(a)} \\ &= \sum_{i=1}^{p_z} \left\{ f_{con,j,i}[\mathbf{z}^*(a), a] \times \partial_a z_i^*(a) \right\} + f_{con,j,a}[\mathbf{z}^*(a), a] \\ \Rightarrow & f_{con,j,a}(\mathbf{z}^*(a), a) = -\sum_{i=1}^{p_z} f_{con,j,i}[\mathbf{z}^*(a), a] \times \partial_a z_i^*(a). \end{aligned}$$

This establishes equation (9.4).

With the Lagrangian

$$\begin{aligned} LA_i\left[\mathbf{z}, a, \mathbf{LME}, \mathbf{LMI}\right] &= \partial_{z_i} LA\left(\mathbf{z}, a, \mathbf{LME}, \mathbf{LMI}\right) \\ &= \partial_{z_i} f_0(\mathbf{z}, a) + \sum_{j \in CON_{eq}} LME_j\ \partial_{z_i} f_{con,j}(\mathbf{z}, a) \\ &\quad + \sum_{j \in CON_{in}} LMI_j\ \partial_{z_i} f_{con,j}(\mathbf{z}, a) \\ &= f_{0,i}(\mathbf{z}, a) + \sum_{j \in CON_{eq}} LME_j\ f_{con,j,i}(\mathbf{z}, a) \\ &\quad + \sum_{j \in CON_{in}} LMI_j\ \partial_{z_i} f_{con,j}(\mathbf{z}, a). \end{aligned}$$

With the *KKT* conditions, $LMI_j^*(a) = 0$, and so

$$\begin{aligned} & 0 = LA_i\left[\mathbf{z}^*(a), a, \mathbf{LME}^*(a), \mathbf{LMI}^*(a)\right] \\ \Rightarrow\quad & f_{0,i}[\mathbf{z}^*(a), a] = -\sum_{j \in CON_{eq}} LME_j^*(a)\ f_{con,j,i}[\mathbf{z}^*(a), a]. \end{aligned}$$

Next, we (a) take a derivative of the objective function at the optimum, then (b) insert the expression for $f_{0,i}[\mathbf{z}^*(a), a]$ and finally (c) utilize equation (9.4). This yields

$$
\begin{aligned}
\partial_a f_0[\mathbf{z}^*(a), a] &= \sum_{i=1}^{p_z} \left\{ \partial_{z_i} f_0(\mathbf{z}, a)|_{\mathbf{z}=\mathbf{z}^*(a)} \times \partial_a z_i^*(a) \right\} + \partial_a f_0(\mathbf{z}, a)|_{\mathbf{z}=\mathbf{z}^*(a)} \\
&= \sum_{i=1}^{p_z} \{ f_{0,i}[\mathbf{z}^*(a), a] \times \partial_a z_i^*(a) \} + f_{0,a}[\mathbf{z}^*(a), a] \\
&= -\sum_{i=1}^{p_z} \left\{ \left[\sum_{j \in CON_{eq}} LME_j^*(a)\ f_{con,j,i}[\mathbf{z}^*(a), a] \right] \times \partial_a z_i^*(a) \right\} \\
&\qquad + f_{0,a}[\mathbf{z}^*(a), a] \\
&= -\sum_{j \in CON_{eq}} LME_j^*(a) \left\{ \sum_{i=1}^{p_z} f_{con,j,i}[\mathbf{z}^*(a), a] \times \partial_a z_i^*(a) \right\} \\
&\qquad + f_{0,a}[\mathbf{z}^*(a), a] \\
&= \sum_{j \in CON_{eq}} LME_j^*(a)\ f_{con,j,a}[\mathbf{z}^*(a), a] + f_{0,a}[\mathbf{z}^*(a), a],
\end{aligned}
$$

as in equation (9.3).

10

Sensitivity and Data Uncertainty

Chapter Preview. In Chapter 9, we explored how optimal values of retained risk measures, our objective functions, change as assumptions vary. Although this understanding is helpful, it is the retention parameters, and hence decision variables, that are contractually documented and likely to be scrutinized carefully. This raises the question: to what extent will the optimal values of the decision variables change as assumptions vary?

Perturbation Sensitivity Proposition. To address this question, this chapter develops an extension of the Envelope theorem that provides explicit expressions for optimal decision variable sensitivities based on small perturbations of an auxiliary variable. We refer this extension, presented in Section 10.1, as the *Perturbation Sensitivity Proposition.* This extension is derived by taking additional derivatives and thus assumes additional smoothness on the objective and constraint functions. Although complex, with this extension we are able to compute risk retention sensitivities *directly* in many cases without the need of additional optimization runs. To demonstrate the usefulness of this proposition, we revisit some of the examples from Chapter 9 and show how to supplement previous results with information about the decision variable sensitivity.

In addition to the examples already discussed, risk managers are eager to understand the robustness of results due to variations in the risk distribution parameters assumptions. However, a marginal distribution parameter affects both risk transfer costs and risk measures. Because this is an instance when the objective and constraint functions both depend on an auxiliary variable, insights from Chapter 9 results alone are difficult to come by. However, as will be seen in Section 10.2, we will be readily be able to handle this type of auxiliary variable with the Perturbation Sensitivity Proposition.

Curvature. To explain why decision variable uncertainty is important for some risk retention problems but not others, Section 10.3 summarizes an investigation into the curvature of the associated Lagrangian. This section shows how to use a quantity known as a *condition number* from numerical analysis to describe the curvature. From this investigation, we learn that excess of loss retention programs are robust in contrast to quota share programs, which are more susceptible to decision variable uncertainty.

Data Uncertainty. Variations in risk distribution parameters may be due to vagueness of the analyst's perception of the distribution but may also be due to the uncertainty that arises from the variability in a data generating process. Here, *data uncertainty* refers to the observation that small changes in the risk distribution parameters can lead to large material changes in optimal risk portfolio coefficients. Section 10.4 provides an introduction to data uncertainty and shows how to calibrate it with what we call a *stochastic sensitivity.*

DOI: 10.1201/9781003469568-10

10.1 Decision Variable Sensitivities

This section introduces a proposition that can be used to calibrate the sensitivity of the optimal decision variables to changes in exogenous (non-decision) variables. To this end, let us revisit the basic problem in Display (9.2) which introduces a variable, "a", as auxiliary into the optimization problem and represents the assumption we wish to vary. In applications developed in Section 10.2, a primarily represents an element of the risk distribution parameter vector $\boldsymbol{\gamma}$ but can also be, for example, the level of an acceptable budget.

Table 10.1 summarizes the notation introduced in Section 9.2 and provides some extensions needed for this development. In particular, we use the notation $f_0(\mathbf{z}, a)$ and $f_{con,j}(\mathbf{z}, a)$ to emphasize that both the objective and constraint functions may depend on a. As before, we require smoothness of the objective and constraint functions in the auxiliary variable, with partial derivatives as displayed in Table 10.1. Recall that there are p_z decision variables, and we only utilize the m active constraints at the optimum.

To economize on notation, we collect the constraints using matrix notation as $f_{con,\bullet,a}(\mathbf{z}, a) = [f_{con,1,a}(\mathbf{z}, a), \ldots, f_{con,m,a}(\mathbf{z}, a)]'$, a $m \times 1$ vector, and $f_{con,\bullet,\mathbf{z}}(\mathbf{z}, a) = [f_{con,1,\mathbf{z}}(\mathbf{z}, a), \ldots, f_{con,m,\mathbf{z}}(\mathbf{z}, a)]$, a $p_z \times m$ matrix.

To present results, recall the shortened version of the Lagrangian $SLA(\mathbf{z}, a) = f_0(\mathbf{z}, a) + \sum_{j=1}^m LME_j \ f_{con,j}(\mathbf{z}, a)$, where LME_j are Lagrange multipliers, that considers only those constraints that are active at the optimum. Table 10.1 introduces the Hessian of the shortened Lagrangian $SLA_{\mathbf{zz}'}(\mathbf{z}, a, \mathbf{LME})$, a $p_z \times p_z$ matrix. In the same way, taking mixed second-order partial derivatives yields the $p_z \times 1$ vector $SLA_{\mathbf{z}a}(\mathbf{z}, a, \mathbf{LME})$.

At the optimum, $\mathbf{LME}^*(a) = \mathbf{LME}^*$ and $\mathbf{z}^*(a) = \mathbf{z}^*$ are values of the Lagrange multipliers and the decision variables (both of which depend on a). Our interest is in the set of *sensitivities* that we now express as vectors

TABLE 10.1
Perturbation Sensitivity Notation Summary.

Description	Mathematical Expression
Objective Function	$f_0(\mathbf{z}, a)$
Partial Derivatives *wrt* decision variables	$f_{0,\mathbf{z}}(\mathbf{z}, a) = \partial_{\mathbf{z}} f_0(\mathbf{z}, a)$
Partial Derivative *wrt* auxiliary variable	$f_{0,a}(\mathbf{z}, a) = \partial_a f_0(\mathbf{z}, a)$
Partial Derivatives *wrt* decision and auxiliary variables	$f_{0,\mathbf{z}a}(\mathbf{z}, a) = \partial_{\mathbf{z}} \partial_a f_0(\mathbf{z}, a)$
Hessian of the Objective Function	$f_{0,\mathbf{zz}'}(\mathbf{z}, a) = \partial_{\mathbf{zz}'} f_0(\mathbf{z}, a)$
Constraint Functions	$f_{con,j}(\mathbf{z}, a)$
Partial Derivatives *wrt* decision variables	$f_{con,j,\mathbf{z}}(\mathbf{z}, a) = \partial_{\mathbf{z}} f_{con,j}(\mathbf{z}, a)$
Partial Derivatives *wrt* auxiliary variable	$f_{con,j,a}(\mathbf{z}, a) = \partial_a f_{con,j}(\mathbf{z}, a)$
Partial Derivative *wrt* auxiliary variable	$f_{con,j,\mathbf{z}}(\mathbf{z}, a) = \partial_{\mathbf{z}} f_{con,j}(\mathbf{z}, a)$
Shortened Lagrangian	$SLA(\mathbf{z}, a, \mathbf{LME}) = f_0(\mathbf{z}, a) + \sum_{j=1}^m LME_j \ f_{con,j}(\mathbf{z}, a)$
Partial Derivatives *wrt* decision and auxiliary variables	$SLA_{\mathbf{z}a}(\mathbf{z}, a, \mathbf{LME}) = \partial_{\mathbf{z}} \partial_a SLA(\mathbf{z}, a, \mathbf{LME})$
Hessian of the Shortened Lagrangian	$SLA_{\mathbf{zz}'}(\mathbf{z}, a, \mathbf{LME}) = \partial_{\mathbf{z}} \partial_{\mathbf{z}'} SLA(\mathbf{z}, a, \mathbf{LME})$

Note: wrt means "with respect to"

$$\partial_a \mathbf{z}^* = \begin{pmatrix} \partial_a z_1^*(a) \\ \vdots \\ \partial_a z_{p_z}^*(a) \end{pmatrix} \quad \text{and} \quad \partial_a \mathbf{LME}^* = \begin{pmatrix} \partial_a LME_1^*(a) \\ \vdots \\ \partial_a LME_m^*(a) \end{pmatrix}.$$

All functions are evaluated at the optimum, denoted using the asterisk * notation as $f^*_{con,\bullet,a} = f_{con,\bullet,a}(\mathbf{z}^*, a)$, $f^*_{con,\bullet,\mathbf{z}} = f_{con,\bullet,\mathbf{z}}(\mathbf{z}^*, a)$, $SLA^*_{\mathbf{z}a} = SLA_{\mathbf{z}a}(\mathbf{z}^*, a, \mathbf{LME}^*)$, and $SLA^*_{\mathbf{zz}'} = SLA_{\mathbf{zz}'}(\mathbf{z}^*, a, \mathbf{LME}^*)$.

To evaluate the sensitivities, we use the

Perturbation Sensitivity Proposition. Assume that the objective and constraint functions are sufficiently smooth and that the Hessian $SLA^*_{\mathbf{zz}'}$ is invertible. Then, we have

$$\begin{aligned} \partial_a \mathbf{LME}^* &= \left\{ f^*_{con,\bullet,\mathbf{z}'} SLA^{*-1}_{\mathbf{zz}'} \; f^*_{con,\bullet,\mathbf{z}} \right\}^{-1} \\ &\quad \left\{ f^*_{con,\bullet,a} - f^*_{con,\bullet,\mathbf{z}'} \; SLA^{*-1}_{\mathbf{zz}'} \; SLA^*_{\mathbf{z}a} \right\} \end{aligned} \tag{10.1}$$

and

$$\partial_a \; \mathbf{z}^* = -SLA^{*-1}_{\mathbf{zz}'} \left\{ SLA^*_{\mathbf{z}a} + f^*_{con,\bullet,\mathbf{z}} \; \partial_a \mathbf{LME}^* \right\}. \tag{10.2}$$

Equations (10.1) and (10.2) give explicit expressions for the sensitivities.

Discussion. The proofs of equations (10.1) and (10.2) are in Appendix Section 10.5.4 where detailed discussions of precise statements of "sufficiently smooth" may be found based on work in Fiacco (1976) and Fiacco (1983). We also refer to Robinson (1974) and Fiacco and McCormick (1990), Theorem 6, that provide alternative smoothness conditions.

To see how these results may be applied, consider the following special cases.

Example 10.1. RTC_{max} as an Auxiliary Variable in the ANU Case. Let us continue with the set up in Example 9.3. Here, there is a single active constraint, $f_{con,1}(\mathbf{z}) = RTC(\mathbf{z}) - RTC_{max}$ and we take as the auxiliary variable $a = RTC_{max}$. Then, $f_{con,1,a}(\mathbf{z}) = -1$. Further, $SLA_{\mathbf{z}a}(\mathbf{z}, a, LME_1) = f_{0,\mathbf{z}a}(\mathbf{z}, a) + LME_1 \;\; \partial_{\mathbf{z}a} f_{con,1}(\mathbf{z}, a) = 0$. Thus, equations (10.1) and (10.2) reduce to

$$\partial_{RTC_{max}} LME_1^* = -\left\{ f^*_{con,1,\mathbf{z}'} SLA^{*-1}_{\mathbf{zz}'} \; f^*_{con,1,\mathbf{z}} \right\}^{-1}$$

and

$$\partial_{RTC_{max}} \; \mathbf{z}^* = -SLA^{*-1}_{\mathbf{zz}'} f^*_{con,1,\mathbf{z}} \;\; \partial_{RTC_{max}} LME_1^*.$$

After calculations, it turns out that $\partial_{RTC_{max}} LME_1^*$ = -0.003 and the first three decision variable sensitivities are $\partial_{RTC_{max}} z_1^*$ = -1.552, $\partial_{RTC_{max}} z_2^*$ = -0.972, and $\partial_{RTC_{max}} z_3^*$ = -1.019. For example, by increasing RTC_{max} by one unit, we anticipate the optimal value of the upper limit for the first risk, General and Products liability, to decrease by 1.552.

Example 10.2. Confidence Level α as an Auxiliary Variable in the ANU Case. Let us continue with the set up in Example 9.5. As in Example 10.1, there is a single active constraint, $f_{con,1}(\mathbf{z}) = RTC(\mathbf{z}) - RTC_{max}$ but now we take as the auxiliary variable $a = \alpha$. Then, $f_{con,1,\alpha}(\mathbf{z}) = 0$. Further, $SLA_{\mathbf{z}a}(\mathbf{z}, \alpha, LME_1) = f_{0,\mathbf{z}\alpha}(\mathbf{z}, \alpha)$. Thus, equations (10.1) and (10.2) reduce to

$$\partial_\alpha LME_1^* = -f^*_{con,1,\mathbf{z}'}\ SLA^{*-1}_{\mathbf{zz}'}\ f^*_{0,\mathbf{z}\alpha}(\mathbf{z},\alpha)/\left\{f^*_{con,1,\mathbf{z}'} SLA^{*-1}_{\mathbf{zz}'}\ f^*_{con,1,\mathbf{z}}\right\}$$

and

$$\partial_\alpha\ \mathbf{z}^* = -SLA^{*-1}_{\mathbf{zz}'}\left\{f^*_{0,\mathbf{z}\alpha}(\mathbf{z},\alpha) + f^*_{con,1,\mathbf{z}}\ \partial_\alpha LME_1^*\right\}.$$

Recall from Example 9.4, we established $f_{0\alpha}(\mathbf{z}^*,\alpha) = \frac{1}{1-\alpha}\{ES^* - VaR^*\}$ that simplifies calculations. With the calculations, it turns out that $\partial_\alpha LME_1^* = 7.565$. This partial derivative is per unit change in α; it is easier to divide by 100 and think about changes per percentage unit of α. With this scale, the first three decision variable sensitivities are $\partial_\alpha z_1^* = 0.02356$, $\partial_\alpha z_2^* = 0.21466$, and $\partial_\alpha z_3^* = 0.21271$. For example, by increasing α by one percentage point, we anticipate the optimal value of the upper limit for the first risk, General and Products liability, to increase by 0.02356. From a practical perspective, these changes are small suggesting that risk managers need not be overly concerned when selecting the confidence level.

Linear Constraint Corollary. Suppose that the jth constraint is linear in the decision variables and is of the form $f_{con,j}(\mathbf{z}) = \mathbf{p}_j'\mathbf{z} - p_0 \le 0$, for a constant p_0 and vector $\mathbf{p}_j$. Then, under the conditions of the Perturbation Sensitivity Proposition, we have

$$\mathbf{p}_j'\ \partial_a \mathbf{z}^* = 0.$$

Remark. In particular, note that if an edge constraint such as $z_j \ge 0$ or $z_j \le 1$ is active (as will be the case in the subsequent quota share examples), this means that the corresponding sensitivity is zero.

10.2 Risk Retention Sensitivities

Following up on Examples 10.1 and 10.2, this section shows how to use the perturbation sensitivity proposition for the risk retention problem with an expected shortfall objective function

$$f_0(\mathbf{z},a) = f_0(z_0,\boldsymbol{\theta},a) = z_0 + \frac{1}{1-\alpha}\left\{\mathrm{E}[g(\mathbf{X};\boldsymbol{\theta}) - z_0]_+\right\}.$$

The constraints include a budget constraint $f_{con,1}(\mathbf{z},a) = RTC(\boldsymbol{\theta},a) - RTC_{max} \le 0$ and additional active edge constraints $f_{con,j}(\mathbf{z},a)$, $j = 2,\ldots,m$, of the form $\mathbf{P}\boldsymbol{\theta} \le \mathbf{p}_0$.

10.2.1 Hessian of the Active Set Lagrangian

Although analysts may not need second derivatives for routine constrained optimization problems, we have seen that it is a key component of the perturbation sensitivity proposition. In addition, Section 10.3 will show that we can get additional insights into the decision variable uncertainty by examining the Hessian of the Lagrangian at the optimum. Recall that the solution to many constrained optimization problems can be determined by minimizing the Lagrangian and so naturally one is interested in the curvature of this function at the optimum. The Hessian describes this curvature (see, for example, Luenberger et al. (2016), page 192, or Boyd and Vandenberghe (2004), page 71). The sharper this curvature, the more reliable are results that may be subject to decision variable uncertainty.

Appendix Section 10.5.3 develops the Hessian of the shortened, or active set, Lagrangian as

$$SLA_{\mathbf{zz}'}(\mathbf{z}, \mathbf{LME}) = \tilde{\mathbf{X}}'\tilde{\mathbf{X}} + \sum_{j=1}^{m} LME_j \ \partial_{\mathbf{z}}\partial_{\mathbf{z}'} f_{con,j}(\mathbf{z}), \tag{10.3}$$

with $\tilde{\mathbf{X}} = (\ \sqrt{\mathbf{W_z}}\ \mathbf{1} \quad -\sqrt{\mathbf{W_z}}\ \mathbf{G}\ )$, the gradients $\mathbf{G} = [\partial_{\boldsymbol{\theta}} g(\mathbf{X}_1;\boldsymbol{\theta}), \ldots, \partial_{\boldsymbol{\theta}} g(\mathbf{X}_R;\boldsymbol{\theta})]'$, and a diagonal matrix, $\sqrt{\mathbf{W_z}} = diag(\sqrt{w_1(\mathbf{z})}, \ldots, \sqrt{w_R(\mathbf{z})}\ )$ having weights $w_r(\mathbf{z}) = k\left(\frac{z_0 - g(\mathbf{X}_r;\boldsymbol{\theta})}{b}\right)/[(1-\alpha)Rb]$. The first matrix on the right hand side of equation (10.3), $\tilde{\mathbf{X}}'\tilde{\mathbf{X}}$, is the Hessian of the objective function, $f_{0,\mathbf{zz}'}(\mathbf{z})$.

What is the role of the constraint functions? The edge constraints of the form $\mathbf{P}\boldsymbol{\theta} \leq \mathbf{p}_0$ are linear in the risk retention/decision variables. Hence, their second derivative is zero and they do not appear in the expression for $SLA_{\mathbf{zz}'}$. Thus, one only needs to consider the budget constraint for which we have $\partial_{\mathbf{z}}\partial_{\mathbf{z}'} f_{con,1}(\mathbf{z}) = \partial_{\mathbf{z}}\partial_{\mathbf{z}'} RTC(\boldsymbol{\theta}, a)$. From this, it is easy to see that for the basic quota share problem, the second term on the right-hand side of equation (10.3) is a matrix of zeros. In contrast, for the excess of loss problem, this term is a diagonal matrix $LME_1\ \partial_{\mathbf{z}}\partial_{\mathbf{z}'} f_{con,1}(\mathbf{z}) = LME_1\ \boldsymbol{\Lambda}$, where $\boldsymbol{\Lambda} = diag\,[0, f_1(\theta_1), \ldots, f_p(\theta_p)]$.

The expression in equation (10.3) provides a handy computational form that was already used in Examples 10.1 and 10.2 and will be used again in examples that follow.

10.2.2 Risk Distribution Parameter Sensitivities - Bivariate Excess of Loss

Example 10.3. Parameter Sensitivities for Excess of Loss with Two Risks. Let us continue from Example 7.2, a bivariate excess of loss problem where X_1 follows a gamma distribution and X_2 follows a Pareto distribution. Refer to Example 7.2 for specific parameter values of the distributions.

This problem has been solved four times with different objective functions: one for standard deviation and three for the expected shortfall with confidence levels $\alpha = 0.95, 0.85, 0.75$. Table 10.2 summarizes results, with each objective function presented in a separate column. As before, once an objective function has been selected and optimal values of retention parameters have been determined, all other risk measures can be evaluated at the optimum. This allows for comparison of results from different objective functions.

In this example, given the long tail of the Pareto distribution, we initially investigate the sensitivity to the scale of the Pareto distribution. Thus, we represent the second risk as

TABLE 10.2
Excess of Loss: Baseline Results from Four Optimization Problems.

	Std Dev	$ES:0.95$	$ES:0.85$	$ES:0.75$
Optimal u_1	5064.60	5480.40	5364.56	5092.53
Optimal u_2	1782.48	1210.94	1336.12	1730.78
Optimal LME_1	1.13	4.14	3.66	2.95
Std Dev	1906.24	1927.92	1918.27	1906.35
$ES:0.95$	6847.07	6691.33	6700.68	6823.31
$ES:0.85$	6700.66	6655.38	6648.79	6690.92
$ES:0.75$	6375.39	6429.27	6402.87	6375.06

$X_2(a) = X_2(1 + \frac{a}{100})$. This formulation allows us to consider small changes in the auxiliary variable a in percentage terms.

All other aspects of the problem remain unchanged. For bivariate excess of loss, the risk owner retains $S(u_1, u_2, a) = X_1 \wedge u_1 + X_2(a) \wedge u_2$. The budget requirement (assumed to be binding) is represented as $f_{con,1}(u_1, u_2, a) = RTC(u_1, u_2, a) - RTC_{max} \leq 0$ where, for simplicity, we use a fair risk transfer cost $RTC(u_1, u_2, a) = \mathrm{E}[S(\infty, \infty, a)] - \mathrm{E}[S(u_1, u_2, a)]$. A maximal risk transfer cost $RTC_{max} = 0.20 \times \{\mathrm{E}(X_1) + \mathrm{E}(X_2(0)\} = 1,000$ is used in this example.

For each of the four optimization problems, Table 10.3 shows the risk retention sensitivities $\partial_a u_1^*(a)$ and $\partial_a u_2^*(a)$, as well as the Lagrange multiplier sensitivity $\partial_a LME^*(a)$. Despite the example being designed to only affect small changes in the scale of the Pareto distribution (the second distribution), we still observe changes in the first risk retention parameter u_1^* and the Lagrange multiplier.

The results were obtained through a relatively straightforward implementation of `R` code. Following the coding of the objective and budget constraint functions, numerical evaluation of the derivatives and matrix of second derivatives was conducted using the functions `grad` and `hessian`, respectively, from the `numDeriv` package.

Table 10.4 shows similar results when varying the scale parameter for the first risk (Scale 1).

Exercise 10.5 outlines follow-up analyses of this very interesting example.

10.2.3 Wisconsin Property Fund Sensitivities

Section 8.1.2 demonstrated a comprehensive optimization analysis for a specific fund member (number 2), presenting optimal risk retention parameters, risk transfer costs, and risk measures across a range of maximal risk transfer costs; see Table 8.3. Supplementary graphs provide an intuitive understanding for the trade-off between risk transfer costs and uncertainty of retained risks.

Naturally, it is possible to present many alternative "what-if" scenarios, examining each assumption underpinning the analysis. However, there were over one thousand fund members when these data were analyzed. An analyst would prefer a set of techniques that can be

TABLE 10.3
Excess of Loss Sensitivities with Auxiliary a = Pareto Scale.

	Std Dev	ES : 0.95	ES : 0.85	ES : 0.75
Sensitivity u_1	10.974	11.540	9.160	9.524
Sensitivity u_2	15.896	16.634	19.333	18.781
Sensitivity LME_1	0.002	0.018	0.005	0.003

TABLE 10.4
Excess of Loss Sensitivities with Auxiliary a = Gamma Scale.

	Std Dev	ES : 0.95	ES : 0.85	ES : 0.75
Sensitivity u_1	67.529	73.008	66.799	63.307
Sensitivity u_2	16.671	7.137	14.355	23.919
Sensitivity LME_1	0.004	0.028	0.010	0.009

TABLE 10.5
Fund Member 2 Excess of Loss: Scale Parameter Sensitivities.

	u_{BC}	u_{IM}	u_{CN}	u_{CO}	u_{PN}	u_{PO}	LME
Base	51.672	17.676	26.377	35.479	15.097	9.668	3.664
$Scale_{BC}$	1.378	0.188	0.148	0.229	0.091	0.083	-0.108
$Scale_{IM}$	0.140	0.223	0.026	0.040	0.016	0.014	-0.019
$Scale_{CN}$	0.098	0.023	0.287	0.028	0.011	0.010	-0.013
$Scale_{CO}$	0.243	0.057	0.044	0.397	0.027	0.025	-0.032
$Scale_{PN}$	0.055	0.013	0.010	0.016	0.150	0.006	-0.007
$Scale_{PO}$	0.070	0.016	0.013	0.020	0.008	0.101	-0.009

readily scaled up to multiple analyses, aligning with one of the objectives of the Envelope theorem approach developed here.

To maintain focus, we assume that fund member 2 aims for a maximal risk transfer cost of $RTC_{max} = 66$, in thousands of dollars. According to Table 8.3, this is near the center of potential costs, representing 55 percent of maximal costs.

Similar to the Section 10.2.2 example, here we investigate the sensitivity to the scale of each variable by employing alternatives $X_j(a) = X_j(1 + \frac{a}{100})$, $j = 1, \ldots, 6$. Table 10.5 shows the results. The first row provides the basic outcomes from the optimization run, without considering alternatives. For the second row, we examine sensitivities for the scale of the first variable. For instance, a 10% increase in the first scale parameter would lead us to anticipate an increase of 13.78 (in thousands) in the optimal u_{BC}. The table also shows we expect the optimal u_{IM} to increase by 1.88, and similarly for other columns and other rows.

10.2.4 ANU Case Sensitivities

Examples 9.5 and 9.6 discussed the sensitivity of ANU case results to the assumptions of the confidence level and the form of the risk measure. This section follows up these examples by examining the sensitivity of results to the scale assumptions of each risk.

To be specific, we focus on the middle value of the maximal risk transfer cost that is $RTC_{max} = 722$, corresponding to the sixth row of Table 8.8.

The sensitivities for scale parameters are given in Table 10.6. This table provides a fairly complete picture; for a partial change of each of the 14 scale parameters, the table summarizes the change in each of the 14 upper limit parameters. As one would expect, changes along the diagonal are generally the largest.

How reliable are these estimates of the sensitivities? One way to assess this is via bootstrapping as shown in Table 10.7. For readers interested in more detail of these analyses and corresponding code, see Frees and Shi (2024).

10.3 Curvature of the Active Set Lagrangian

As outlined in Section 10.2.1, for excess of loss and quota share insurable risk portfolios, one can express the Hessian of the active, or shortened, Lagrangian as

TABLE 10.6
ANU Excess of Loss: Scale Parameter Sensitivities.

	u_2	u_3	u_4	u_5	u_6	u_7	u_8	u_9	u_{10}	u_{11}	u_{12}	u_{13}	u_{14}	u_{15}
Base	334.15	346.73	261.60	246.07	98.32	124.28	185.00	289.59	200.69	191.09	245.61	415.37	183.23	212.82
$Scale_2$	17.01	5.46	6.04	4.42	0.68	0.00	0.02	5.60	2.48	1.07	0.61	0.10	0.32	0.80
$Scale_3$	2.21	4.39	1.00	0.73	0.11	0.00	0.00	0.93	0.41	0.18	0.10	0.02	0.05	0.13
$Scale_4$	2.56	1.05	3.98	0.85	0.13	0.00	0.01	1.08	0.48	0.21	0.12	0.02	0.06	0.15
$Scale_5$	2.25	0.92	1.02	2.91	0.11	0.00	0.00	0.95	0.42	0.18	0.10	0.02	0.05	0.13
$Scale_6$	0.11	0.04	0.05	0.04	0.93	0.00	0.00	0.05	0.02	0.01	0.00	0.00	0.00	0.01
$Scale_7$	0.00	0.00	0.00	0.00	0.00	0.00	0.00	0.00	0.00	0.00	0.00	0.00	0.00	0.00
$Scale_8$	0.00	0.00	0.00	0.00	0.00	0.00	1.26	0.00	0.00	0.00	0.00	0.00	0.00	0.00
$Scale_9$	2.55	1.04	1.16	0.85	0.13	0.00	0.00	3.84	0.47	0.21	0.12	0.02	0.06	0.15
$Scale_{10}$	0.40	0.16	0.18	0.13	0.02	0.00	0.00	0.17	2.14	0.03	0.02	0.00	0.01	0.02
$Scale_{11}$	0.19	0.08	0.09	0.06	0.01	0.00	0.00	0.08	0.04	1.84	0.01	0.00	0.00	0.01
$Scale_{12}$	0.00	0.00	0.00	0.00	0.00	0.00	0.00	0.00	0.00	0.00	2.35	0.00	0.00	0.00
$Scale_{13}$	0.00	0.00	0.00	0.00	0.00	0.00	0.00	0.00	0.00	0.00	0.00	4.15	0.00	0.00
$Scale_{14}$	0.01	0.00	0.00	0.00	0.00	0.00	0.00	0.00	0.00	0.00	0.00	0.00	1.81	0.00
$Scale_{15}$	0.03	0.01	0.01	0.01	0.00	0.00	0.00	0.01	0.01	0.00	0.00	0.00	0.00	2.00

TABLE 10.7
Bootstrap Standard Errors of Scale Parameter Sensitivities.

	u_2	u_3	u_4	u_5	u_6	u_7	u_8	u_9	u_{10}	u_{11}	u_{12}	u_{13}	u_{14}	u_{15}
$Scale_2$	0.85	0.73	0.83	0.38	0.08	0	0.10	0.78	0.30	0.09	0.13	0.03	0.06	0.16
$Scale_3$	0.15	0.29	0.14	0.07	0.01	0	0.01	0.14	0.05	0.02	0.02	0.00	0.01	0.03
$Scale_4$	0.21	0.15	0.32	0.09	0.02	0	0.02	0.14	0.06	0.01	0.02	0.01	0.01	0.03
$Scale_5$	0.14	0.13	0.12	0.12	0.01	0	0.01	0.14	0.05	0.01	0.02	0.00	0.01	0.03
$Scale_6$	0.01	0.01	0.01	0.00	0.09	0	0.00	0.01	0.00	0.00	0.00	0.00	0.00	0.00
$Scale_7$	0.00	0.00	0.00	0.00	0.00	0	0.01	0.00	0.00	0.00	0.00	0.00	0.00	0.00
$Scale_8$	0.00	0.00	0.00	0.00	0.00	0	0.57	0.00	0.00	0.00	0.00	0.00	0.00	0.00
$Scale_9$	0.16	0.16	0.17	0.08	0.02	0	0.01	0.32	0.06	0.02	0.02	0.01	0.01	0.03
$Scale_{10}$	0.03	0.03	0.04	0.01	0.00	0	0.00	0.02	0.18	0.00	0.00	0.00	0.00	0.01
$Scale_{11}$	0.02	0.01	0.01	0.01	0.00	0	0.00	0.01	0.00	0.12	0.00	0.00	0.00	0.00
$Scale_{12}$	0.00	0.00	0.00	0.00	0.00	0	0.01	0.00	0.00	0.00	0.39	0.00	0.00	0.00
$Scale_{13}$	0.00	0.00	0.00	0.00	0.00	0	0.01	0.00	0.00	0.00	0.00	0.64	0.00	0.00
$Scale_{14}$	0.00	0.00	0.00	0.00	0.00	0	0.00	0.00	0.00	0.00	0.00	0.00	0.16	0.00
$Scale_{15}$	0.01	0.00	0.00	0.00	0.00	0	0.00	0.00	0.00	0.00	0.00	0.00	0.00	0.29

$$SLA_{\mathbf{zz'}}(\mathbf{z}, \mathbf{LME}) = \tilde{\mathbf{X}}'\tilde{\mathbf{X}} + LME_1 \left\{ \begin{array}{ll} \mathbf{0} & \text{Quota Share} \\ \mathbf{\Lambda} & \text{Excess of Loss} \end{array} \right. . \tag{10.4}$$

See also Appendix Section 10.5.3. Equation (10.4) provides not only a handy computational formula for the Hessian of the active set Lagrangian but also one that is amenable to an informative interpretation.

This Hessian can be viewed as a balance between an ill-conditioned matrix ($f_{0,\mathbf{zz'}}(\mathbf{z}) = \tilde{\mathbf{X}}'\tilde{\mathbf{X}}$) and a well-conditioned one ($\mathbf{\Lambda}$); this balance is a feature of many related problems. Many analysts may draw analogies between a comparison of ordinary least squares and ridge regression (see, for example, Hastie et al. (2009), Section 3.4.1).

To explore why excess of loss contracts do not suffer from decision variable uncertainty, we revisit the discussion in Section 10.2.1. There, we posited that the curvative of the Hessian of the Lagrangian yield insights into the decision variable uncertainty problem.

Condition Number. There are numerous ways to summarize the curvature of the Hessian, but arguably the most natural choice is the condition number. Based on equations (10.3) and (10.4), we observe that the Hessians we work with are typically symmetric and nonnegative definite. In this scenario, one may interpret the condition number as the ratio of the largest to the smallest eigenvalues. Each eigenvalue summarizes the curvature in the direction of its corresponding eigenvector. Hence, the condition number of the Hessian can be understood as the ratio of the largest to the smallest curvatures at the optimum (see for example, Bierlaire (2018), Section 2.5).

Example 10.4. Property Fund Portfolio Condition Numbers. Frees and Shi (2024) examined the Wisconsin Property Fund similarly to Section 8.1.3, but imposed retention limits at the portfolio level, rather than in a proportional manner as demonstrated in that section. For this application, Table 10.8 displays the condition numbers of the Hessian of the excess of loss objective function as well as the active set Lagrangian, both at the optimum. We report the condition number using the infinity norm. We observe that incorporating

TABLE 10.8
Comparison of Excess of Loss Condition Numbers.

Excess of Loss f_{0zz}	8.23
Excess of Loss $SLAzz$	1.73

the extra term for the excess of loss in equation (10.3) substantially reduces the condition number, implying that analysts can anticipate greater curvature, and thus greater reliability, of the results.

Investments and Quota Share Portfolios Condition Numbers. In the investments portfolio problem, analysts use shrinkage estimators to estimate the variance-covariance of returns, employing a combination of (ill-conditioned) data-driven estimators and a more stable estimator such as a market model (see, for example, Ledoit and Wolf (2003), Candelon et al. (2012), and Schäfer and Strimmer (2005)). Moreover, in the investments portfolio problem, DeMiguel et al. (2009) suggested adding an additional constraint that the norm of the portfolio-weight vector be smaller than a given threshold. The motivation underpinning this suggestion was to enhance the out-of-sample performance of portfolio performance (see also Ban et al. (2018)). As demonstrated in the following, this type of additional constraint can sharpen the curvature of the Lagrangian, reducing the influence of decision variable uncertainty on risk retention uncertainty.

Building on Example 10.4, Table 10.9 displays the condition number of the Hessian of the active set Lagrangian for the quota share contract. Comparing this to the excess of loss contract in Table 10.8, the quota share condition numbers are substantially larger. This can be explained by examining the vector on which the Hessian is based, $\tilde{\mathbf{X}}$, from Section 10.2.1. For quota share, the rth row of this vector is $\sqrt{w_r(\mathbf{z}^*)}(1, -\mathbf{X}_r')$, with weight $w_r(\mathbf{z}^*) = k\left[(z_0^* - \boldsymbol{\theta}^{*'}\mathbf{X}_r)/b\right]/[(1-\alpha)Rb]$. By analyzing these weights, we observe that only observations that are nearly linear about the optimum receive substantive weight. Consequently, when examining the columns of $\tilde{\mathbf{X}}$, one can anticipate them to be nearly a linear combination of one another. Columns that are nearly a linear combination of one another naturally result in an ill-conditioned matrix, creating decision variable uncertainty.

Achieving a balance between ill- and well-conditioned matrices in equation (10.4) involves introducing nonlinear constraint functions. This suggests the consideration of retaining a linear mechanism for sharing risks yet introducing a nonlinear risk transfer cost for pricing. We pursued this approach, following the work of DeMiguel et al. (2009). Specifically, for the quota share problem, we imposed the quadratic constraint $f_{con}(\mathbf{z}) = f_{con}(z_0, \boldsymbol{\theta}) = \boldsymbol{\theta}'\boldsymbol{\theta} - \delta$. With the choice of $\delta = 4$, this constraint was found to be binding at the optimum and was therefore included in the shortened Lagrangian. The bottom two rows of Table 10.9 demonstrate that by incorporating this constraint, one can dramatically sharpen the curvature of the Lagrangian which mitigates the decision variable uncertainty problem.

TABLE 10.9
Comparison of Quota Share Condition Numbers (in Millions).

Quota Share $SLA_{zz} = f_{0zz}$	824.836
Quota Share Quad Constraint f_{0zz}	958.152
Quota Share Quad Constraint SLA_{zz}	0.351

10.4 Data Uncertainty and Stochastic Sensitivity

The preceding sections of this chapter established sensitivities based on perturbations of a scalar auxiliary variable a. It is evident that one can extend these sensitivities to an auxiliary *vector* $\mathbf{a}$ - after all, a partial derivative with respect a vector, a gradient, is simply a vector of partial derivatives. For example, an interesting case when $\mathbf{a} = \gamma$, that is, the auxiliary vector is a vector of marginal risk parameters, such as those that account for the scale and shape of a distribution. Chapter 12 will delve into another interesting choice where $\mathbf{a}$ represents a collection of dependence parameters.

Like all other assumptions, the optimal decision variables depend on the set of marginal risk parameters, denoted as $\mathbf{z}^*(\gamma)$. Associated with each parameter, one may use equations (10.1) and (10.2) to compute perturbation sensitivities $\nabla \mathbf{z}^*(\gamma) = \partial_\gamma\ \mathbf{z}^*(\gamma)$, a matrix of partial derivatives.

Data Uncertainty. Risk distribution parameters may be imprecise due to either simulation uncertainty or sampling (of data) uncertainty, although other sources are clearly possible. In applications, the full distribution may be unknown, but it is common to have available a sample of $data = \{\mathbf{X}_1, \ldots, \mathbf{X}_n\}$ from which one computes distribution parameter estimates $\hat{\gamma} = \gamma(data)$. From this, one computes the optimal decision variables $\hat{\mathbf{z}}^* = \mathbf{z}^*(\hat{\gamma})$. The difficulty arises because small changes in distribution parameters $\hat{\gamma}$ can lead to significant changes in decision variables $\hat{\mathbf{z}}^*$; I refer to this difficulty as *data uncertainty.*

To illustrate, one could consider a size n sample of risks from a parametric distribution. With the usual estimators of parameters, e.g., maximum likelihood, we can rely on standard methods to quantify uncertainty in the distribution parameter estimates. As is customary, we can invoke the central limit theorem and assume that $\sqrt{n}\,(\hat{\gamma} - \gamma)$ has an approximate normal distribution with vector mean $\mathbf{0}$ and variance-covariance matrix, say, $\mathbf{\Sigma}_\gamma$. With this, we can quantify the data uncertainty in the optimal decision variables using a Taylor-series approximation.

Stochastic Sensitivity Corollary. Suppose that $\sqrt{n}\,(\hat{\gamma} - \gamma) \to_D\ N(\mathbf{0}, \mathbf{\Sigma}_\gamma)$. Then, under mild smoothness conditions,

$$\sqrt{n}\,[\mathbf{z}^*(\hat{\gamma}) - \mathbf{z}^*(\gamma)] \to_D\ N(\mathbf{0},\ \nabla \mathbf{z}^*(\gamma)\mathbf{\Sigma}_\gamma \nabla \mathbf{z}^*(\gamma)'). \tag{10.5}$$

With the asymptotic distribution, one can readily compute the standard errors of (estimated) optimal decision variables. This provides another way to interpret the sensitivities.

Example 10.5. Property Fund Portfolio Stochastic Sensitivities. This is a continuation from Example 10.4. In the work of Frees and Shi (2024), a single sample of size $n = 50$ was taken, and standard parametric likelihood analysis was used to estimate the parameters as well as their asymptotic variance. From the estimated parameters, the optimal risk retention parameters were determined and are presented in the first column of Table 10.10. To assess their reliability, the stochastic sensitivity corollary in equation (10.5) was employed to calculate the corresponding standard errors, which are given in the second column of Table 10.10.

To evaluate the accuracy of this asymptotic approximation, a summary of a bootstrap analysis is provided in columns three and four of Table 10.10. For these columns, 100 bootstrap resamples of the single sample were taken. For each resample, an optimal retention portfolio was fitted. Columns three and four of Table 10.10 provide the mean and standard deviations of this procedure. The bootstrap analysis relies on fewer assumptions at the

TABLE 10.10
Stochastic Sensitivities for Excess of Loss.

	Stochastic Sens		Bootstrap	
	Dec Var	Std Error	Mean	Std Dev
VaR	8611.9	143.2	8604.8	124.4
u_1^*	3988.3	83.8	3979.4	86.0
u_2^*	1201.1	41.0	1209.4	72.3
u_3^*	913.1	24.7	916.3	30.8
u_4^*	1628.0	55.1	1628.4	46.6
u_5^*	676.3	23.4	692.4	51.0
u_6^*	582.9	21.9	582.6	19.6

Source: Frees and Shi (2024)

expense of additional computation time. For this application, the results are consistent with one another, instilling confidence in risk portfolio managers regarding their reliability.

10.5 Supplemental Materials

10.5.1 Further Resources and Reading

This chapter is based on the work in Frees and Shi (2024).

10.5.2 Exercises

See the *Online Supplement* for selected exercise solutions.

Section 10.1 Exercises

Exercise 10.1. Quota Share. Use the Perturbation Sensitivity Proposition to establish the results in Exercise 9.10.

Exercise 10.2. Quota Share . Use the same set-up as in Exercise 10.1 except now take the auxiliary variable is $\sigma = a$, where σ is an element of $\boldsymbol{\Sigma}$. Use the Perturbation Sensitivity Proposition to establish the results in Exercise 9.11.

Exercise 10.3. Asset Allocation. Use the Perturbation Sensitivity Proposition to establish the results in Section 9.4.1.

Exercise 10.4. Asset Allocation . As a follow-up to Exercise 10.3, now the interest is in sensitivity due to the uncertainty parameters σ. Use the Perturbation Sensitivity Proposition to establish the results in Exercise 9.12.

Section 10.2 Exercises

Exercise 10.5. Excess of Loss VaR Sensitivity. Continue with Example 10.3.

a. For some new baseline optimizations, determine the optimal upper limits use VaR as a risk measure, for $\alpha = 0.95, 0.85, 0.75$. Provide code to replicate the work in Table 10.11.

b. Using $nsim = 2,000,000$ simulated realizations of losses, determine the risk retention and Lagrange multiplier sensitivities for each of the seven optimization problems in part (a). Your results should be comparable to those in Table 10.12.

c. You note that the results for the VaR optimization problems seem to be a bit counter-intuitive. To check their accuracy, repeat part (b) 10 times and calculate simulation standard errors. Summarize the mean, or average, of these 10 replications as in Table 10.13 and the standard error as in Table 10.14.

d. One can use the part (b) strategy to investigate the sensitivity of shape parameters. For example, suppose that the shape parameter for the first risk is the auxiliary variable of interest. Your results should be comparable to those in Table 10.15.

10.5.3 Appendix. Kernel Smoothing of Expectations, with Derivatives

We start with a function $h(\cdot)$ and use a kernel density estimator $k(\cdot)$ to define $\mathrm{E}_{Rk}[h(Y)]$. Then with a change of variables $y = (\tilde{y} - Y_r)/b$. Thus, we can approximate $\mathrm{E}[h(Y)]$ using

$$
\begin{aligned}
\mathrm{E}_{Rk}[h(Y)] &= \int h(\tilde{y}) f_{Rk}(\tilde{y}) d\tilde{y} \\
&= \frac{1}{R\,b} \sum_{r=1}^{R} \int h(\tilde{y}) k\left(\frac{\tilde{y} - Y_r}{b}\right) d\tilde{y} \\
&= \int \left\{ \frac{1}{R} \sum_{r=1}^{R} h[Y_r + by] \right\} k(y) dy \\
&= \int \{\mathrm{E}_R h[Y_r + by]\} \; k(y) dy.
\end{aligned}
$$

Expected Shortfall. Applying this result to the expected shortfall auxiliary function, we get

$$
\begin{aligned}
ES1_{Rk}(z_0, \boldsymbol{\theta}) &= z_0 + \frac{1}{1-\alpha} \{\mathrm{E}_{Rk}[g(\mathbf{X}; \boldsymbol{\theta})] - \mathrm{E}_{Rk}[g(\mathbf{X}; \boldsymbol{\theta}) \wedge z_0]\} \\
&= z_0 + \frac{1}{1-\alpha} \{\mathrm{E}_{Rk}[g(\mathbf{X}; \boldsymbol{\theta}) - z_0]_+\} \\
&= z_0 + \frac{1}{1-\alpha} \frac{1}{R} \sum_{r=1}^{R} \int [g(\mathbf{X}_r; \boldsymbol{\theta}) + by - z_0]_+ \; k(y) dy \\
&= z_0 + \frac{1}{1-\alpha} \int \left\{ \frac{1}{R} \sum_{r=1}^{R} [g(\mathbf{X}_r; \boldsymbol{\theta}) + by - z_0]_+ \right\} k(y) dy.
\end{aligned}
$$

Expected Shortfall Gradients. Now note that

$$
\partial_{z_0} \; [g(\mathbf{X}_r; \boldsymbol{\theta}) + by - z_0]_+ = -I\left(y > [z_0 - g(\mathbf{X}_r; \boldsymbol{\theta})]/b\right)
$$

and

$$
\partial_{\boldsymbol{\theta}} \; [g(\mathbf{X}_r; \boldsymbol{\theta}) + by - z_0]_+ = I\left(y > [z_0 - g(\mathbf{X}_r; \boldsymbol{\theta})]/b\right) \partial_{\boldsymbol{\theta}} g(\mathbf{X}_r; \boldsymbol{\theta}).
$$

TABLE 10.11
Excess of Loss: Baseline Results from Seven Optimization Problems.

	Std Dev	$VaR : 0.95$	$VaR : 0.85$	$VaR : 0.75$	$ES : 0.95$	$ES : 0.85$	$ES : 0.75$
Optimal u_1	5064.60	5480.40	4985.57	4974.50	5480.40	5364.56	5092.53
Optimal u_2	1782.48	1210.94	1944.56	1969.36	1210.94	1336.12	1730.78
Optimal LME	1.13	4.14	2.56	0.00	4.14	3.66	2.95
Std Dev	1906.24	1927.92	1907.35	1907.70	1927.92	1918.27	1906.35
$VaR : 0.95$	6847.07	6691.33	6930.12	6943.86	6691.33	6700.68	6823.31
$VaR : 0.85$	6208.36	6400.16	6174.11	6169.40	6400.16	6344.77	6220.65
$VaR : 0.75$	5615.22	5819.67	5573.30	5567.30	5819.67	5765.06	5629.62
$ES : 0.95$	6847.07	6691.33	6930.12	6943.86	6691.33	6700.68	6823.31
$ES : 0.85$	6700.66	6655.38	6734.60	6740.14	6655.38	6648.79	6690.92
$ES : 0.75$	6375.39	6429.27	6379.93	6381.02	6429.27	6402.87	6375.06

TABLE 10.12
Excess of Loss Sensitivities with Auxiliary a = Scale 2.

	Std Dev	$VaR : 0.95$	$VaR : 0.85$	$VaR : 0.75$	$ES : 0.95$	$ES : 0.85$	$ES : 0.75$
Sensitivity u_1	10.919	11.540	14.202	-167.924	11.540	9.472	9.466
Sensitivity u_2	16.002	16.634	7.704	420.143	16.634	18.968	18.884
Sensitivity LME	0.002	0.018	-0.008	0.168	0.018	0.005	0.003

TABLE 10.13
Mean Excess of Loss Sensitivities with Auxiliary a = Scale 2.

	Std Dev	$VaR : 0.95$	$VaR : 0.85$	$VaR : 0.75$	$ES : 0.95$	$ES : 0.85$	$ES : 0.75$
Sensitivity u_1	10.75	11.54	-2.54	16.26	11.54	9.11	9.57
Sensitivity u_2	16.32	16.63	44.79	2.69	16.63	19.39	18.69
Sensitivity LME	0.00	0.02	0.01	-0.01	0.02	0.00	0.00

TABLE 10.14
Standard Error Excess of Loss Sensitivities with Auxiliary a = Scale 2.

	Std Dev	$VaR : 0.95$	$VaR : 0.85$	$VaR : 0.75$	$ES : 0.95$	$ES : 0.85$	$ES : 0.75$
Sensitivity u_1	0.1890	0	12.5151	20.5931	0	0.1698	0.2093
Sensitivity u_2	0.3588	0	27.7274	46.6745	0	0.1985	0.3776
Sensitivity LME	0.0001	0	0.0608	0.0490	0	0.0003	0.0002

TABLE 10.15
Excess of Loss Sensitivities with Auxiliary a = Shape 1.

	Std Dev	ES : 0.95	ES : 0.85	ES : 0.75
Sensitivity u_1	-6.581	1119.928	-257.704	-1.605
Sensitivity u_2	49.234	-1090.854	331.324	38.856
Sensitivity LME	0.006	0.027	1.422	-0.004

Using the observation that $\int I(y > c)k(y)dy = 1 - K(c)$ for a constant c, we have the partial derivatives

$$\partial z_0 \ ES1_{Rk}(z_0, \boldsymbol{\theta}) = 1 - \frac{1}{1-\alpha}\frac{1}{R}\sum_{r=1}^{R}\left\{1 - K\left(\frac{z_0 - g(\mathbf{X}_r;\boldsymbol{\theta})}{b}\right)\right\}$$

and

$$\partial_{\boldsymbol{\theta}} \ ES1_{Rk}(z_0, \boldsymbol{\theta}) = \frac{1}{1-\alpha}\frac{1}{R}\sum_{r=1}^{R}\left\{1 - K\left(\frac{z_0 - g(\mathbf{X}_r;\boldsymbol{\theta})}{b}\right)\right\}\partial_{\boldsymbol{\theta}} g(\mathbf{X}_r;\boldsymbol{\theta}),$$

which is sufficient for the result.

Expected Shortfall Hessian. The Hessian can be written as

$$\partial_{\mathbf{z}}\partial_{\mathbf{z}'} \ ES1_{Rk}(\mathbf{z}) = \left(\begin{array}{cc} \partial_{z_0}\partial_{z_0} \ ES1_{Rk}(z_0, \boldsymbol{\theta}) & \partial_{z_0}\partial_{\boldsymbol{\theta}'} \ ES1_{Rk}(z_0, \boldsymbol{\theta}) \\ \partial_{\boldsymbol{\theta}}\partial_{z_0} \ ES1_{Rk}(z_0, \boldsymbol{\theta}) & \partial_{\boldsymbol{\theta}}\partial_{\boldsymbol{\theta}'} \ ES1_{Rk}(z_0, \boldsymbol{\theta}) \end{array}\right).$$

For second derivatives, we have

$$\begin{aligned}
\partial_{z_0 z_0} \ ES1_{Rk}(z_0) &= \frac{1}{1-\alpha}\frac{1}{Rb}\sum_{r=1}^{R}\left\{k\left(\frac{z_0 - g(\mathbf{X}_r;\boldsymbol{\theta})}{b}\right)\right\} \\
\partial_{z_0 \boldsymbol{\theta}} \ ES1_{Rk}(z_0) &= -\frac{1}{1-\alpha}\frac{1}{Rb}\sum_{r=1}^{R}\left\{k\left(\frac{z_0 - g(\mathbf{X}_r;\boldsymbol{\theta})}{b}\right)\partial_{\boldsymbol{\theta}} g(\mathbf{X}_r;\boldsymbol{\theta})\right\} \\
\partial_{\boldsymbol{\theta}\boldsymbol{\theta}'} \ ES1_{Rk}(z_0) &= \frac{1}{1-\alpha}\frac{1}{R}\sum_{r=1}^{R}\left[\left\{1 - K\left(\frac{z_0 - g(\mathbf{X}_r;\boldsymbol{\theta})}{b}\right)\right\}\partial_{\boldsymbol{\theta}\boldsymbol{\theta}'} g(\mathbf{X}_r;\boldsymbol{\theta})\right. \\
&\quad \left. + \frac{1}{b}\, k\left(\frac{z_0 - g(\mathbf{X}_r;\boldsymbol{\theta})}{b}\right)\partial_{\boldsymbol{\theta}} g(\mathbf{X}_r;\boldsymbol{\theta}) \ \partial_{\boldsymbol{\theta}'} g(\mathbf{X}_r;\boldsymbol{\theta})\right].
\end{aligned}$$

Note that for both the excess of loss and quota share special cases, we have $\partial_{\boldsymbol{\theta}\boldsymbol{\theta}'} g(\mathbf{X}_r;\boldsymbol{\theta}) = \mathbf{0}$.

To simplify these expressions, define the weight function $w_r(\mathbf{z}) = k\left(\frac{z_0 - g(\mathbf{X}_r;\boldsymbol{\theta})}{b}\right)/[(1-\alpha)Rb]$. Assuming $\partial_{\boldsymbol{\theta}\boldsymbol{\theta}'} g(\mathbf{X}_r;\boldsymbol{\theta}) = \mathbf{0}$, we may write

$$f_{0,\mathbf{z}\mathbf{z}'}(\mathbf{z}) = \sum_{r=1}^{R} w_r(\mathbf{z})\left(\begin{array}{cc} 1 & -\partial_{\boldsymbol{\theta}'} g(\mathbf{X}_r;\boldsymbol{\theta}) \\ -\partial_{\boldsymbol{\theta}} g(\mathbf{X}_r;\boldsymbol{\theta}) & \partial_{\boldsymbol{\theta}} g(\mathbf{X}_r;\boldsymbol{\theta})\partial_{\boldsymbol{\theta}'} g(\mathbf{X}_r;\boldsymbol{\theta}) \end{array}\right).$$

Further define $\mathbf{W_z} = diag(w_1(\mathbf{z}), \ldots, w_R(\mathbf{z}))$ and $\mathbf{G} = [\partial_{\boldsymbol{\theta}} g(\mathbf{X}_1;\boldsymbol{\theta}), \ldots, \partial_{\boldsymbol{\theta}} g(\mathbf{X}_R;\boldsymbol{\theta})]'$. With this notation, we may write

$$f_{0,\mathbf{z}\mathbf{z}'}(\mathbf{z}) = \left(\begin{array}{cc} \mathbf{1}'\mathbf{W_z}\mathbf{1} & -\mathbf{1}'\mathbf{W_z}\mathbf{G} \\ -\mathbf{G}'\mathbf{W_z}\mathbf{1} & \mathbf{G}'\mathbf{W_z}\mathbf{G} \end{array}\right).$$

From this, we may write the Hessian of the active set Lagrangian as

$$SLA_{\mathbf{zz}'}(\mathbf{z},\mathbf{LME}) = \begin{pmatrix} \mathbf{1}'\mathbf{W_z 1} & -\mathbf{1}'\mathbf{W_z G} \\ -\mathbf{G}'\mathbf{W_z 1} & \mathbf{G}'\mathbf{W_z G} \end{pmatrix} + LME_1 \begin{cases} \mathbf{0} & \text{Quota Share} \\ \mathbf{\Lambda} & \text{Excess of Loss} \end{cases},$$

where $\mathbf{\Lambda} = diag\,[0, f_1(\theta_1), \ldots, f_p(\theta_p)]$.

To interpret the curvature of the Hessian, it is also helpful to define

$$\tilde{\mathbf{X}} = \begin{pmatrix} \sqrt{\mathbf{W_z}}\,\mathbf{1} & -\sqrt{\mathbf{W_z}}\,\mathbf{G} \end{pmatrix}$$

where we define $\sqrt{\mathbf{W_z}} = diag(\sqrt{w_1(\mathbf{z})}, \ldots, \sqrt{w_R(\mathbf{z})})$. With this, one can write

$$SLA_{\mathbf{zz}'}(\mathbf{z},\mathbf{LME}) = \tilde{\mathbf{X}}'\tilde{\mathbf{X}} + LME_1 \begin{cases} \mathbf{0} & \text{Quota Share} \\ \mathbf{\Lambda} & \text{Excess of Loss} \end{cases}.$$

This verifies equation (10.3).

10.5.4 Appendix. Proof of the Perturbation Sensitivity Proposition

The proof is based on results from Fiacco (1976) and Fiacco (1983). To keep the presentation self-contained, we begin by restating relevant results from these works using the notation of this paper. For simplicity, we ignore inequality constraints that are not binding at the optimum and use only constraints that are binding at the optimum.

At this stage, we consider a general objective and constraint functions that need not be related to risk retention. The constrained optimization problem is

$$\boxed{\begin{array}{ll} \text{minimize}_{\mathbf{z}} & f_0(\mathbf{z}, a) \\ \text{subject to} & f_{con,j}(\mathbf{z}, a) \le 0 \quad j = 1, \ldots, m. \end{array}} \qquad (10.6)$$

In this formulation, the auxiliary variable a is perturbed about 0. The corresponding (shortened) Lagrangian is

$$SLA(\mathbf{z}, a) = f_0(\mathbf{z}, a) + \sum_{j=1}^{m} LME_j f_{con,j}(\mathbf{z}, a).$$

We need the following two sets of conditions that appear in Fiacco (1976), Lemma 2.1.

First Order Karush-Kuhn-Tucker Conditions.

$$\boxed{\begin{array}{ccl} SLA^*_{\mathbf{z}} & = f^*_{0,\mathbf{z}} + \sum_{j=1}^{m} LME^*_j\ f^*_{con,j,\mathbf{z}} & \\ \partial_{z_j} f^*_{con,j} & \le 0 & (j = 1, \ldots, m) \\ LME^*_j\ \partial_{z_j} f^*_{con,j} & = 0 & (j = 1, \ldots, m). \end{array}}$$

Second Order Conditions.

$$\boxed{\begin{array}{cll} \mathbf{y}'\ SLA^*_{\mathbf{zz}'}\mathbf{y} & > 0 & \text{for all } \mathbf{y} \ne \mathbf{0} \text{ such that} \\ \mathbf{y}' f^*_{con,j,\mathbf{z}} & \le 0 & (j = 1, \ldots, m) \quad \text{where } f^*_{con,j} = 0 \\ \mathbf{y}' f^*_{con,j,\mathbf{z}} & = 0 & (j = 1, \ldots, m) \quad \text{where } LME^*_j > 0. \end{array}}$$

The following result provides first-order sensitivity results for a second order local minimizing point.

Theorem 2.1 and Corollary 2.1 of Fiacco (1976). Suppose

- the functions $f_0(\mathbf{z}, a)$ and $f_{con,j}(\mathbf{z}, a)$, $j = 1, \ldots, m$ are twice continuously differentiable in $(\mathbf{z}, a)$ in a neighborhood of $(\mathbf{z}^*, 0)$
- the *Lemma 2.1* first order KKT and second order conditions hold,
- for a near 0, the gradients $\partial_{z_j} f_{con,j}(\mathbf{x}^*, a)$ are linearly independent, and
- $LME_j^* > 0$ when $\partial_{z_j} f^*_{con,j} = 0$, $j = 1, \ldots, m$ (strict complementary slackness).

Then,

$$\begin{pmatrix} \partial_a \mathbf{z}^* \\ \partial_a \mathbf{LME}^* \end{pmatrix} = \begin{pmatrix} SLA^*_{\mathbf{zz}'} & f^*_{con,1,\mathbf{z}} & \cdots & f^*_{con,m,\mathbf{z}} \\ -LME_1^* \, f^*_{con,1,\mathbf{z}'} & 0 & \cdots & 0 \\ \vdots & & \ddots & \vdots \\ -LME_m^* \, f^*_{con,m,\mathbf{z}'} & 0 & \cdots & 0 \end{pmatrix}^{-1} \times \begin{pmatrix} -SLA^*_{\mathbf{z},a} \\ LME_1^* \, f^*_{con,1,a} \\ \vdots \\ LME_m^* \, f^*_{con,m,a} \end{pmatrix}. \tag{10.7}$$

Proof of the Perturbation Sensitivity Proposition. We now use equation (10.7) to establish the perturbation sensitivity results. As in Fiacco (1983) (Section 4.2), we define the $m \times m$ diagonal matrix $\mathbf{D_{LME}}^* = diag(LME_1^*, \ldots, LME_m^*)$. We also recall $f_{con,\bullet,\mathbf{z}}(\mathbf{z}, a)$ $= [f_{con,1,\mathbf{z}}(\mathbf{z}, a), \ldots, f_{con,m,\mathbf{z}}(\mathbf{z}, a)]$, a $p_z \times m$ gradient matrix, that when evaluated at the optimum is $f^*_{con,\bullet,\mathbf{z}}$. With this notation, equation (10.7) yields

$$\begin{pmatrix} \partial_a \mathbf{z}^* \\ \partial_a \mathbf{LME}^* \end{pmatrix} = \left[\begin{pmatrix} \mathbf{I}_{p_z} & \mathbf{0} \\ \mathbf{0} & \mathbf{D_{LME}}^* \end{pmatrix} \begin{pmatrix} SLA^*_{\mathbf{zz}'} & f^*_{con,1,\mathbf{z}} & \cdots & f^*_{con,1,\mathbf{z}} \\ -f^*_{con,1,\mathbf{z}'} & 0 & \cdots & 0 \\ \vdots & & \ddots & \vdots \\ -f^*_{con,m,\mathbf{z}'} & 0 & \cdots & 0 \end{pmatrix} \right]^{-1} \times \begin{pmatrix} \mathbf{I}_{p_z} & \mathbf{0} \\ \mathbf{0} & \mathbf{D_{LME}}^* \end{pmatrix} \begin{pmatrix} -SLA^*_{\mathbf{z},a} \\ f^*_{con,1,a} \\ \vdots \\ f^*_{con,m,a} \end{pmatrix}$$

$$= \begin{bmatrix} SLA^*_{\mathbf{zz}'} & f^*_{con,\bullet,\mathbf{z}} \\ -f_{con,\bullet,\mathbf{z}'}(\mathbf{z}^*, a) & \mathbf{0} \end{bmatrix}^{-1} \begin{pmatrix} -SLA^*_{\mathbf{z},a} \\ f^*_{con,\bullet,a} \end{pmatrix}. \tag{10.8}$$

From a standard result on block matrices, we use

$$\begin{pmatrix} \mathbf{A}_{11} & \mathbf{A}_{12} \\ \mathbf{A}_{21} & \mathbf{A}_{22} \end{pmatrix}^{-1} = \begin{pmatrix} \mathbf{A}_{11}^{-1} + \mathbf{A}_{11}^{-1}\mathbf{A}_{12}\mathbf{A}_{22\cdot 11}^{-1}\mathbf{A}_{21}\mathbf{A}_{11}^{-1} & -\mathbf{A}_{11}^{-1}\mathbf{A}_{12}\mathbf{A}_{22\cdot 11}^{-1} \\ -\mathbf{A}_{22\cdot 11}^{-1}\mathbf{A}_{21}\mathbf{A}_{11}^{-1} & \mathbf{A}_{22\cdot 11}^{-1} \end{pmatrix},$$

that requires $\mathbf{A}_{11}$ to be invertible but not $\mathbf{A}_{22}$. Here, $\mathbf{A}_{22\cdot 11} = \mathbf{A}_{22} - \mathbf{A}_{21}\mathbf{A}_{11}^{-1}\mathbf{A}_{12}$.

Use $SLA^*_{\mathbf{zz}'}$ for $\mathbf{A}_{11}$, $f^*_{con,\bullet,\mathbf{z}}$ for $\mathbf{A}_{12}$, $-f^*_{con,\bullet,\mathbf{z}'}$ for $\mathbf{A}_{21}$, and $\mathbf{0}$ for $\mathbf{A}_{22}$. Further define $\mathbf{M} = f^*_{con,\bullet,\mathbf{z}'} \, SLA^{*-1}_{\mathbf{zz}'} \, f^*_{con,\bullet,\mathbf{z}}$ for $\mathbf{A}_{22\cdot 11}$.

From the second row of blocks in Display (10.8), we have

$$\begin{aligned}\partial_a \mathbf{LME}^* &= \mathbf{A}_{22\cdot 11}^{-1}\left\{-\mathbf{A}_{21}\mathbf{A}_{11}^{-1}(-SLA^*_{\mathbf{z},a}) + (f^*_{con,\bullet,a})\right\}\\ &= \mathbf{M}^{-1}\left\{f^*_{con,\bullet,a} - f^*_{con,\bullet,\mathbf{z}}\; SLA^{*-1}_{\mathbf{zz}'}\; SLA^*_{\mathbf{z},a}\right\},\end{aligned}$$

which is sufficient for equation (10.1).

From the first row of blocks in Display (10.8), we have

$$\begin{aligned}\partial_a\ \mathbf{z}^* &= \left\{\mathbf{A}_{11}^{-1} + \mathbf{A}_{11}^{-1}\mathbf{A}_{12}\mathbf{A}_{22\cdot 11}^{-1}\mathbf{A}_{21}\mathbf{A}_{11}^{-1}\right\}(-SLA^*_{\mathbf{z},a}) - \left\{\mathbf{A}_{11}^{-1}\mathbf{A}_{12}\mathbf{A}_{22\cdot 11}^{-1}\right\}(f^*_{con,\bullet,a})\\ &= -SLA^{*-1}_{\mathbf{zz}'}SLA^*_{\mathbf{z},a} + SLA^{*-1}_{\mathbf{zz}'}f^*_{con,\bullet,\mathbf{z}}\mathbf{M}^{-1}f^*_{con,\bullet,\mathbf{z}'}SLA^{*-1}_{\mathbf{zz}'}SLA^*_{\mathbf{z},a}\\ &- SLA^{*-1}_{\mathbf{zz}'}f^*_{con,\bullet,\mathbf{z}}\mathbf{M}^{-1}f^*_{con,\bullet,a}\\ &= SLA^{*-1}_{\mathbf{zz}'}\left\{-SLA^*_{\mathbf{z},a} + f^*_{con,\bullet,\mathbf{z}}\mathbf{M}^{-1}\left[f^*_{con,\bullet,\mathbf{z}}SLA^{*-1}_{\mathbf{zz}'}SLA^*_{\mathbf{z},a} - f^*_{con,\bullet,a}\right]\right\}\\ &= -SLA^{*-1}_{\mathbf{zz}'}\left\{SLA^*_{\mathbf{z},a} + f^*_{con,\bullet,\mathbf{z}}\ \partial_a\mathbf{LME}^*\right\},\end{aligned}$$

which is sufficient for equation (10.2).

Proof of the Linear Constraint Corollary. It is straightforward to multiply each side of equation (10.7) by the inverse of first matrix on the right hand side and, examining the bottom m rows, get

$$-LME^*_j\ f^*_{con,j,\mathbf{z}'}\ \partial_a\mathbf{z}^* = LME^*_j\ f^*_{con,j,a},$$

for $j = 1, \ldots, m$.

Now consider a linear constraint of the form $f_{con,j}(\mathbf{z}) = \mathbf{p}_j'\mathbf{z} - p_0 \le 0$, for a constant p_0 and vector $\mathbf{p}_j$. For partial derivatives, we have $f_{con,j,a}(\mathbf{z}) = 0$ and $f_{con,j,\mathbf{z}}(\mathbf{z}) = \mathbf{p}_j$. At the optimum $f_{con,j,a}(\mathbf{z}^*) = 0$ and by the strict complementary slackness, the corresponding Lagrange multiplier is $LME^*_j > 0$.

Thus, with the prior expression we can cancel out LME^*_j, to get

$$\begin{aligned}0 &= -f_{con,j,a}(\mathbf{z}^*) = f_{con,j,\mathbf{z}'}(\mathbf{z}^*)\partial_a\mathbf{z}^*(a)\\ &= \mathbf{p}_j'\ \partial_a\mathbf{z}^*(a),\end{aligned}$$

which is sufficient for the result.

11

Risk Retention Conditions

Chapter Preview. When using the method of constrained optimization to propose an effective risk retention policy, it can be valuable to interpret the conditions necessary for transferring a risk. This chapter introduces the Karush-Kuhn-Tucker, or KKT, conditions that must be satisfied for an optimal risk transfer solution. To underscore their relevance, we initially revisit the framework of Chapter 3 for a single risk and demonstrate how the KKT framework can summarize the detailed work developed in that context.

Moving to the multivariate case, these conditions can verify classical conditions on balance among risks for multivariate excess of loss and extend them to encompass cases of dependence. This framework is subsequently employed to establish conditions to achieve (1) a binding budget, (2) a balance among retention parameters at the optimum, and (3) boundary constraints. All these conditions rely on the risk measure relative marginal change (RM^2). Additionally, because of its significance in multivariate risk retention, special consideration is given to risk retention within a simulation context. Specifically, this framework is utilized to establish the equivalence between quantile regression and ES optimization.

11.1 KKT Conditions

We now return to the general constrained optimization problem in Display (3.9) and the associated Lagrangian in equation (3.10). Suppose that $\mathbf{z}^*$ is a local minimizer on the feasible set. Then, there exist multipliers $\mathbf{LMI}^* = \left(LMI_j^*;\ j \in CON_{in}\right)'$ and $\mathbf{LME}^* = \left(LME_j^*;\ j \in CON_{eq}\right)'$ that satisfy the **Karush-Kuhn-Tucker (KKT) Conditions.**

$$
\boxed{\begin{array}{ll}
\partial_{z_i}\ LA\left(\mathbf{z}, \mathbf{LMI}^*, \mathbf{LME}^*\right)\big|_{\mathbf{z}=\mathbf{z}^*} = 0 & i = 1, \ldots, p_z \\
LMI_j^* f_{con,j}(\mathbf{z}^*) = 0 & j \in CON_{in} \\
LMI_j^* \geq 0 & j \in CON_{in} \\
f_{con,j}(\mathbf{z}^*) \leq 0 & j \in CON_{in} \\
f_{con,j}(\mathbf{z}^*) = 0 & j \in CON_{eq}.
\end{array}} \qquad (11.1)
$$

The conditions in Display (11.1) are known as the *Kuhn-Tucker* or the *Karush-Kuhn-Tucker* (KKT) conditions, named after their originators. Mathematically, they are necessary conditions; that is, if a point is an optimum, then the conditions must hold. Many constrained optimization algorithms can be seen as methods for solving these conditions; they are extensively used in analyses.

The KKT conditions are also referred to as *first-order necessary conditions* as they are based on first derivatives. Being derivative-based, these conditions only ensure a point is a local optimum and do not indicate global behavior. Additional constraints, such as

DOI: 10.1201/9781003469568-11

convexity or conditions on second derivatives, are necessary to guarantee a point is a global minimizer. It is important to note that these are (partial) derivatives with respect to the decision variables. This differs from the sensitivity analysis in Chapter 9, which focused on derivatives with respect to auxiliary parameters.

Example 11.1. Univariate Risk Retention Conditions. Consider a single (univariate) risk X and the risk retention function given in Section 2.2 that has parameters $\boldsymbol{\theta} = (d, c, u)$. For this example, let us take the coinsurance parameter $c = 1$. For the risk measure, consider the expected shortfall $ES_{\alpha}[g(X); \boldsymbol{\theta}]$ given in equation (2.11). For simplicity, use the fair risk transfer cost $RTC(\boldsymbol{\theta}) = \mathrm{E}(X) - \mathrm{E}[g(X; \boldsymbol{\theta})]$, with an expression for $\mathrm{E}[g(X; \boldsymbol{\theta})]$ in equation (2.7). Explicitly, consider the risk retention problem

$$
\boxed{\begin{array}{ll}
\text{minimize}_{d,u} & f_0(d,u) = ES_{\alpha}[g(X); d, u] \\
\text{subject to} & f_{con,1}(d,u) = RTC(d,u) - RTC_{max} \leq 0 \\
 & d \geq 0, \qquad u \geq d.
\end{array}}
$$

As in Section 3.5, we assume that the RTC_{max} is sufficiently small so that the largest feasible deductible is smaller than the α quantile of the risk X, that is, $d_{max} < F_{\alpha}^{-1}$. Under this constraint, one can use the KKT conditions to verify that the optimal deductible is $d^* = 0$.

ⓘ *Under the Hood.* See the *Online Supplement* for the verification that the Optimal $d = 0$.

11.2 Risk Retention Conditions for a Single Risk

I now extend Example 11.1 to incorporate coinsurance along with other risk measures. This section revisits the work that was done in Section 3.5, but presenting it within the KKT framework offers a more concise and elegant approach. The focus remains on a single (univariate) risk X and the risk retention function detailed in Section 2.2 that has parameters $\boldsymbol{\theta} = (d, c, u)$. Initially, let us consider a generic risk measure RM and risk transfer cost RTC. It is worth noting that when $d = u$, no losses are retained. To introduce some risk retention, assume that the maximal risk transfer cost $RTC_{max} < RTC(d, c, u = d)$.

Explicitly, consider the risk retention problem

$$
\boxed{\begin{array}{ll}
\text{minimize} & f_0(\boldsymbol{\theta}) = RM[g(X); \boldsymbol{\theta}] \\
\text{subject to} & f_{con,1}(\boldsymbol{\theta}) = RTC(\boldsymbol{\theta}) - RTC_{max} \leq 0 \\
 & f_{con,2}(\boldsymbol{\theta}) = c - 1 \leq 0 \\
 & d \geq 0, c \geq 0, u \geq d.
\end{array}}
$$

Here, the constraint on $f_{con,2}$ ensures $c \leq 1$.

Consistent with Table 2.2, I also assume:

$$
\boxed{\begin{array}{ll}
\partial_d RM[g(X); \boldsymbol{\theta}] \leq 0 & \partial_d RTC(\boldsymbol{\theta}) > 0 \\
\partial_c RM[g(X); \boldsymbol{\theta}] > 0 & \partial_c RTC(\boldsymbol{\theta}) < 0 \\
\partial_u RM[g(X); \boldsymbol{\theta}] \geq 0 & \partial_u RTC(\boldsymbol{\theta}) < 0.
\end{array}} \qquad (11.2)
$$

For each parameter θ in the vector $\boldsymbol{\theta}$, define the risk measure relative marginal change

$$RM^2_\theta = -\frac{\partial_\theta RM[g(X);\boldsymbol{\theta}]}{\partial_\theta RTC(\boldsymbol{\theta})}, \quad \text{for } \theta = d, c, u,$$

and, using Display (11.2), note that $RM^2_\theta \geq 0$ for each θ.

At the optimum values of $\boldsymbol{\theta}^* = (d^*, c^*, u^*)$, one can use the KKT conditions to provide sufficient conditions so that the optimal values of parameters are on **boundaries**, as follows

$$\text{if} \quad RM^2_{u^*} > RM^2_{d^*}, \quad \text{then } d^* = 0 \tag{11.3}$$

and

$$\text{if} \quad RM^2_{u^*} > RM^2_{c^*}, \quad \text{then } c^* = 1. \tag{11.4}$$

Under the Hood. See the *Online Supplement* for the proof of the single risk retention results.

To show how to apply these results, we explore the VaR and ES measures. The following Table 11.1 summarizes the RM^2 measures, providing an extension of Table 2.2.

For both the VaR and ES risk measures, from Table 11.1 we see that if the optimal $u^* < F^{-1}_\alpha$, then the condition in Display (11.3) holds and the optimal deductible is $d^* = 0$.

For the expected shortfall measure, suppose that the optimal deductible $d^* < F^{-1}_\alpha$. Then, from Table 11.1 we see that the condition in Display (11.3) holds and the optimal deductible is $d^* = 0$. As in Section 3.5, we can assume that the RTC_{max} is sufficiently small so that $d_{max} < F^{-1}_\alpha$, thus ensuring that $d^* < F^{-1}_\alpha$.

For the coinsurance parameter, let us again consider the case where the optimal $u^* < F^{-1}_\alpha$. Then, for both the value at risk and the expected shortfall, the condition in Display (11.4) can be expressed as

$$\frac{1}{1-F(u^*)} = RM^2_{u^*} > RM^2_{c^*} = \frac{u^* - d^*}{\mathrm{E}(X \wedge u^*) - \mathrm{E}(X \wedge d^*)}$$
$$\iff \mathrm{E}(X \wedge u^*) - \mathrm{E}(X \wedge d^*) > (u^* - d^*)[1 - F(u^*)].$$

TABLE 11.1
RM^2 for Value at Risk and Expected Shortfall.

Summary	**Parameter (θ)**			
Measure	d	c	u	**Range of α**
	0	0	0	$\alpha < F(d)$
RM^2_θ for	$\frac{1}{1-F(d)}$	$\frac{F^{-1}_\alpha - d}{\mathrm{E}(X\wedge u)-\mathrm{E}(X\wedge d)}$	0	$F(d) \leq \alpha < F(u)$
VaR	$\frac{1}{1-F(d)}$	$\frac{u-d}{\mathrm{E}(X\wedge u)-\mathrm{E}(X\wedge d)}$	$\frac{1}{1-F(u)}$	$F(u) \leq \alpha$
	$\frac{1}{1-\alpha}$	$\frac{1}{1-\alpha}$	$\frac{1}{1-\alpha}$	$\alpha < F(d)$
RM^2_θ for ES	$\frac{1}{1-F(d)}$	$\frac{1}{1-\alpha}\{\mathrm{E}(X \wedge u) - \mathrm{E}(X \wedge F^{-1}_\alpha) + (1-\alpha)(F^{-1}_\alpha - d)\} / \{\mathrm{E}(X \wedge u) - \mathrm{E}(X \wedge d)\}$	$\frac{1}{1-\alpha}$	$F(d) \leq \alpha < F(u)$
	$\frac{1}{1-F(d)}$	$\frac{u-d}{\mathrm{E}(X\wedge u)-\mathrm{E}(X\wedge d)}$	$\frac{1}{1-F(u)}$	$F(u) \leq \alpha$

Using

$$\begin{aligned}
\mathrm{E}(X \wedge u^*) - \mathrm{E}(X \wedge d^*) &= \int_{d^*}^{u^*} [1 - F(z)]dz \\
&> \int_{d^*}^{u^*} [1 - F(u)]dz = (u^* - d^*)[1 - F(u^*)],
\end{aligned}$$

we see that this condition is satisfied at the optimum so the optimal parameter is $c^* = 1$.

Readers are invited to explore extensions to the range value at risk, $RVaR$.

11.3 Risk Retention Conditions for Multiple Risks

We now use the KKT conditions introduced in Section 11.1 to investigate general conditions required for achieving an optimal risk portfolio.

11.3.1 Risk Retention Conditions

We start by looking at a risk retention problem in a fairly abstract way and then specialize to problems of interest to us. To be explicit, consider a slight modification of risk retention problem in Display (9.1) as follows. For simplicity, I dropped the decision variable z_0 used in expected shortfall and use RM for a risk measure instead of the generic f_0.

Risk Retention Problem

$$\boxed{\begin{array}{ll}
\text{minimize}_{\boldsymbol{\theta}} & RM(\boldsymbol{\theta}) \\
\text{subject to} & RTC(\boldsymbol{\theta}) \leq RTC_{max} \\
& \mathbf{P}\boldsymbol{\theta} \leq \mathbf{p}_0 \quad .
\end{array}} \tag{11.5}$$

To permit detailed analysis, define $\mathbf{P}_i$ to be the ith row of $\mathbf{P}$ and $p_{0,i}$ to be the ith element of $\mathbf{p}_0$. With this notation, the $(i+1)$st constraint is $\mathbf{P}_i\ \boldsymbol{\theta} \leq p_{0,i}$. Thus, we can write the Lagrangian is

$$\begin{aligned}
LA(\boldsymbol{\theta}, \mathbf{LMI}) = RM[g(\mathbf{X}; \boldsymbol{\theta})] &+ LMI_1[RTC(\boldsymbol{\theta}) - RTC_{max}] \\
&+ \sum\nolimits_{i=1}^{m-1} LMI_{i+1} \left(\mathbf{P}_i\ \boldsymbol{\theta} - p_{0,i}\right).
\end{aligned} \tag{11.6}$$

Using Display (11.1), the risk retention problem KKT conditions are

KKT **Conditions for the Risk Retention Problem**

$$\boxed{\begin{array}{ll}
\partial_{\theta_j}\ LA\left(\boldsymbol{\theta}, \mathbf{LMI}^*\right)|_{\boldsymbol{\theta}=\boldsymbol{\theta}^*} = 0 & j = 1, \ldots, p_z \\
LMI_1^*[RTC(\boldsymbol{\theta}^*) - RTC_{max}] = 0 & \\
LMI_{i+1}^*\left[\mathbf{P}_i\ \boldsymbol{\theta} - p_{0,i}\right] = 0 & i = 1, \ldots, m-1 \\
LMI_i^* \geq 0 & i = 1, \ldots, m \\
RTC(\boldsymbol{\theta}^*) \leq RTC_{max} & \\
\mathbf{P}_i\ \boldsymbol{\theta}^* \leq p_{0,i} & i = 1, \ldots, m-1.
\end{array}} \tag{11.7}$$

Our interest is in summarizing behavior when the optimal value of the risk retention parameters are *not* on one of the edges defined by the vector constraint $\mathbf{P}\boldsymbol{\theta} \leq \mathbf{p}_0$. So, for the

jth decision variable, let us consider a typical row, say, the ith one, where θ_j is present. Mathematically, we can express this as $\partial_{\theta_j}\mathbf{P}_i\boldsymbol{\theta} \neq 0$ for some $\boldsymbol{\theta}$. For such a row, to ensure that the jth decision variable is not on an edge, we require $\mathbf{P}_i\ \boldsymbol{\theta}^* < p_{0,i}$. To summarize, we require:

Edge Condition for the jth Decision Variable. For $i = 1, \ldots, m-1$, if $\partial_{\theta_j}\mathbf{P}_i\boldsymbol{\theta} \neq 0$ for some $\boldsymbol{\theta}$, then $\mathbf{P}_i\ \boldsymbol{\theta}^* < p_{0,i}$.

As before, it is helpful to use the risk measure relative marginal change,

$$RM_j^2(\boldsymbol{\theta}) = -\frac{\partial_{\theta_j} RM[g(\mathbf{X};\boldsymbol{\theta})]}{\partial_{\theta_j} RTC(\boldsymbol{\theta})}.$$

Problems of interest to us largely adhere to the following:

- **Condition *RR1*.** $\partial_{\theta_j}\ \ RTC\left(\boldsymbol{\theta}\right)|_{\boldsymbol{\theta}=\boldsymbol{\theta}^*} \neq 0$.
- **Condition *RR2*.** $\partial_{\theta_j}\ \ RM[g(\mathbf{X};\boldsymbol{\theta})]|_{\boldsymbol{\theta}=\boldsymbol{\theta}^*} \neq 0$.

When the edge condition for the jth decision variables holds, there are opportunities to consider small changes in the parameter. When this happens, we want these small changes to affect both the risk transfer cost (Condition *RR1*) and the risk measure (Condition *RR2*). It is worth noting that the requirement $0 < RM_j^2(\boldsymbol{\theta}^*) < \infty$ is sufficient for Conditions *RR1* and *RR2*. We typically use this simpler condition in applications.

Binding Budget Constraint

Assume that the edge condition for the jth decision variable and the corresponding Conditions *RR1* and *RR2* hold. Then, $RTC(\boldsymbol{\theta}^*) = RTC_{max}$.

ⓘ *Under the Hood.* See the *Online Supplement* for the confirmation of the budget constraint to be binding.

If the budget constraint is binding, this reduces the feasible region where one searches for the optimal parameter values. If we have good reason to suspect that an algorithm will converge where one of the parameters has a positive RM^2 (so that both Conditions *RR1* and *RR2* hold), then we might get faster convergence by assuming the risk transfer constraint is binding.

Moreover, from the proof, one can observe that the optimal value of the Lagrange multiplier is $RM_j^2(\boldsymbol{\theta}^*)$. It is interesting to note that this holds for *all* retention parameters where the edge condition on the decision variables hold. This leads to the following.

Balance Among Retention Parameters at the Optimum

Assume that the edge conditions for the ith and jth decision variables and the corresponding Conditions *RR1* and *RR2* hold. Then,

$$RM_i^2(\boldsymbol{\theta}^*) = RM_j^2(\boldsymbol{\theta}^*). \tag{11.8}$$

This result was previously hinted at in Section 2.3 where, in the simpler context of a single risk with only a few parameters, we described the natural interpretation of the risk measure relative marginal change. Recall that RM^2 can be interpreted as measuring the marginal (negative) change in the risk measure per unit marginal change in the risk transfer cost. In other words, if a change in the jth parameter causes a unit change in the risk transfer cost, the RM_j^2 quantifies the amount of change in the measure of uncertainty. It seems reasonable that, at the optimum, we would strive for a balance among the parameter settings. Without this balance, a change in one parameter would result in a greater (or lesser) change in the objective function, given the same change in the risk transfer cost.

Example 11.2. Multivariate Excess of Loss with Variance as a Risk Measure. For the variance as a risk measure, we have

$$\begin{aligned}\partial_{\theta_j} RM[g(\mathbf{X};\boldsymbol{\theta})] &= \partial_{\theta_j} \frac{1}{2}\mathrm{Var}[g(\mathbf{X};\boldsymbol{\theta})] \\ &= \frac{1}{2}\partial_{\theta_j} \mathrm{E}[g(\mathbf{X};\boldsymbol{\theta})]^2 - \mathrm{E}[g(\mathbf{X};\boldsymbol{\theta})]\{\partial_{\theta_j} \mathrm{E}[g(\mathbf{X};\boldsymbol{\theta})]\} \\ &= \mathrm{E}[g(\mathbf{X};\boldsymbol{\theta})\partial_{\theta_j} g(\mathbf{x};\boldsymbol{\theta})] - \mathrm{E}[g(\mathbf{X};\boldsymbol{\theta})]\{\mathrm{E}[\partial_{\theta_j} g(\mathbf{x};\boldsymbol{\theta})]\}.\end{aligned}$$

For excess of loss, we require that upper limit parameters u be nonnegative. So, the constraint $u_j \geq 0$ is equivalent to $\mathbf{P}_j\ \boldsymbol{\theta} \leq p_{0,j}$ by taking $\mathbf{u} = \boldsymbol{\theta}$, $\mathbf{P}_j = -\mathbf{1}_j'$, and $p_{0,j} = 0$.

For the multivariate excess of loss, we take $g(\mathbf{X};\mathbf{u}) = X_1 \wedge u_1 + \cdots + X_p \wedge u_p = S(\mathbf{u})$. With this, we have $\partial_{u_j} g(\mathbf{x};\mathbf{u}) = I(X_j > u_j)$. Thus,

$$\begin{aligned}RM_j^2(\mathbf{u}) &= -\frac{\partial_{u_j} RM[S(\mathbf{u})]}{\partial_{u_j} RTC(\mathbf{u})} \\ &= \mathrm{E}[S(\mathbf{u})|X_j > u_j] - \mathrm{E}[S(\mathbf{u})] \qquad (11.9) \\ &= u_j - \mathrm{E}[X_j \wedge u_j] + \sum\nolimits_{i \neq j}^{p} \{\mathrm{E}[X_i \wedge u_i | X_j > u_j] - \mathrm{E}[X_i \wedge u_i]\}.\end{aligned}$$

In the case of independence, this reduces to $RM_j^2 = u_j - \mathrm{E}[X_j \wedge u_j]$, the result for the classical problem that we developed in Section 4.1.2.

Under the Hood. See the *Online Supplement* for the verification of the RM^2.

At the optimum, if for some risk $u_j^* > 0$ and $0 < RM_j^2(\mathbf{u}^*) < \infty$, then the budget constraint is binding. In addition, if for another risk $u_i^* > 0$ and the risk measure relative marginal change is positive and finite, then we have a balance $RM_i^2(\mathbf{u}^*) = RM_j^2(\mathbf{u}^*)$.

Example 11.3. Bivariate Excess of Loss with VaR as a Risk Measure. For excess of loss with two risks, the retained losses are the limited sum $g(\mathbf{X};\boldsymbol{\theta}) = S(u_1,u_2) = X_1 \wedge u_1 + X_2 \wedge u_2$ where $\boldsymbol{\theta} = (u_1,u_2)'$. For simplicity, we assume a fair risk transfer cost.

Under the Hood. See the *Online Supplement* for the verification of the VaR balance.

Using equation (11.8), the balance among retention parameters can be expressed as

$$\begin{aligned}&I(u_1^* < z_0^*) f_2(z_0^* - u_1^*) \frac{1 - C_1[F_2(z_0^* - u_1^*), F_1(u_1^*)]}{1 - F_1(u_1^*)} \\ &= I(u_2^* < z_0^*) f_1(z_0^* - u_2^*) \frac{1 - C_1[F_1(z_0^* - u_2^*), F_2(u_2^*)]}{1 - F_2(u_2^*)}.\end{aligned}$$

at the optimal value of $z_0^* = F^{-1}_{S(u_1^*,u_2^*)}(\alpha)$. Here, C_1 is a copula derivative defined in Appendix Section 14.1. From this expression, we see that the balance depends on the copula and hence on the dependence between risks. Further, both optimal retention levels u_1^* and u_2^* must be less than the optimal value $F^{-1}_{S(u_1^*,u_2^*)}(\alpha)$. Yet we need $u_1^* + u_2^* > F^{-1}_{S(u_1^*,u_2^*)}(\alpha)$. In addition:

- **Special Case of Independence.** To get further insights, it is also of interest to consider the special case where the risks are independent. Here, $C_1(u,v) = v$ and we have

$$I(u_1^* < z_0^*) f_2(z_0^* - u_1^*) = I(u_2^* < z_0^*) f_1(z_0^* - u_2^*).$$

- **Special Case of Identical Distributions.** As another interesting case, if the distributions are equal, then we have

$$\begin{aligned} I(u_1^* < z_0^*) f(z_0^* - u_1^*) & \frac{1 - C_1[F(z_0^* - u_1^*), F(u_1^*)]}{1 - F(u_1^*)} \\ = I(u_2^* < z_0^*) f(z_0^* - u_2^*) & \frac{1 - C_1[F(z_0^* - u_2^*), F(u_2^*)]}{1 - F(u_2^*)}. \end{aligned}$$

This naturally holds true if $u_1^* = u_2^*$ *regardless of the dependence.*

Example 11.4. Separable Contracts and Risk Measures. Now consider the case where the portfolio risk transfer costs can be subdivided into separate contracts; see equation (7.10) of Section 7.5.1 where we wrote $g(\mathbf{X};\boldsymbol{\theta}) = \sum_{j=1}^{p} g_j(X_j;\boldsymbol{\theta}_j)$. From this, let us assume that one can write the risk transfer cost as $RTC(\boldsymbol{\theta}) = \sum_{j=1}^{p} RTC_j(\boldsymbol{\theta}_j)$. In the same way, we also assume that the risk measure can be subdivided additively as $RM(\boldsymbol{\theta}) = \sum_{j=1}^{p} RM_j(\boldsymbol{\theta}_j)$. With these assumptions, we may write $RM_j^2 = -\partial_{\theta_j} RM_j(\theta_j)/\partial_{\theta_j} RTC_j(\theta_j)$.

Suppose that the jth contract has an upper limit form (excess of loss). Then, from Table 11.1,

$$RM_j^2 = \begin{cases} \frac{1}{1-F_j(u_j)} I[F_j(u_j-) \le \alpha] & \text{for } VaR \\ \min\left(\frac{1}{1-\alpha}, \frac{1}{1-F_j(u_j)}\right) & \text{for } ES. \end{cases}$$

Under the mild conditions required for equation (11.8), this suggests that $F_j(u_j^*)$ is the same. That is, at the optimum, all positive upper limits have the same quantile.

The assumption of separable risk measures in Example 11.4 is unlikely to hold in general. However, it is true when using, for example, VaR and ES, as risk measures and when risks are comonotonic (see Section 9.1 or, for a broader introduction, Denuit et al. (2006), Chapter 2). Comonotonicity represents extreme positive dependence and so is unlikely to hold in practice. However, it may be useful to make this assumption to quickly generate upper limit values and then use these as **starting values** for general situations.

11.3.2 Risk Retention Boundary Conditions

To simplify the presentation, this subsection only considers edge conditions of the form $\theta_j \ge 0$. In this context, one might wonder under what conditions is one of the retention parameters equal to 0? To illustrate, suppose that it is known that $\theta_1^* > 0$ and we would like conditions that lead to $\theta_2^* = 0$. To this end, from equation (11.6)

$$\begin{aligned} \partial_{\theta_2} LA(\boldsymbol{\theta}, \mathbf{LMI}) &= \partial_{\theta_2} RM(\boldsymbol{\theta}) + LMI_1 \, \partial_{\theta_2} RTC(\boldsymbol{\theta}) - LMI_3 \\ &= \partial_{\theta_2} RTC(\boldsymbol{\theta}) \left(\frac{\partial_{\theta_2} RM(\boldsymbol{\theta})}{\partial_{\theta_2} RTC(\boldsymbol{\theta})} + RM_1^2(\boldsymbol{\theta}^*) \right) - LMI_3 \\ &= \partial_{\theta_2} RTC(\boldsymbol{\theta}) \left(RM_1^2 - RM_2^2 \right) - LMI_3. \end{aligned}$$

So, at the optimum, if $\partial_{\theta_2} RTC(\boldsymbol{\theta}) \left(RM_1^2 - RM_2^2 \right)$ is positive, by the first KKT condition, this means that LMI_3^* is positive. By the third condition, this means that $\theta_2^* = 0$. We summarize this as follows.

Risk Retention Boundary Conditions
Assume $\theta_i^* > 0$ for the ith decision variable and the corresponding Conditions *RR1* and *RR2* hold. In addition, suppose that either

- $\partial_{\theta_j} RTC(\boldsymbol{\theta}) > 0$ and $RM_i^2 > RM_j^2$ or
- $\partial_{\theta_j} RTC(\boldsymbol{\theta}) < 0$ and $RM_i^2 < RM_j^2$,

at $\boldsymbol{\theta} = \boldsymbol{\theta}^*$. Then $\theta_j^* = 0$.

In summary, from equation (11.8), we have a collection of RM^2 ratios that are the same. If a ratio is not equal to the collective number, then there are mild conditions so that the corresponding retention parameter is zero.

In many of our problems, we focus on cases where an increase in a parameter θ means that more risk is retained, such as with upper limits and coinsurances. More risk retained means that as θ increases, we expect risk transfer costs to decrease and our (risk) measure of retained risk to increase. However, for other problems, such as deductibles, an increase in a parameter means that less risk is retained.

Example 11.5. Multivariate Deductible with Variance as a Risk Measure. As in Example 11.2, the partial derivative of the risk measure is

$$\partial_{\theta_j} RM[g(\mathbf{X};\boldsymbol{\theta})] = \mathrm{E}[g(\mathbf{X};\boldsymbol{\theta})\partial_{\theta_j} g_j(\mathbf{X};\boldsymbol{\theta})] - \mathrm{E}[g(\mathbf{X};\boldsymbol{\theta})]\{\mathrm{E}[\partial_{\theta_j} g_j(\mathbf{X};\boldsymbol{\theta})]\}.$$

For the multivariate deductible, we take $d_j = \theta_j$ and

$$\begin{aligned} g(\mathbf{X};\mathbf{d}) &= (X_1 - d_1)_+ + \cdots + (X_p - d_p)_+ \\ &= [X_1 - X_1 \wedge d_1] + \cdots + [X_p - X_p \wedge d_p] = S(\infty) - S(\mathbf{d}). \end{aligned}$$

With this, we have $\partial_{d_j} g(\mathbf{x};\mathbf{d}) = -I(X_j > d_j)$. Further calculations show

$$\begin{aligned} RM_j^2(\mathbf{d}) &= -\frac{\partial_{d_j} RM[S(\mathbf{d})]}{\partial_{d_j} RTC(\mathbf{d})} \\ &= d_j - \mathrm{E}[X_i \wedge d_j] + \sum\nolimits_{i \neq j}^{p} \{\mathrm{E}[X_i \wedge d_i | X_j > d_j] - \mathrm{E}[X_i \wedge d_i]\}. \end{aligned} \tag{11.10}$$

This is the same RM^2 measure as in Example 11.2. The only difference between the two problems is the sign of the partial derivative of the risk transfer cost, $\partial_{\theta_j} RTC(\boldsymbol{\theta})$. So, when we take a look at the *edge condition* for the deductible problem, a small value of $RM_j^2(\mathbf{d}^*)$ at the optimum means that $d_j = 0$, signifying full retention.

ⓘ *Under the Hood.* See the *Online Supplement* for the verification of the RM^2.

Suppose that all of the parameters adhere to the assumption that retained risk increases with a parameter increase. Additionally, if all of the parameters equal 0, that is, if $\theta_1 = \cdots = \theta_p = 0$, then there is no risk retention, corresponding to full transfer (which could be full insurance). This is illustrated in Examples 11.6 - 11.9 of the subsequent Section 11.4.1. In this scenario, we can set the maximal risk transfer cost below the case of no risk retention, $RTC_{max} < RTC(\mathbf{0})$, ensuring that we have some positive retention parameters. Under this assumption, at least one $\theta_j^* > 0$, satisfying one of the basic ingredients for risk retention conditions to hold.

Re-parameterization

The result of a parameter at a boundary has a natural interpretation, such as full transfer for upper limit and coinsurance parameters, and full retention for deductible parameters. In some cases, analysts can re-parameterize (or redefine) parameters to achieve alternative desirable interpretations.

For instance, if the jth risk has upper limit $u_j < \infty$, then we may re-parameterize the deductible as $d_{1j} = (u_j - d_j)_+$, representing the amount that the deductible d_j falls below the upper limit. The new parameter d_{1j} has zero risk retention when $d_{1j} = 0$ and aligns with our objective of taking on more risk as the parameter increases. In such cases, we can interpret the boundary condition result $\theta_j^* = 0$ to indicate no retention, or full transfer, of the jth risk.

The re-parameterization tactic can be applied to identify conditions for full risk retention in other scenarios. For example, in a multivariate excess of loss policy, if an upper limit u_j is associated with a risk, then $u_j = 0$ implies no risk retention, whereas $u_j = \infty$ signifies full retention. Therefore, we might define a new parameter, the reciprocal of the upper limit $u_j^R = 1/u_j$. We could then identify conditions so that the boundary conditions hold so that $u_j^R = 0$, signifying full risk retention.

11.4 Risk Measure Relative Marginal Changes

As seen in Section 11.3, the risk measure relative marginal change, RM_j^2, plays a key role in the investigation of risk transfer conditions. Because practical applications involves simulation of multivariate risks, this section develops RM_j^2 metrics in this context.

Separate Contract

Because the focus of RM_j^2 is on marginal changes, we can extend slightly the idea of separable contracts introduced in Section 7.5.1 to specify that *only* the jth contract is separate. Mathematically, define $\boldsymbol{\theta}_{(j)} = (\theta_1, \ldots, \theta_{j-1}, \theta_{j+1}, \ldots, \theta_p)'$ to be the vector of parameters excluding the jth one and similarly for $\mathbf{X}_{(j)}$. Then, we could write the retention function for the jth risk as

$$\tilde{g}_j(X_j; \theta_j) = g(\mathbf{X}; \boldsymbol{\theta}) - g_{(j)}(\mathbf{X}_{(j)}; \boldsymbol{\theta}_{(j)}),$$

where $g_{(j)}(\cdot)$ is the retention function for all risks except the jth one. Unlike in Section 7.5.1, we do not require that $g_{(j)}(\cdot)$ be additive.

If the jth risk is separate, then determination of partial derivatives of risk retention become easier as they rely on only one parameter. Table 11.2 summarizes this calculation for different types of retentions.

11.4.1 Variance Risk Retention

Using the variance as a risk measure, we are able to utilize the usual empirical simulation estimators of the distribution because the variance is differentiable in the risk retention parameters. Readers will not be surprised that we can get some intuitively pleasing results using the classic assumption of the variance as a risk measure. To recap, here is a summary of the simulation version of this risk retention problem.

TABLE 11.2
Separate Risk Retention Functions and Partial Derivatives.

Retention Type	Retention Function $g(\mathbf{X};\boldsymbol{\theta})$	Partial Derivative $\partial_{\boldsymbol{\theta}}\, g(\mathbf{X};\boldsymbol{\theta})$
Deductible	$g(x;d) = (x-d)_+$	$\partial_d\, g(x;d) = -I(x>d)$
Coinsurance	$g(x;c) = c\, x$	$\partial_c\, g(x;c) = x$
Upper Limit	$g(x;u) = x \wedge u$	$\partial_u\, g(x;u) = I(x>u)$
Upper Limit with Reciprocal Parameter	$g(x;u_R) = x \wedge \frac{1}{u_R}$	$\partial_{u_R}\, g(x;u_R) = \frac{-1}{u_R^2} I(x > \frac{1}{u_R})$

$$
\boxed{\begin{array}{ll} \text{minimize}_{\boldsymbol{\theta}} & RM_{var}(\boldsymbol{\theta}) = \frac{1}{2R}\sum_{r=1}^{R}\left\{g(\mathbf{X}_r;\boldsymbol{\theta})^2 - \overline{g(\mathbf{X}_R;\boldsymbol{\theta})}\right\}^2 \\ \text{subject to} & RTC_R(\boldsymbol{\theta}) \le RTC_{max}. \end{array}} \tag{11.11}
$$

For notation, let $g_{rj} = \partial_{\theta_j} g(\mathbf{X}_r;\boldsymbol{\theta})$ and $\bar{g}_j = \frac{1}{R}\sum_{r=1}^{R} g_{rj}$. Also, let the average risk retained be denoted as $\overline{g(\mathbf{X}_R;\boldsymbol{\theta})}$. With this, we can express the risk measure relative marginal statistic as

$$
RM^2_{var,j}(\boldsymbol{\theta}) = -\frac{\partial_{\theta_j}\, RM_{var}(\boldsymbol{\theta})}{\partial_{\theta_j} RTC_R(\boldsymbol{\theta})} = \overline{g(\mathbf{X}_R;\boldsymbol{\theta})} - \frac{\mathrm{E}_R[g(\mathbf{X}_R;\boldsymbol{\theta})g_{Rj}]}{\bar{g}_j}. \tag{11.12}
$$

Here, the second term on the right-hand side of equation (11.12) is a weighted average of retained risks where the weights are given by the partial derivatives g_{rj}. It is worth noting that if we are comparing RM_i^2 to RM_j^2 to determine the balance in the system, only the second term is important as the first term, the average retained risk, to the same for the ith and jth measures.

ⓘ *Under the Hood.* See the *Online Supplement* for the verification of the RM^2.

Interpretations of this result are best seen in the context of some special cases.

Example 11.6. Coinsurance. For the jth risk, take $c_j = \theta_j$. Thus, from Table 11.2 we have $g_{rj} = X_{rj}$. With equation (11.12), in this case we can express the risk measure relative marginal change as

$$
RM^2_{var,j} = \overline{g(\mathbf{X}_R;\boldsymbol{\theta})} - \frac{\mathrm{E}_R[g(\mathbf{X}_R;\boldsymbol{\theta})X_{rj}]}{\overline{X}_{Rj}}.
$$

Thus, the jth risk measure relative marginal change is the average retained risk minus a weighted average retained risk where the weights are given by the size of the jth risk.

Example 11.7. Excess of Loss. Now let $u_j = \theta_j$. Using Table 11.2, we have $g_{rj} = I(X_{rj} > u_j)$. Then, equation (11.12) becomes

$$
\begin{aligned}
RM^2_{var,j} &= \overline{g(\mathbf{X}_R;\boldsymbol{\theta})} - \frac{\mathrm{E}_R[g(\mathbf{X}_R;\boldsymbol{\theta})g_{Rj}]}{\bar{g}_j} \\
&= \overline{g(\mathbf{X}_R;\boldsymbol{\theta})} - \frac{\sum_{r=1}^{R} g(\mathbf{X}_r;\boldsymbol{\theta})I(X_{rj} > u_j)]}{\sum_{r=1}^{R} I(X_{rj} > u_j)}.
\end{aligned}
$$

The jth risk measure relative marginal change can be expressed as the overall average minus average portfolio risk of those losses that exceeds the upper limit.

11.4.2 Quantile-Based Risk Retention

Using the KKT conditions yields desirable properties that, when applied to risk retention problems, can be described in terms of the risk measure relative marginal change, RM^2. However, both the KKT conditions and RM^2 metric are based on marginal derivatives that rely on smoothness in terms of risk retention parameters. However, basic simulation techniques assign a weight of $1/R$ to each simulated outcome, and when used with typical check or indicator functions, the result is that simulated approximations are no longer smooth.

Unlike the variance, the usual empirical simulation approximations of quantile-based risk measures such as the VaR and the ES are not differentiable in the risk retention parameters. Therefore, technically, we need to use smooth estimators of the distribution introduced in Section 7.3, with properties developed in Section 10.5.3. Recall the density, distribution function, and general expectations from those sections as given in Table 11.3.

From these expressions, one can determine the quantile, or value at risk, in the usual way and is denoted as VaR_{Rk}. In addition, the above display provides a partial derivative of the distribution function with respect to a retention parameter which is needed in the following.

For the VaR, recall that in Section 5.3.1 we developed an expression for the quantile sensitivity but that development assumed smoothness in the argument of the distribution function as well as the risk retention parameters. This result in equation (5.5), with a smoothed empirical estimator of the distribution, can be expressed as

$$\partial_{\theta_j} VaR_{Rk}(\boldsymbol{\theta}) = \frac{-1}{f_{Rk}[VaR_{Rk}]} F_{Rk,\theta_j}[VaR_{Rk}].$$

We also need partial derivatives with respect to the risk transfer cost. From the general expectation for retained risks, one can see that $\partial_{\theta_j}\ RTC_R(\boldsymbol{\theta}) = -\bar{g}_j$.

ⓘ *Under the Hood.* See the *Online Supplement* to confirm the partial derivative of RTC.

Thus, risk measure relative marginal change for the value at risk is

$$\begin{aligned} RM^2_{VaR,j}(\boldsymbol{\theta}) &= \frac{\partial_{\theta_j}\ VaR_R(\boldsymbol{\theta})}{\partial_{\theta_j} RTC_R(\boldsymbol{\theta})} \\ &= \frac{1}{\bar{g}_j\ f_{Rk}[VaR_{Rk}]} F_{Rk,\theta_j}[VaR_{Rk}]. \end{aligned}$$

TABLE 11.3
Kernel Smoothed Expressions.

Term	Symbol	Expression
Density	$f_{Rk}(y)$	$= \frac{1}{R\,b} \sum_{r=1}^R k\left(\frac{y-g(\mathbf{X}_r;\boldsymbol{\theta})}{b}\right)$
Distribution function	$F_{Rk}(y)$	$= \frac{1}{R} \sum_{r=1}^R K\left(\frac{y-g(\mathbf{X}_r;\boldsymbol{\theta})}{b}\right)$ $= \mathrm{E}_R\left\{K\left(\frac{y-g(\mathbf{X}_R;\boldsymbol{\theta})}{b}\right)\right\}$
General expectation of retained risk	$\mathrm{E}_{Rk}\{h[g(\mathbf{X}_R;\boldsymbol{\theta})]\}$	$= \int \mathrm{E}_R\left\{h[g(\mathbf{X};\boldsymbol{\theta}) + bz]\right\}\ k(z)dz$
Partial derivative of the distribution function *wrt* a retention parameter	$F_{Rk,\theta_j}(y)$	$= \frac{1}{R} \sum_{r=1}^R \partial_{\theta_j} K\left(\frac{y-g(\mathbf{X}_r;\boldsymbol{\theta})}{b}\right)$ $= \frac{-1}{b} \mathrm{E}_R\left\{k\left(\frac{y-g(\mathbf{X}_R;\boldsymbol{\theta})}{b}\right) g_{Rj}\right\}$

For the expected shortfall, we use the ES sensitivity in equation (5.7). With this, we have

$$\partial_{\theta_j} ES_{Rk}[g(\mathbf{X};\theta)] = \frac{1}{1-\alpha}\left\{\partial_{\theta_j}\mathrm{E}_{Rk}[g(X;\boldsymbol{\theta})] - \partial_{\theta_j}\mathrm{E}_{Rk}[g(X;\boldsymbol{\theta})\wedge z_0]\big|_{z_0=VaR_{Rk}}\right\}$$
$$+[1-\frac{1}{1-\alpha}\{1-F_{Rk}(VaR_{Rk})\}]\times\partial_{\theta_j}VaR_R(\boldsymbol{\theta}).$$

Because the distribution function is smooth, in many cases there is no discreteness at the quantile, so $F_{Rk}(VaR_{Rk}) = \alpha$, and the second term on the right-hand side is zero. Nonetheless, sometimes discreteness is induced by the risk retention parameters, as seen in Example 2.6. With no discreteness, the risk measure relative marginal change for the expected shortfall can be written as

$$RM^2_{ES,j}(\boldsymbol{\theta}) = \frac{1}{(1-\alpha)}\frac{\mathrm{E}_R\left\{\tilde{k}_{R\theta}\ g_{Rj}\right\}}{\bar{g}_j}, \qquad (11.12\text{ES})$$

where $\tilde{k}_{R\theta} = 1 - K\left([VaR_{Rk} - g(\mathbf{X}_R;\boldsymbol{\theta})]/b\right)$. For the rth simulated value, $\tilde{k}_{r\theta}$ represents the probability that the reference (such as the normal) distribution exceeds the argument $[VaR_{Rk} - g(\mathbf{X}_r;\boldsymbol{\theta})]/b$. Thus, large values of $g(\mathbf{X}_r;\boldsymbol{\theta})$ mean that $\tilde{k}_{r\theta}$ is large, so it can be interpreted as a measure of the size of the retained risk.

ⓘ *Under the Hood.* See the *Online Supplement* the Expected Shortfall RM^2.

Example 11.8. Coinsurance. As in Example 11.6, take $c_j = \theta_j$ and $g_{rj} = X_{rj}$ for the jth risk. In this case, equation (11.12ES) becomes

$$RM^2_{ES,j}(\boldsymbol{\theta}) = \frac{1}{(1-\alpha)}\frac{1}{\bar{g}_j}\mathrm{E}_R\left\{\tilde{k}_{R\theta}\ g_{Rj}\right\},$$
$$= \frac{1}{(1-\alpha)}\frac{\sum_{r=1}^R \tilde{k}_{r\theta}\ X_{rj}}{\sum_{r=1}^R X_{rj}}.$$

This is proportional to the weighted average of the measure of size of retained risk, where the weight is given by the jth loss.

Example 11.9. Upper Limit. As in Example 11.7, take $u_j = \theta_j$ and $g_{rj} = I(X_{rj} > u_j)$ for the jth risk. Now, equation (11.12ES) becomes

$$RM^2_{ES,j}(\boldsymbol{\theta}) = \frac{1}{(1-\alpha)}\frac{\sum_{r=1}^R \tilde{k}_{r\theta}\ I(X_{rj} > u_j)}{\sum_{r=1}^R I(X_{rj} > u_j)}.$$

This is proportional to the average of the measure of size of retained risk, where the average is taken over large values of the jth risk, those exceeding the upper limit u_j.

Example 11.10. Varying the Cyber Risk Premium. This example continues from Exercise 8.7, which analyzed the ANU Excess of Loss problem with a market loading *Cyber Load* given in Table 11.4. In Exercise 8.7, readers were tasked with minimizing

TABLE 11.4
ANU Cyber Risk Retention.

Cyber Load	VaR	u_3: Cyber	RM_3^2	$se(RM_3^2)$	u_2	u_4
0.5	6093	0	0.395	0.005	175	288
1.0	6226	432	0.327	0.005	186	252
1.5	6357	1506	0.295	0.008	198	144
2.0	6361	5633	0.338	0.055	204	145
2.5	6364	5848	0.386	0.014	204	147
3.0	6364	5848	0.321	0.011	204	147

the expected shortfall and determining optimal upper limits from which risk measures were computed.

Table 11.4 displays selected optimal retention levels. As in Exercise 8.7, the results are pleasing. A lower value of *Cyber Load* (at half the fair cost) suggests full insurance, indicated by the optimal upper limit $u_3^* = 0$. Conversely, a high value of *Cyber Load* (at double the fair cost) extends the optimal upper limit u_3^* well beyond the 99th risk percentile, suggesting full retention. In this example, we supplement this analysis by computing the ES risk measure relative marginal change (RM^2).

To assess the simulation accuracy of the RM^2, note that it can be expressed as the ratio of two averages. The theory underpinning the standard errors are sketched out in Exercise 11.3. For this example, the standard errors that appear in Table 11.4 (with $R = 100000$) enable the assessment of accuracy.

In addition, Table 11.5 shows risk measure relative marginal changes for other risks RM_j^2. Corresponding standard errors were also computed although not displayed, they showed approximately the same level as for the cyber risk displayed in Table 11.4. One can see a reasonable balance among RM^2 metrics for most risks (although not the 13th and 15th) for the *Cyber Load* 1.0 and 1.5. For other values of the *Cyber Load* as well as risks 13 and 15 (corresponding to Motor Vehicle and Marine Hull), there is a lack of balance suggesting that the corresponding upper limits may be in doubt.

11.5 Supplemental Materials

11.5.1 Further Resources and Reading

See, for example, Boyd and Vandenberghe (2004) and Simon and Blume (1994), for fascinating introductions to constrained optimization and more details about the Lagrange multiplier method.

11.5.2 Exercises

See the *Online Supplement* for selected exercise solutions.

Exercise 11.1. de Finetti Optimal Retention Proportions. Consider the quota share agreement described in Section 4.3.2. Here, the insurer's portion of the portfolio risk is $Y_{insurer} = \sum_{i=1}^{n} c_i X_i = \mathbf{c}'\mathbf{X}$. We seek to find those values of c_i that minimize $\mathrm{Var}(Y_{insurer})$

TABLE 11.5
ANU Cyber Risk Measure Relative Marginal Change.

Cyber Load	RM_1^2	RM_2^2	RM_3^2: Cyber	RM_4^2	RM_5^2	RM_6^2	RM_9^2	RM_{10}^2	RM_{13}^2	RM_{15}^2
0.5	0.233	0.303	0.395	0.325	0.317	0.310	0.328	0.329	0.267	0.336
1.0	0.234	0.303	0.327	0.319	0.317	0.310	0.324	0.330	0.245	0.386
1.5	0.235	0.290	0.295	0.309	0.306	0.304	0.309	0.319	0.290	0.353
2.0	0.233	0.291	0.338	0.308	0.307	0.304	0.308	0.300	0.290	0.361
2.5	0.233	0.291	0.386	0.309	0.307	0.303	0.308	0.300	0.254	0.353
3.0	0.233	0.291	0.321	0.308	0.307	0.303	0.309	0.299	0.258	0.353

subject to the constraint that $\mathrm{E}(Y_{reinsurer}) = RTC_{max}$. Subject to this budget constraint, the insurer wishes to minimize the uncertainty of the retained risks as measured by the variance. We now further impose constraints that the sharing coefficients are bounded between 0 and 1, as considered by De Finetti (1940) (although he assumed independence among risks).

See also Glineur and Walhin (2006) and Pressacco et al. (2011) for additional background.

$$\boxed{\begin{array}{ll} \text{minimize}_{c_1,\ldots,c_n} & \frac{1}{2}\mathrm{Var}(Y_{insurer}) = \frac{1}{2}\mathbf{c}'\mathbf{\Sigma}\mathbf{c} \\ \text{subject to} & \sum_i (1-c_i)\mathrm{E}(X_i) \le RTC_{max} \\ & 0 \le c_i \le 1, \qquad i = 1,\ldots,n, \end{array}}$$

a. Put the problem into standard form and develop the Lagrangian. Henceforth, assume zero correlations among risks and write the variance of the ith risk as σ_i^2. Verify that the KKT conditions can be expressed as:

$$\boxed{\begin{array}{cl} \sigma_i^2 c_i^* - LMI_1^*\,[\mathrm{E}\,X_i] + LMI_{2i}^* - LMI_{3i}^* = 0 & i = 1,\ldots,n \\ LMI_{2i}^* \ge 0, LMI_{3i}^* \ge 0 & i = 1,\ldots,n \\ LMI_{2i}^*(c_i^* - 1) = 0, LMI_{3i}^* c_i^* = 0 & i = 1,\ldots,n \\ 0 \le c_i^* \le 1 & i = 1,\ldots,n \\ LMI_1^* \ge 0, \sum_i (1-c_i^*)\mathrm{E}(X_i) \le RTC_{max} & \\ LMI_1^*\,(\sum_i (1-c_i^*)\mathrm{E}(X_i) - RTC_{max}) = 0 & \end{array}} \tag{11.13}$$

b. To get some practice, let us make some assumptions about the optimal coefficients and use the KKT conditions to verify how we expect the Lagrange multipliers to behave. Verify:

Suppose	KKT Results	
$0 < c_i^* < 1$	$LMI_{2i}^* = LMI_{3i}^* = 0$	$\sigma_i^2 c_i^* - LMI_1^*\,[\mathrm{E}\,X_i] = 0$
$c_i^* = 0$	$LMI_{2i}^* = LMI_{3i}^* = 0$	$LMI_1^* = 0$
$c_i^* = 1$	$LMI_{3i}^* = 0$	$LMI_{2i}^* = LMI_1^*\,[\mathrm{E}\,X_i] - \sigma_i^2$.

(11.14)

c. For another approach, let us make some assumptions about the Lagrange multipliers and use the KKT conditions to verify how we expect the optimal coefficients to behave. Verify:

Suppose		KKT Results	
$LMI_{3i}^* > 0$		$c_i^* = 0$	$LMI_{2i}^* = 0$
$LMI_{2i}^* > 0$		$c_i^* = 1$	$LMI_{3i}^* = 0$
$LMI_{2i}^* = LMI_{3i}^* = 0$		$\sigma_i^2 c_i^* - LMI_1^*\,[\mathrm{E}\,X_i] = 0$	
$LMI_1^* = 0$		$LMI_{2i}^* = LMI_{3i}^* = 0$	$c_i^* = 0$
$LMI_1^* > 0$	$LMI_{3i}^* > 0$	violates conditions	
$LMI_1^* > 0$	$LMI_{2i}^* > 0$	$c_i^* = 1$	$\sigma_i^2 - LMI_1^*\,[\mathrm{E}\,X_i] + LMI_{2i}^* = 0$
$LMI_1^* > 0$	$LMI_{2i}^* = LMI_{3i}^* = 0$	$\sigma_i^2 c_i^* - LMI_1^*\,[\mathrm{E}\,X_i] = 0$	

(11.15)

d. Check the following. To determine the optimal coefficients, we first find the value of Lagrange multiplier LMI_1^* as the solution of this equation:

$$\sum_i \mathrm{E}(X_i) - \sum_i \min\left\{1, \frac{LMI_1^*\,[\mathrm{E}\,X_i]}{\sigma_i^2}\right\}\mathrm{E}(X_i) = RTC_{max} \tag{11.16}$$

and then determine the optimal coefficients as

$$c_i^* = \min\left\{1, \frac{LMI_1^* \left[\mathrm{E}\ X_i\right]}{\sigma_i^2}\right\}. \tag{11.17}$$

Exercise 11.2. Lasso Regression Conditions. To get additional practice with the *KKT* conditions, we now apply them to the lasso regression problem introduced in Exercise 3.1. Similar to Exercise 3.1, we consider the problem

$$\boxed{\begin{array}{ll} \text{minimize}_{\boldsymbol{\beta}} & f_{0,ls}(\boldsymbol{\beta}) = \frac{1}{2}\sum_{i=1}^{n}(y_i - \mathbf{x}_i'\boldsymbol{\beta})^2 \\ \text{subject to} & \sum_{j=1}^{p}|\beta_j| \leq c_{lasso}, \end{array}}$$

with observations $(\mathbf{x}_i, y_i)$, for $i = 1, \ldots, n$. If c_{lasso} is large enough, then the constraint has no effect and the resulting estimator is $\hat{\boldsymbol{\beta}}^{ols} = [\mathbf{X}'\mathbf{X}]^{-1}\mathbf{X}'\mathbf{y}$. where $\mathbf{y}$ is a vector of responses and $\mathbf{X} = (\mathbf{x}_1', \ldots, \mathbf{x}_n')'$ is a matrix of explanatory variables. So, we choose c_{lasso} to be smaller than $\sum_{j=1}^{p}|\hat{\beta}_j^{ols}|$ so the constraint has some effect.

As is common for problems involving absolute values, we can linearize the constraints by taking positive and negative parts. Specifically, define the positive part $\beta_j^+ = \max(0, \beta_j)$ and the negative part $\beta_j^- = \max(0, -\beta_j)$. Thus, the problem becomes

$$\boxed{\begin{array}{ll} \text{minimize}_{\boldsymbol{\beta}^+,\boldsymbol{\beta}^-} & f_{0,ls}(\boldsymbol{\beta}^+, \boldsymbol{\beta}^-) = \frac{1}{2}\sum_{i=1}^{n}[y_i - \mathbf{x}_i'(\boldsymbol{\beta}^+ - \boldsymbol{\beta}^-)]^2 \\ \text{subject to} & \sum_{j=1}^{p}(\beta_j^+ + \beta_j^-) \leq c_{lasso}. \end{array}}$$

a. Confirm that the Lagrangian can be expressed as

$$\begin{aligned} LA(\boldsymbol{\beta}^+, \boldsymbol{\beta}^-) &= f_{0,ls}(\boldsymbol{\beta}^+, \boldsymbol{\beta}^-) + LMI_1\left[\sum\nolimits_{j=1}^{p}(\beta_j^+ + \beta_j^-) - c_{lasso}\right] \\ &\quad - \sum\nolimits_{j=1}^{p} LMI_{1+j}^+\ \beta_j^+ - \sum\nolimits_{j=1}^{p} LMI_{1+j}^-\ \beta_j^-. \end{aligned}$$

b. Show that

$$\partial_{\boldsymbol{\beta}^+}\ f_{0,ls}(\boldsymbol{\beta}^+, \boldsymbol{\beta}^-) = -\mathbf{X}'\mathbf{e} = -\partial_{\boldsymbol{\beta}^-}\ f_{0,ls}(\boldsymbol{\beta}^+, \boldsymbol{\beta}^-),$$

where $\mathbf{e} = \mathbf{y} - \mathbf{X}\boldsymbol{\beta}$ is a vector of residuals.

c. At the optimum, show that $\beta_j^+ > 0$ implies $\beta_j^- = 0$ and vice-versa, that $\beta_j^- > 0$ implies $\beta_j^+ = 0$. Thus, there is no contradiction, as one would hope.

d. Let $\mathbf{x}^{(j)}$ be the jth column of $\mathbf{X}$. At the optimum, show that $\mathbf{x}^{(j)'}\mathbf{e} = 0$ implies $\beta_j = 0$.

Exercise 11.3. Linearity of Lasso Regression Conditions. This exercise is motivated by Exercise 3.27 of Hastie et al. (2009). Similar to Exercises 3.1 and 11.2, we consider an equivalent way of writing the problem is using the penalized version

$$\text{minimize}_{\boldsymbol{\beta}^+,\boldsymbol{\beta}^-} \quad f_{0,ls}(\boldsymbol{\beta}^+, \boldsymbol{\beta}^-) + \lambda\sum\nolimits_{j=1}^{p}(\beta_j^+ + \beta_j^-).$$

The Lagrangian has essentially the same form as in Exercise 11.2

$$LA(\boldsymbol{\beta}^+, \boldsymbol{\beta}^-) = f_{0,ls}(\boldsymbol{\beta}^+, \boldsymbol{\beta}^-) + \lambda \sum\nolimits_{j=1}^{p} (\beta_j^+ + \beta_j^-)$$
$$- \sum\nolimits_{j=1}^{p} LMI_{1+j}^+ \, \beta_j^+ - \sum\nolimits_{j=1}^{p} LMI_{1+j}^- \, \beta_j^-,$$

and so the solution set is the same.

a. At the optimum, show that $\beta_j \neq 0$ implies $\lambda = -sgn(\beta_j)\, \mathbf{x}^{(j)\prime} \, [\mathbf{y} - \mathbf{X}\boldsymbol{\beta}]$.

b. At the optimum, show that the Hessian is

$$\nabla^2 LA \;= \begin{pmatrix} \mathbf{X}'\mathbf{X} & -\mathbf{X}'\mathbf{X} \\ -\mathbf{X}'\mathbf{X} & \mathbf{X}'\mathbf{X} \end{pmatrix}.$$

c. Suppose that we perturb the penalty parameter λ slightly and solve the problem for $\lambda_\delta = \lambda + \delta$. Call the perturbed solution $\boldsymbol{\beta}_\delta$. Then, if $sgn(\beta_j) > 0$ show that

$$\delta \;= \mathbf{x}^{(j)\prime}\mathbf{X}\,[\boldsymbol{\beta}_\delta - \boldsymbol{\beta}]$$

d. At the optimum, use the Perturbation Sensitivity Proposition from Section 10.1 to show that

$$-\mathbf{1}_A = [\nabla^2 LA]_A \; \partial_\lambda(\boldsymbol{\beta}^+, \boldsymbol{\beta}^-)_A,$$

where the subscript A means that we restrict the rows and columns to be those in the active set (where the corresponding parameters are not zero). Thus, the vector of derivatives $\partial_\lambda(\boldsymbol{\beta}^+, \boldsymbol{\beta}^-)_A$ is constant in the penalty parameter λ, meaning that the solution is locally linear. (See Rosset and Zhu (2007) for extensions of this idea to more general settings.)

Exercise 11.4. Standard Errors of Ratios. To determine the accuracy of the RM^2 in Example 11.10, note that it can be expressed as the ratio of two averages. Especially with simulation applications, it can be useful to understand the reliability of this type of statistic.

To this end, let $\{(x_{11}, x_{21}), \ldots, (x_{1R}, x_{2R})$ be a random sample of size R. Summarize this sample with means $\bar{x}_1$ and $\bar{x}_2$, sample variances s_1^2 and s_2^2, coefficient of variations $CV_1 = s_1/\bar{x}_1$ and $CV_2 = s_2/\bar{x}_2$ and correlation coefficient $r_{12} = [(R-1)s_1\, s_2]^{-1}\left[\sum_{r=1}^{R}(x_{1r} - \bar{x}_1)(x_{2r} - \bar{x}_2)\right]$. It is known from the mathematical statistic literature (see, for example, Levy and Lemeshow (2013)), that a desirable approximate standard error for the ratio of the means is

$$se\left(\frac{\bar{x}_1}{\bar{x}_2}\right) = \frac{(\bar{x}_1/\bar{x}_2)}{\sqrt{R}}\sqrt{CV_1^2 + CV_2^2 - 2r_{12}CV_1CV_2}.$$

a. For the risk measure relative marginal change in equation (11.12ES), identify the two variables x_1 and x_2 that you would use to compute the standard error.

b. Suppose that both x_1 and x_2 are binary variables (as in the upper limit special case). Show that $s_j^2 = \bar{x}_j(1-\bar{x}_j)/R$, for $j = 1, 2$, and the correlation coefficient reduces to

$$r_{12} \approx \frac{\bar{x}_{12} - \bar{x}_1\bar{x}_2}{\sqrt{\bar{x}_1(1-\bar{x}_1)\;\bar{x}_2(1-\bar{x}_2)}},$$

where $\bar{x}_{12} = (\sum_{r=1}^{R} x_{1r}x_{2r})/R$.

c. Suppose further that $x_{1r} = 0 \implies x_{2r} = 0$. Then show that the correlation coefficient can be expressed in an *odds ratio* form

$$r_{12} = \sqrt{\frac{(1-\bar{x}_1)/\bar{x}_1}{(1-\bar{x}_2)/\bar{x}_2}}.$$

Exercise 11.5. Quantile Regression Approach. This is a follow-up to Exercise 7.1 where we explored the relationship between *quantile regression* and the auxiliary version of expected shorted fall. In this exercise, we defined the objective functions

$$\begin{aligned} f_{QR,0}(z_0,\boldsymbol{\theta}) &= \mathrm{E}_R\ \phi_\alpha[g(\mathbf{X};\boldsymbol{\theta}) - z_0] \\ f_{ES,0}(z_0,\boldsymbol{\theta}) &= z_0 + \frac{1}{(1-\alpha)}\mathrm{E}_R[g(\mathbf{X};\boldsymbol{\theta}) - z_0]_+ \end{aligned}$$

and showed their relationship through the expression

$$f_{ES,0}(z_0,\boldsymbol{\theta}) = \mathrm{E}_R\ g(\mathbf{X};\boldsymbol{\theta}) + \frac{1}{1-\alpha} f_{QR,0}(z_0,\boldsymbol{\theta}). \qquad (7.12)$$

Now, assume that we only have a budget constraint of the form $RTC(\boldsymbol{\theta}) \leq RTC_{max}$.

a. Show that the Lagrangians derivatives of the two problems can be expressed as the following.

$$\begin{aligned} \partial_{\theta_j} LA_{QR} &= \partial_{\theta_j} f_{QR,0}(z_0,\boldsymbol{\theta}) - \bar{g}_j \quad LMI_{1,QR} \\ \partial_{\theta_j} LA_{ES} &= \frac{1}{1-\alpha}\left\{\partial_{\theta_j} f_{QR,0}(z_0,\boldsymbol{\theta}) - \bar{g}_j(1-\alpha)(LMI_{1,ES}-1)\right\}. \end{aligned}$$

b. Interpret this to mean that the two algorithms will converge to the same solutions, so analysts need only implement one of them. Moreover, it will be the case that generally the two approaches will have the same level of computational difficulty.

11.5.3 Appendix. Establishing the KKT Conditions

A proof to establish the *KKT* conditions for a general setting is cumbersome and not very insightful. Fortunately, there are several good resources where readers can access the details, including Boyd and Vandenberghe (2004), Simon and Blume (1994), and Nocedal and Wright (2006).

In the following, we provide sketches of proofs in two important special cases, of a single equality and inequality constraints. These sketches provide some useful intuition into the underpinnings of the *KKT* conditions.

Single Equality Constraint

We now restrict considerations to a single equality constraint so that the problem in Display (3.9) reduces to

minimize	$f_0(\mathbf{z})$
subject to	$f_{con,1}(\mathbf{z}) = 0$

and corresponding Lagrangian

$$LA\left(\mathbf{z}, LME_1\right) = f_0(\mathbf{z}) + LME_1\ f_{con,1}(\mathbf{z}).$$

In this setting, the Display (11.1) reduces to

$$\boxed{\begin{array}{ll} \partial_{\theta_i} \; LA(\mathbf{z}, LME_1^*)|_{\mathbf{z}=\mathbf{z}^*} = 0 & i = 1, \ldots, p \\ f_{con,1}(\mathbf{z}^*) = 0 & . \end{array}}$$

These are the classical conditions that appear in many beginning economics and calculus courses.

Sketch of Proof:
Consider a point $\mathbf{z}$ that is in the feasible set. Now, suppose that it is not optimal so that $\mathbf{z} \neq \mathbf{z}^*$. Then we can find many directions s so that $f_0(\mathbf{z}+s) < f_0(\mathbf{z})$. Choose a direction s that is small and use Taylor series expansion to write

$$f_0(\mathbf{z}+s) \approx f_0(\mathbf{z}) + \nabla f_0(\mathbf{z})' s \Rightarrow \nabla f_0(\mathbf{z})' s < 0.$$

Restrict considerations to those directions s so that $\mathbf{z}+s$ is also in the feasible set. In the same way, with a Taylor series expansion, we have

$$0 = f_{con,1}(\mathbf{z}+s) \approx f_{con,1}(\mathbf{z}) + \nabla f_{con,1}(\mathbf{z})' s \Rightarrow \nabla f_{con,1}(\mathbf{z})' s = 0.$$

Is such a restriction sensible? Yes, for example, we might choose

$$s = -\epsilon \left(\mathbf{I} - \frac{\nabla f_{con,1}(\mathbf{z}) \nabla f'_{con,1}(\mathbf{z})}{||\nabla f_{con,1}(\mathbf{z})||^2} \right) \nabla f'_0(\mathbf{z}),$$

for a small positive constant ϵ. With this choice, one can check that we still have $\nabla f_{con,1}(\mathbf{z})' s = 0$ and $\nabla f_0(\mathbf{z})' s < 0$.

This establishes that a given feasible point $\mathbf{z}$ is not optimal if we can find a small step s that both retains feasibility and decreases the objective function f to first order.

Thus, we cannot find a constant LME_1 such that $\nabla f_0(\mathbf{z}) + LME_1 \nabla f_{con,1}(\mathbf{z}) = 0$ (that is, the gradients are proportional to one another). If this were true, then

$$\left[\nabla f_0(\mathbf{z}) + LME_1 \nabla f_{con,1}(\mathbf{z})\right]' s = 0,$$

contradicting prior arguments. That is, if we can find a constant, say LME_1 so that $\nabla f_0(\mathbf{z}) + LME_1 \nabla f_{con,1}(\mathbf{z}) = 0$, then this point is optimal $\mathbf{z} = \mathbf{z}^*$.

Single Inequality Constraint

We now restrict considerations to a single inequality constraint so that the problem in Display (3.9) reduces to

$$\boxed{\begin{array}{ll} \text{minimize} & f_0(\mathbf{z}) \\ \text{subject to} & f_{con,1}(\mathbf{z}) \leq 0 \end{array}}$$

and corresponding Lagrangian

$$LA(\mathbf{z}, LMI) = f_0(\mathbf{z}) + LMI \; f_{con,1}(\mathbf{z}). \tag{11.18}$$

In this setting, the Display (11.1) reduces to

$$\boxed{\begin{array}{l} \partial_{\theta_i} \ LA(\mathbf{z}, LMI^*)|_{\mathbf{z}=\mathbf{z}^*} = 0 \quad i = 1, \ldots, p \\ LMI^* \times f_{con,1}(\mathbf{z}^*) = 0 \\ LMI^* \geq 0 \\ f_{con,1}(\mathbf{z}^*) \leq 0. \end{array}}$$

Sketch of Proof:
As in the Section 11.5.3 sketch of the proof, we first seek conditions so that a small step s decreases the objective function while retaining feasibility. As before, for decreasing the objective function, we have $f_0(\mathbf{z}+s) < f_0(\mathbf{z})$. With a Taylor series expansion, this means

$$f_0(\mathbf{z}+s) \approx f_0(\mathbf{z}) + \nabla f_0(\mathbf{z})'s \Rightarrow \nabla f_0(\mathbf{z})'s < 0.$$

For feasibility, we now require that $f_{con,1}(\mathbf{z}) \leq 0$ and $f_{con,1}(\mathbf{z}+s) \leq 0$. Using a Taylor series expansion, we have

$$f_{con,1}(\mathbf{z}+s) \approx f_{con,1}(\mathbf{z}) + \nabla f_{con,1}(\mathbf{z})'s.$$

Taken together, we seek

$$\nabla f_0(\mathbf{z})'s < 0, \quad f_{con,1}(\mathbf{z}) + \nabla f_{con,1}(\mathbf{z})'s \leq 0. \tag{11.19}$$

Case 1. Begin by assuming that the constraint is inactive so that $f_{con,1}(\mathbf{z}) < 0$. From this, any direction s that is sufficiently small satisfies the feasibility constraint

$$f_{con,1}(\mathbf{z}) + \nabla f_{con,1}(\mathbf{z})'s \leq 0.$$

Thus, if $\mathbf{z}$ is an optimal point then $\nabla f_0(\mathbf{z})' = 0$. Otherwise, the point could be improved upon and would not be optimal. A proportionality constant is not required and so we may define $LMI = 0$ without imposing any restrictions.

Case 2. Now assume that the constraint is active at $\mathbf{z}$ so that $f_{con,1}(\mathbf{z}) = 0$. Then, equation (11.19) becomes

$$\nabla f_0(\mathbf{z})' \ s < 0, \quad \nabla f_{con,1}(\mathbf{z})' \ s \leq 0. \tag{11.20}$$

Suppose that we cannot find a s such that equation (11.20) holds (assuming an active constraint). Then, as in the Section 11.5.3 proof sketch, $\mathbf{z}$ is an optimal point, $\mathbf{z} = \mathbf{z}^*$, and this can only occur when $\nabla f_0(\mathbf{z})$ is proportional to $\nabla f_{con,1}(\mathbf{z})$. Writing this proportionality constraint as

$$\nabla f_0(\mathbf{z}) + LMI \ \nabla f_{con,1}(\mathbf{z}) = 0,$$

for some constant LMI, we see that we now also require that $LMI > 0$. If $LMI < 0$, then we could choose a direction s so that both the objective function and the constraint function would decrease without bound (both $\nabla f_0(\mathbf{z})'s < 0$ and $\nabla f_{con,1}(\mathbf{z})' \ s < 0$).

Using these constructions for LMI, we can define the Lagrangian as in equation (11.18). With this,

- if the constraint is inactive, then $LMI = 0$ and $\nabla LA(\mathbf{z}) = \nabla f_0(\mathbf{z}) = 0$ at the optimum. Thus, the KKT conditions are satisfied.
- if the constraint is active, then $f_{con,1}(\mathbf{z}) = 0$, $\nabla LA(\mathbf{z}) = 0$, and $LMI > 0$. Thus, the KKT conditions are satisfied.

This is sufficient for the result.

12

The Role of Dependence in Managing Insurable Risks

Chapter Preview. Managers of risk portfolios understand that dependence among risks can be critical when making decisions to manage those risks. In extreme cases where all risks in a portfolio are strongly related, such as due to earthquake and flood events, opportunities for risk sharing and diversification benefits are limited. More broadly, dependence influences decisions to take on risks, how much risk to retain, the pricing of risks, and even whether a private marketplace for insurance is available.

The chapter begins with Section 12.1 which describes the ubiquity of dependence when considering insurable risks. To provide guidance in assessing its impact, the section categorizes dependence based on the number of risks and whether one should focus on small ("local") versus broad ("global") impacts of dependence.

Following a brief discussion on approaches for assessing broad impacts of dependence, Section 12.2 emphasizes local impacts by introducing a tool called *portfolio expectation dependence sensitivity.* Building on sensitivity work done in prior chapters, this tool enables analysts to assess the impact of small changes in dependence.

With the structure developed in the first two sections, Section 12.3 describes the impact of dependence on small collection of risks that this chapter calls a *portfolio.* Specifically, the discussion addresses the impact on risk transfer costs, the uncertainty of a portfolio, and the risk retention parameters obtained when constructing an optimal insurable risk portfolio.

Section 12.4 continues with a discussion for a large collection of risks that this chapter calls a *pool.* In modeling pools of many entities with several risks, it is common to allow for dependencies among different risk types yet assuming independence among entities. However, these commonly used assumptions may not be a sensible approximation to reality. In particular, risk managers are concerned with effects of *contagion*, where there is an association or spread of loss from one entity to another within a portfolio. Contagion can be most naturally interpreted in the context of a natural disaster or other catastrophic event common to many entities within the portfolio. Contagion can eradicate any benefits achieved from pooling, and this section provides approaches for assessing effects of contagion.

12.1 Role of Dependence

As we learned in Chapter 6, managing insurable risks begins with identifying and mitigating important risks, determining which are insurable or partially insurable, and introducing risk controls and prevention techniques to mitigate their frequency and severity. Despite extensive efforts to control and prevent undesirable outcomes, risk owners will still have a portfolio of risks with a financial impact that can be managed by risk transference, the subject of this book. But what is the role of dependence among risks?

 DOI: 10.1201/9781003469568-12

Dependence is common among insurable risks. Any risk owner, whether a person, corporation, or an insurer, responsible for several insurable risks can expect dependency among them. For example, a person's unobserved attitude towards risk has a common influence on outcomes of automobile, liability, and health risks. For a corporation, management's approach towards undertaking risky projects, personnel and their training in mitigating risky outcomes, as well as common geographical influences, all suggest that risky outcomes can depend upon one another in unforeseen ways. For an insurer, overall market conditions, the corporate philosophy for underwriting standards, and the industries in which they underwrite risks have common latent effects on all risks, inducing dependencies. Dependence is the norm when considering insurable risks, not the exception. Beginning with the Chapter 4 introduction of multiple (multivariate) risks, this book immediately introduced a structure to allow for dependence in a flexible manner - the copula. Any framework designed to manage insurable risks portfolios needs to incorporate a dependency structure at its foundations.

Dependency, whether weak, strong, or extreme, is a common feature of collections of insurable risks. This chapter argues that dependency is critical for insurable risk management for some purposes. Nonetheless, as will be seen, other business decisions are relatively robust to the effects of dependency. Dependency can be difficult to quantify and so it worthwhile to understand which business decisions critically rely upon dependence and which are relatively robust to its effects.

Portfolio versus Pool. When the number of risks is large, even small dependencies among risks can lead to a breakdown in the usual asymptotic results such as the law of large numbers. Inspired by these mathematical results, it is useful to think about situations in which the number of insurable risks can be categorized as "small" versus "large." This chapter refers to a small collection of risks as a *portfolio* and a large number of risks as a *pool*. For instance, a portfolio could consist of 2, 6, or 15 risks, as seen in previous examples in this book, and a pool is for a larger number of risks, typically in the hundreds, thousands, or more. Admittedly, this distinction is vague and there is a continuum over this size measure, but this categorization will help us appreciate potential impacts of dependency in different scenarios.

Local versus Global. Another distinction to be made when discussing dependency is whether the impact is "local" or "global." In some instances, the extent of dependence is well understood either from empirical evidence or knowledge of the situation. In these cases, local measures that quantify the impact of small changes in dependence are useful. Sensitivities for general auxiliary variables have already been introduced, and this framework will be applied to dependence parameters in this chapter. Conversely, in other situations, the extent of dependence is not well understood. More global assessment is needed in these instances, informing risk managers of the best and worst likely outcomes over a range of dependencies.

Why is dependency difficult to measure? In part this is because of this meaning of the word "dependency" in insurance risk modeling. Managers not accustomed to the subtleties of stochastic modeling think about one variable affecting another as a type of dependency. In contrast, for an insurance analyst, an observed variable affecting an outcome can be thought of as a regression-type effect which is relatively straight-forward to model using standard tools. That is, an observed variable, sometimes known as a *covariate*, can be used to "explain" another variable in a regression setting. *Dependency modeling* is something more; after known covariate effects have been accounted for in the model, "dependencies" measure other relationships not present in known observable covariates. That is, dependencies account for latent, unobserved, relationships that persist.

Despite the well-appreciated effects of dependence in insurance practice, the scope of quantitative tools for assessing the impact of dependence has been limited in the actuarial and insurance literature. This is especially surprising given its prominence in finance, a closely related field that also deals with financial risks. As just one example, students of finance learn about the capital asset pricing model (CAPM) that employs the covariance between a stock and a market index. Historically, finance has profitably used basic tools such as covariance and regression to explain relationships because of the linearity among risk types. In contrast, insurance risk applications typically deal with highly non-linear relationships meaning that these basic tools are less helpful.

Another reason for the lack of development of insurance risk management tools for assessing dependency is that dependence is typically not a decision variable representing an action that a portfolio manager might take. A portfolio manager might choose the upper limit of a risk, whether to take on a risk, or set a price to take on a risk. Each of these actions may be influenced by dependencies but a portfolio manager cannot, for example, set the correlation of a risk within a portfolio to be negative (although this could be desirable for diversification purposes).

Some Motivating Examples. Before presenting some general results, let us first examine two examples where it can be helpful to understand the impact of dependence.

Example 12.1. Reinsurance Expected Payments. Suppose that X_1 represents a general liability claim and X_2 represents expenses associated with the claims. Specifically, this is known as an *allocated loss adjustment expense* (ALEA), a type of insurance company expense that is attributable to the settlement of individual claims such as lawyers' fees and claims investigation expenses. In Frees and Valdez (1998), a sample of 1,500 claims and expenses from then Insurance Services Office (now Verisk Analytics) were used to calibrate a joint distribution with Pareto marginal distributions for claims and expenses. Fitting a normal copula model, it turns out that the fitted association parameter is 0.473.

In addition, Frees and Valdez (1998) calculated a reinsurer's expected payment as $\mathrm{E}\{\min(X_1, 10000) + X_2 \times [1 - 5000/\min(X_1, 10000)]\}$. With the fitted normal copula and Pareto marginals, this turns out to be 8,183. Actuarial analysts are quite comfortable fitting distributions, such as Pareto, to marginal risks. Yet, how sensitive is this result to the dependence between claims and expenses? Techniques to be introduced in this chapter show that for every one percent change (increase) in the dependence, one expects the reinsurance premium to increase by 19. Put another way, a one percent increase in the dependence leads to a 19 / 8183 = 0.0023 proportional change in expected payments. This is a relatively small proportional change, suggesting that this expected payment is robust to dependence effects.

Example 12.2. Value at Risk for an Insurance Portfolio. We consider a portfolio of Wisconsin Property Fund analyzed in Section 8.1.3. Recall that there are $p = 6$ lines of business, building and contents (BC), contractor's equipment (known as "inland marine", IM, for historical reasons), and four types of commercial auto. For simplicity, this example looks to a portfolio before any risk transference. The interest is in quantifying the value at risk (at the 98 % level) for a portfolio of 1,098 policyholders in a held-out dataset.

As described in Frees et al. (2016), a longitudinal/panel data set allows one to calibrate the joint distribution. For this example, we use Tweedie marginal distributions and a normal copula among the six lines. Because of covariate information, each policyholder has a different distribution. This example assumes independence among policyholders yet allow for dependence among lines of business (in Section 12.4, this assumption will be weakened). To be specific, Table 12.1 shows the dependence parameters estimated from the longitudinal/panel dataset. To emphasize, these dependence parameters summarize associations among lines *after* having removed (via the Tweedie regressions) the effects of covariates (such as exposures) that may induce dependencies.

Figure 12.1 shows the distribution of outcomes based on 2,000 simulations of the portfolio. Because of the number of independent policyholders, it is not surprising that the distribution of the portfolio is approximately normal due to the Central Limit Theorem even though individual risks are very non-normal and non-identical. The value at risk turns out to be 29,293, in thousands.

How sensitive is this value at risk to the dependence parameters? Using the techniques in this chapter, results shows that a one percent increase in the dependence between BC and IM, from 0.21 to 0.22, induces a decrease in the value at risk by 25.94, in thousands. On the one hand, this change is small relative to the total value at risk for the portfolio, 29,293. On the other hand, this is only one of 15 dependence parameters in Table 12.1 that concerns the portfolio manager.

TABLE 12.1
Dependence Parameters for Six Commercial Lines of a Portfolio.

	BC	IM	PN	PO	CN	CO
BC	1.000	0.210	0.276	0.359	0.265	0.417
IM	0.210	1.000	0.362	0.410	0.266	0.358
PN	0.276	0.362	1.000	0.552	0.553	0.490
PO	0.359	0.410	0.552	1.000	0.329	0.564
CN	0.265	0.266	0.553	0.329	1.000	0.533
CO	0.417	0.358	0.490	0.564	0.533	1.000

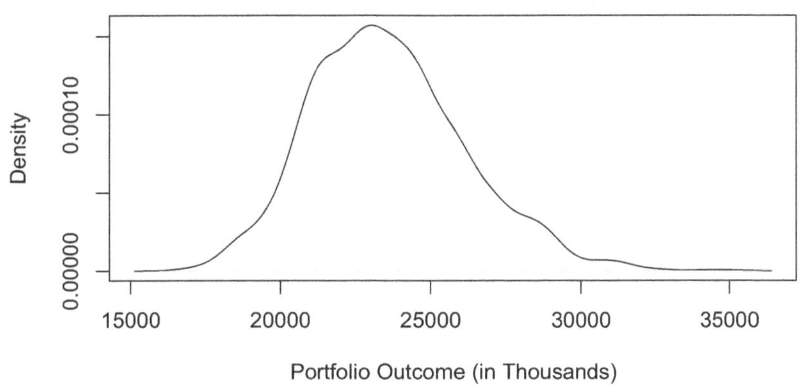

FIGURE 12.1
Simulated Portfolio Distribution.

12.2 Measuring Global and Local Impacts

12.2.1 Measuring Global Impact

Because dependence can be difficult to quantify, it is useful in some applications to provide a global assessment of the best and worst likely outcomes over a range of dependencies. For the worst most likely outcome, we have already utilized the concept of comonotonicity among risks, discussed in Section 9.1, that can be interpreted as perfect (positive) dependence. As also discussed in that section, the concept of the best most likely outcome in a multivariate setting is elusive, as perfect negative dependence can be vague for more than $p = 2$ risks. A model of independence among risks serves as a useful proxy for the best most likely outcome. Although it is possible to have scenarios even more favorable than independence, this is an approximation that most risk managers would gladly accept. As described in Section 9.1, a mixture model between these two bookend scenarios is a useful approach for quantifying the global impact of dependencies.

In the special case of investment portfolios, Section 9.4.2 introduced a special type of *robustness* where analysts aim to maximize penalized worst-case portfolio returns. The introduction in Section 9.4.2 focused on uncertainty sets for mean parameters over which one considers worst-case scenarios. In principle, this approach could be extended to insurable risks looking at uncertainty sets for dependence. Robustness has found favor among analysts in portfolio analysis; it may be that one day researchers will develop comparable tools for applications with insurable risk portfolios.

12.2.2 Measuring Local Impact

When dependencies are reasonably well understood, analysts require local measures that quantify the impact of small changes in dependence. For small changes, this book has featured a differential approach termed *sensitivities.* We learned how to determine risk retention sensitivities in Section 5.3 for the important cases of the value at risk and expected shortfall. In that investigation, consideration was limited to partial derivatives with respect to risk retention variables. This section shows how to use the same approach for dependencies, subject to the requisite smoothness assumptions needed for differentiation.

Additionally, sensitivities in the more complex setting of portfolio optimization were calculated in Chapter 9. There, a generic auxiliary variable a was introduced, which might represent a level of confidence α or maximal risk transfer cost RTC_{max}. Further, in Chapter 10, a was even used to represent a parameter associated with a marginal distribution, such as the scale. In this chapter, the auxiliary variable is a dependence parameter.

12.2.3 Portfolio Expectation Dependence Sensitivity

This section provides a tool for calibrating the sensitivity of dependence in measures of interest in actuarial and insurance applications. Here, "measures of interest" include premiums and risk measures such as value at risk and expected shortfall.

Specifically, recall that for a collection of risks, $\mathbf{\Sigma}$ represents a matrix of association parameters, and let σ be a generic element from this matrix. Further, let $h(\cdot)$ represent some summary measure of all the risks in the portfolio. It may be a risk transfer function or some other function, such as reinsurance expected payments in Example 12.1. This section shows how to compute the partial derivative $\partial_{\sigma}\mathrm{E}[h(\mathbf{X})]$, referred to here as a *portfolio expectation dependence sensitivity.*

Why is this tool useful? Because of their complexity, baseline portfolio expectations and risk measures are typically determined via simulation for a given dependence matrix $\boldsymbol{\Sigma}$. Using a "brute force" method, one would repeat this over several variations of $\boldsymbol{\Sigma}$ to obtain a numerical approximation of the sensitivities. In many complex situations of interest, this approach is computationally prohibitive. This section provides an alternative approach for a large class of dependence models, elliptical copulas.

Portfolio Expectation Dependence Sensitivity. To promote interpretation, the main body provides results for the Gaussian (normal) copula. In the Appendix Section 14.2, results are extended to general elliptical copulas.

Based on a Gaussian copula, the portfolio expectation dependence sensitivity can be written as

$$\partial_\sigma \mathrm{E}[h(\mathbf{X})] = \frac{1}{2}\mathrm{Cov}\left[h(\mathbf{X}),\ W_\sigma(\mathbf{X})\right], \tag{12.1}$$

where $\mathrm{Cov}[\cdot,\cdot]$ is the usual covariance operator. Further, define the weights

$$W_\sigma(\mathbf{X}) = \mathbf{ns}'\ \boldsymbol{\Sigma}^{-1}(\partial_\sigma \boldsymbol{\Sigma})\boldsymbol{\Sigma}^{-1}\ \mathbf{ns}, \tag{12.2}$$

where $\mathbf{ns}' = (\Phi^{-1}[V_1], \ldots, \Phi^{-1}[V_p])$ and $(V_1, \ldots, V_p)$ is a vector of uniform random numbers that has the same copula as $(X_1, \ldots, X_p)$. Here, $\Phi^{-1}[V]$ is simply a normal score associated with each outcome. As described further in Appendix Section 14.2, in the case where distribution of outcomes are continuous (as in the following Example 12.3), one may simply use $F_j(X_j)$ for V_j. However, for other instances with outcomes that have jumps (such as Example 12.2), resorting to more basic uniform numbers is needed. The weights in equation (12.2) appear complex but can readily be computed via simulation. To see this, consider the following example.

Example 12.3. Reinsurance Expected Payments. This is a continuation of Example 12.1. To illustrate, consider a reinsurer's expected payment on a policy with limit u and insurer's retention d. Then, assuming a pro-rata sharing of expenses, we have

$$h(X_1, X_2) = \begin{cases} 0 & \text{if } X_1 < d \\ X_1 - d + \frac{X_1 - d}{X_1} X_2 & \text{if } d \le X_1 < u \\ u - d + \frac{u-d}{u} X_2 & \text{if } X_1 \ge u. \end{cases}$$

Table 12.2 shows the maximum likelihood estimates of the fitted distribution, both under the joint copula model and assuming independence of losses and expenses.

Maximum likelihood estimates can be used to determine reinsurance expected payments, as in Table 12.3. Here, the rows are for different upper limits (u) and the columns are for

TABLE 12.2
Maximum Likelihood Estimation Results.

	Estimate	Std Error	Univariate Estimate	Univariate Std Error
Loss shape	1.1882	0.0699	1.1349	0.066
Loss scale	15.5303	1.4895	14.4433	1.3865
ALAE shape	2.376	0.1873	2.3524	0.1957
ALAE scale	16.5625	1.7649	15.8936	1.789
Dependence	0.4734	0.0192	NA	NA

TABLE 12.3
Simulation-Based Reinsurance Expected Payments by Upper Limit u and Retention per Limit (d/u).

Limit (u)	Retention per Limit (d/u)				
	0.0	0.25	0.50	0.75	0.95
10,000	19401	13137	8183	3876	747
100,000	37984	16124	8244	3446	617
500,000	51863	13046	6213	2510	440
1,000,000	57005	11497	5427	2175	380

TABLE 12.4
Standard Errors of Simulation-Based Reinsurance Premium.

Limit (u)	Retention per Limit (d/u)				
	0.0	0.25	0.50	0.75	0.95
10,000	13	10	7	3	1
100,000	21	16	10	5	1
500,000	42	28	17	8	2
1,000,000	57	37	22	10	2

different deductibles as a fraction of the upper limit (d/u). Table 12.4 gives the corresponding simulation standard errors, suggesting that using five million simulations used here is adequate for our purposes.

Table 12.5 gives estimates of the dependence sensitivities. These measure how much the reinsurance expected payments change per a one percent change in the association parameter, e.g., moving from 0.4734 to 0.4834. Table 12.6 gives the corresponding information but as a proportion to the reinsurance expected payments. From this table, we see that a one percent change in the association parameter means less than a one percent change in reinsurance expected payments for all levels of limits u and retention d. These results, and the follow-up investigation outlined in Exercise 12.5, suggest that the dependence plays a minor role in this example. In contrast, several examples are given in Dong et al. (2022) where dependence plays an important role.

TABLE 12.5
Dependence Sensitivity by Upper Limit u and Retention per Limit (d/u).

Limit (u)	Retention per Limit (d/u)				
	0.0	0.25	0.50	0.75	0.95
10,000	0.48	21.25	18.97	10.81	2.30
100,000	0.80	43.45	26.77	12.42	2.37
500,000	1.16	24.79	12.44	5.24	0.96
1,000,000	1.88	17.03	8.45	3.75	0.71

TABLE 12.6
Dependence Sensitivity per Reinsurance Expected Payments.

Limit (u)	Retention per Limit (d/u)				
	0.0	0.25	0.50	0.75	0.95
10,000	0	0.0016	0.0023	0.0028	0.0031
100,000	0	0.0027	0.0032	0.0036	0.0039
500,000	0	0.0019	0.0020	0.0021	0.0022
1,000,000	0	0.0015	0.0016	0.0017	0.0019

12.3 Managing Portfolios

The impact of dependence varies according to the modeling purpose or, put another way, the type of business decision. This section focuses on three types of decisions associated with collections of insurable risks: (i) questions of risk transfer costs, (ii) quantifying overall portfolio uncertainty, and (iii) determining optimal risk retention parameters. As considerations differ by the number of risks, this section focuses on collections with a smaller number of risks, referred to here as a *portfolio*.

12.3.1 Risk Transfer Costs

When modeling collections of risks for pricing purposes or understanding appropriate risk transfer costs, issues of dependence are generally not at the forefront. For instance, when persons or corporations are risk owners, there are often many potential parties willing to take on a risk. A lack of coordination can be expected when different parties take on different risks, or even when different units within the same risk-taking firm are responsible for different risks. For example, a risk manager for a large corporation may purchase general and products liability, medical malpractice, and workers' compensation, all from different vendors. For purposes of determining risk transfer costs to the corporation, prices of different risks can be treated as unrelated and independent from one another.

For another instance, consider an insurer pricing a portfolio of risks for a firm or corporation. Suppose that $X_1, \ldots, X_p$ represents different risks and $S = X_1 + \cdots + X_p$ represents the portfolio sum. Technical pricing often starts with a representation such as $\mathrm{E}(S) = \mathrm{E}(X_1) + \cdots + \mathrm{E}(X_p)$; that is, the expectation of the sum is the sum of expectations. This mathematical relationship holds regardless of the dependence and reduces the pricing of a complex multivariate portfolio to the pricing of each marginal risk. Before prices are determined, insurers will rescale technical prices through adjustments based on their own portfolio of risks as well as market conditions, cf. Mildenhall and Major (2022). Dependence does come into play into these adjustments and can be important in many situations.

Naturally, there are situations when dependence affects the determination of technical prices. Example 12.1 illustrated one such situation.

12.3.2 Portfolio Uncertainty

Unlike the case for risk transfer costs, dependencies typically have a major impact on metrics that quantify portfolio uncertainty. For example, consider the case of $p = 2$ risks, X_1 and X_2, that form the portfolio $S = X_1 + X_2$. In an extreme case, where $X_1 = -X_2$ is known as *counter-monotonic*, we have $S = 0$ resulting in no uncertainty. Another extreme case, where $X_1 = X_2$ that we have already identified as *comonotonic*, results in no diversification benefits when forming a portfolio. These are situations where even small changes in dependence are likely to be important to risk managers.

More generally, consider a collection of p risks $\mathbf{X}$ whose dependence can be expressed in terms of a matrix of dependence parameters $\mathbf{\Sigma}$. From the multivariate set of risks, one forms a portfolio $h(\mathbf{X}) = h(\mathbf{X}; \boldsymbol{\theta})$. Note that the risk retention parameters in $\boldsymbol{\theta}$ are dropped, as we will not be optimizing over these variables in this subsection. Now consider several cases of $RM(\mathbf{X})$, a risk measure associated with $h(\mathbf{X})$. The interest here is in computing the partial derivative $\partial_\sigma RM(\mathbf{X})$.

Value at Risk Dependence Sensitivity

Risk retention sensitivities for the important cases of value at risk and expected shortfall were covered in Section 5.3. This framework can also be applied to dependence sensitivities. Specifically, let F be the distribution function associated with the random variable $h(\mathbf{X})$. Consider only functions h so that the distribution function is smooth (differentiable) in potential outcomes and in a dependence parameter.

The quantile, or value at risk, is $F^{-1}(\alpha; \mathbf{\Sigma}) = VaR(\mathbf{\Sigma}) = VaR$ at confidence level α. This forms the basis of a *value at risk dependence sensitivity,*

$$\partial_\sigma \, VaR(\mathbf{\Sigma}) = -\frac{F_\sigma[VaR; \mathbf{\Sigma}]}{F_y[VaR; \mathbf{\Sigma}]}. \tag{12.3}$$

This style of calculation was previously seen in Section 5.3, equation (5.5), in the context of risk retention parameters $\boldsymbol{\theta}$. A short review of this calculation is provided to appreciate the smoothness requirements.

Under the Hood. See the *Online Supplement* for confirmation of this result.

As described in the prior section, many measures of interest in actuarial applications can be expressed as an expectation of a portfolio function. Moreover, this result can readily be utilized when considering quantiles. For problems of interest, evaluating the value at risk VaR and the density function $F_y(\cdot)$ are typically straightforward using basic simulation techniques. The difficult term is $F_\sigma(\cdot)$

For the portfolio distribution function, we can consider $I[h(\mathbf{X}) \leq y]$ for some argument y. With this, the expected portfolio function $\mathrm{E}\{I[h(\mathbf{X}) \leq y]\} = \Pr[h(\mathbf{X}) \leq y] = F(y)$ is a distribution function. Taking a partial derivative of each side yields $\partial_\sigma \mathrm{E}\{I[h(\mathbf{X}) \leq y]\} = \partial_\sigma F(y) = F_\sigma(y)$, for a given value of y. From equation (12.1), we know how to compute this as a portfolio expectation dependence sensitivity, which is

$$F_\sigma(y) = \frac{1}{2} \, \mathrm{Cov}\left\{I[h(\mathbf{X}) \leq y], \; W_\sigma(\mathbf{X})\right\}, \tag{12.4}$$

where the weight $W_\sigma(\mathbf{X})$ is defined in equation (12.2).

As with the portfolio expectation dependence sensitivity in equation (12.1), using equation (12.4) to determine the value at risk sensitivity appears complex but can readily be computed via simulation.

Example 12.4. Value at Risk for an Insurance Portfolio (Continued). Using the structure from Example 12.2, Table 12.7 provides the value at risk dependence sensitivities, $\partial_\sigma \, VaR$. Here, there are 15 sensitivities, one for each parameter in Table 12.1. For example, a 1% increase in the association parameter between the BC and IM lines imply a decrease of 26 in the value at risk. To assess the magnitude of these sensitivities, Table 12.8 provides

TABLE 12.7
***VaR* Dependence Sensitivity (in Percentages).**

	BC	IM	PN	PO	CN	CO
BC	0	-26	2	51	3	33
IM	-26	0	39	52	31	-55
PN	2	39	0	40	84	-25
PO	51	52	40	0	-9	52
CN	3	31	84	-9	0	25
CO	33	-55	-25	52	25	0

TABLE 12.8
***VaR* Dependence Sensitivity (in Percentages) as a Proportion of *VaR*.**

	BC	IM	PN	PO	CN	CO
BC	0.000	-0.001	0.000	0.002	0.000	0.001
IM	-0.001	0.000	0.001	0.002	0.001	-0.002
PN	0.000	0.001	0.000	0.001	0.003	-0.001
PO	0.002	0.002	0.001	0.000	0.000	0.002
CN	0.000	0.001	0.003	0.000	0.000	0.001
CO	0.001	-0.002	-0.001	0.002	0.001	0.000

the same information but as a proportion of the value at risk, $(\partial_\sigma\ VaR)/VaR$. The results are not surprising. Small changes in dependence among risks within a portfolio imply only small changes in the value at risk.

Note. Technical Note on the Computation of Weights.

For the Wisconsin Property Fund, there are 1098 fund members each of which has up to six risks. Many of the fund members only have a subset of coverages for the six risks so that the number of risks is unbalanced. Further, this example assumes independence among fund members.

Thus, let $\mathbf{\Sigma}_j$ be the matrix of dependence matrix associated with the jth fund member. Then, $\mathbf{\Sigma} = blkdiag(\mathbf{\Sigma}_1, \ldots, \mathbf{\Sigma}_{1098})$.

From equation (12.2), the weight can be expressed as

$$\begin{aligned} W_\sigma(\mathbf{X}) &= \mathbf{ns}'\ \mathbf{\Sigma}^{-1}(\partial_\sigma \mathbf{\Sigma})\mathbf{\Sigma}^{-1}\mathbf{ns} \\ &= \Sigma_{j=1}^{1098}\mathbf{ns}_j'\ \mathbf{\Sigma}_j^{-1}(\partial_\sigma \mathbf{\Sigma}_j)\mathbf{\Sigma}_j^{-1}\ \mathbf{ns}_j, \end{aligned}$$

where $\mathbf{ns}_j$ is the sub-vector of $\mathbf{ns}$ associated with the jth fund member.

ES and other Quantile-Based Risk Measure Dependence Sensitivity

The expected shortfall parameter sensitivity follows in the same fashion. Assuming smoothness in potential outcomes and the dependence parameter, one can modify the risk retention sensitivity from equation (5.7) to write

$$\begin{aligned} \partial_\sigma\ ES[h(\mathbf{X})] &= \frac{1}{1-\alpha}\left\{\partial_\sigma \mathrm{E}[h(\mathbf{X})] - \partial_\sigma \mathrm{E}[h(\mathbf{X}) \wedge y]|_{y=VaR}\right\} \\ &\quad + [1 - \frac{1}{1-\alpha}\{1 - F[VaR]\}] \times \partial_\sigma VaR. \end{aligned} \tag{12.5}$$

Computing the risk measure dependence sensitivities in equations (12.3) and (12.5) can be difficult. To underscore this difficulty, Exercise 12.2 shows the calculations by restricting considerations to $p = 2$ risks.

It is not difficult to extend this approach to general quantile-based risk measures. To be specific, let us consider the range value at risk introduced in Section 2.1. Recall from equation (2.4) that this is defined as

$$RVaR_{(\alpha,\beta)} = \begin{cases} F^{-1}(\alpha) & \text{if } \beta = 0 \\ \frac{1}{\beta}\int_{\alpha}^{\alpha+\beta} F^{-1}(a)\, da & \text{if } 0 < \beta \leq 1-\alpha. \end{cases}$$

In the case $\beta = 1 - \alpha$, the $RVaR$ reduces to the expected shortfall.

This yields the derivative of the range value at risk, another type of *dependence sensitivity*,

$$\partial_\sigma \, RVaR_{(\alpha,\beta)} = \begin{cases} \partial_\sigma F^{-1}(\alpha) & \text{if } \beta = 0 \\ \frac{-1}{\beta}\int_{F^{-1}(\alpha)}^{F^{-1}(\alpha+\beta)} F_\sigma(z)\, dz & \text{if } 0 < \beta \leq 1-\alpha, \end{cases} \tag{12.6}$$

with $\partial_\sigma F^{-1}(\alpha)$ given in equation (12.3).

ⓘ *Under the Hood.* See the *Online Supplement* for confirmation of this result.

Example 12.5. Expected Shortfall for an Insurance Portfolio. This is a continuation of Example 12.4. Take $\beta = 1 - \alpha$, so that $RVaR$ becomes ES. To evaluate $\partial_\sigma ES$, use equations (12.4) and (12.6) to establish

$$\partial_\sigma \, ES = \frac{1}{2(1-\alpha)} \mathrm{Cov}\left(W_\sigma(\mathbf{X}), \{[h(\mathbf{X}) - F^{-1}(\alpha)]\ I[h(\mathbf{X}) > F^{-1}(\alpha)]\}\right).$$

This form is easy to compute using simulation.

ⓘ *Under the Hood.* See the *Online Supplement* for confirmation of this result.

Table 12.9 provides the expected shortfall dependence sensitivities, $\partial_\sigma \, ES$, and Table 12.10 provides the same information but as a proportion of the expected shortfall, $(\partial_\sigma \, ES)/ES$. The results are not surprising. Consistent with what we saw for the value at risk in Example 12.4, small changes in dependence among risks in a portfolio imply only small changes in the expected shortfall.

12.3.3 Constructing Insurable Risk Portfolios

Based on what has been learned so far, when modeling collections of risks for the purpose of constructing insurable risk portfolios, issues of dependence do not appear to be at the forefront.

TABLE 12.9
***ES* Dependence Sensitivity (in Percentages).**

	BC	IM	PN	PO	CN	CO
BC	0	-30	26	80	-5	98
IM	-30	0	60	75	34	-91
PN	26	60	0	119	118	-1
PO	80	75	119	0	12	116
CN	-5	34	118	12	0	28
CO	98	-91	-1	116	28	0

TABLE 12.10
***ES* Dependence Sensitivity (in Percentages) as a Proportion of *ES*.**

	BC	IM	PN	PO	CN	CO
BC	0.000	-0.001	0.001	0.003	0.000	0.003
IM	-0.001	0.000	0.002	0.002	0.001	-0.003
PN	0.001	0.002	0.000	0.004	0.004	0.000
PO	0.003	0.002	0.004	0.000	0.000	0.004
CN	0.000	0.001	0.004	0.000	0.000	0.001
CO	0.003	-0.003	0.000	0.004	0.001	0.000

From a global perspective, Section 9.1 introduced the mixture approach that investigated optimal portfolio construction over a wide range of dependencies. At the optimum, levels of retained risk uncertainty varied as anticipated, with strongly dependent risks exhibiting high levels of uncertainty compared with lesser dependencies. However, the optimal retention parameters were relatively stable. As the retention parameters represent decision variables of interest to risk managers, this suggests that optimal insurable risk portfolios are robust to the effects of dependence.

Also from the global perspective, there is little information about applications of distributionally robust techniques introduced in Section 9.4. Hopefully, future research will shed more light on this approach.

From the local perspective, we saw how, for special cases of linear quota share and asset investment problems, one could compute closed-form expressions for dependence sensitivities in Exercises 9.11 and 9.12. These results were expressed for general insurable risk portfolio problems through the Section 9.2 Envelope Theorem and the Section 10.1 Perturbation Sensitivity Proposition. Specifically, one can use $a = \sigma$ for the auxiliary variable to apply these results.

With the portfolio expectation dependence sensitivity and results from Appendix Section 10.5.3, derivatives of these quantities with respect to association parameters can be evaluated as

$$
\begin{aligned}
\partial_\sigma \, \partial_{z_0} \, ES1(z_0, \boldsymbol{\theta}) &= \frac{1}{2(1-\alpha)} \, \mathrm{E}_R \left\{ W_\sigma(\mathbf{X}) K\left(\frac{z_0 - g(\mathbf{X};\boldsymbol{\theta})}{b}\right)\right\} \\
&\quad - \frac{1}{2(1-\alpha)} \mathrm{tr}(\boldsymbol{\Sigma}^{-1}\partial_\sigma \boldsymbol{\Sigma}) \, \mathrm{E}_R \left\{ K\left(\frac{z_0 - g(\mathbf{X};\boldsymbol{\theta})}{b}\right)\right\}
\end{aligned}
$$

$$
\begin{aligned}
\partial_\sigma \, \partial_{\boldsymbol{\theta}} \, ES1(z_0, \boldsymbol{\theta}) &= \frac{1}{2(1-\alpha)} \, \mathrm{E}_R \left\{ W_\sigma(\mathbf{X}) \left[1 - K\left(\frac{z_0 - g(\mathbf{X};\boldsymbol{\theta})}{b}\right)\right] \partial_{\boldsymbol{\theta}} g(\mathbf{X};\boldsymbol{\theta}) \right\}. \\
&\quad - \frac{1}{2(1-\alpha)} \mathrm{tr}(\boldsymbol{\Sigma}^{-1}\partial_\sigma \boldsymbol{\Sigma}) \, \mathrm{E}_R \left\{ \left[1 - K\left(\frac{z_0 - g(\mathbf{X};\boldsymbol{\theta})}{b}\right)\right] \partial_{\boldsymbol{\theta}} g(\mathbf{X};\boldsymbol{\theta}) \right\}.
\end{aligned}
$$

Example 12.6. Property Fund Portfolio. This is a continuation from Examples 10.4 and 10.5. Here, we examine perturbation sensitivities due to changes in the association, or dependency, parameters.

For a change of 0.01 in the association parameter, Table 12.11 shows the change in each decision variable at the optimum. For example, a 0.01 increase in the first dependence parameter $\rho_{12} \approx -0.11$ leads to a 0.91 increase in the optimal value of VaR, a 0.28 decrease in the optimal value of θ_1, and so forth. Changes in the association parameters have only a small impact on the optimal values of the decision variables.

TABLE 12.11
Association/Dependency Sensitivities for Excess of Loss.

Dep Param	ρ_{12}	ρ_{13}	ρ_{14}	ρ_{15}	ρ_{16}	ρ_{23}	ρ_{24}	ρ_{25}
VaR	0.91	0.54	0.8	0.4	0.36	0.32	0.66	0.18
θ_1	-0.28	-0.06	-0.05	-0.16	0.04	0.1	0.16	0.07
θ_2	-0.83	0.12	0.2	0.06	0.14	-0.21	-0.22	-0.3
θ_3	0.19	-0.52	0.21	0.07	0.09	-0.35	0.21	0.06
θ_4	0.38	0.25	-0.47	0.14	0.16	0.23	-0.38	0.15
θ_5	0.1	0.09	0.15	-0.52	0.08	0.05	0.12	-0.33
θ_6	0.24	0.11	0.17	0.09	-0.41	0.09	0.17	0.06
Dep Est	0.06	0.02	0.05	0.1	0.17	0.08	0.15	0.11
Dep Param	ρ_{26}	ρ_{34}	ρ_{35}	ρ_{36}	ρ_{45}	ρ_{46}	ρ_{56}	
VaR	0.17	0.41	0.11	0.11	0.22	0.25	0.08	
θ_1	0.11	0.11	0.07	0.07	0.1	0.11	0.06	
θ_2	-0.06	0.16	0.04	0.08	0.13	0.12	0.05	
θ_3	0.08	-0.21	-0.2	-0.09	0.09	0.08	0.04	
θ_4	0.15	-0.16	0.09	0.09	-0.26	-0.07	0.07	
θ_5	0.05	0.08	-0.13	0.04	-0.18	0.06	-0.06	
θ_6	-0.31	0.09	0.04	-0.15	0.07	-0.19	-0.17	
Dep Est	0.09	0.43	0.46	0.18	0.4	0.35	0.4	

Note: Sensitivities are per 0.01 change in association parameters.

12.4 Managing Pools

When the number of risks in a collection is large, dependencies among risks, or inter-dependencies, can affect everything. From a minimalist perspective, this can be seen simply by looking to probability theory where basic results hinge on the extent of dependence. For instance, Stout (1974) describes mild dependency conditions where the law of large number continues to hold. Another fundamental source, Billingsley (2013), describes central limit theorems that hold even in the presence of certain types of mild dependencies. However, these basic results can fail to hold in the presence of different types of stronger dependencies.

To illustrate, consider an extreme example of a perfectly dependent pool of risks such that $X_1 = \cdots = X_n$. In this case, the average pool risk exactly equals any one of the risks, say X_1; it does not converge to the mean of the distribution as typically seen in a law of large numbers. This failure holds regardless of the risk distribution, whether it is long-tailed, standard normal, or any other distribution. When the number of risks is large, the effects of dependence are very important.

12.4.1 Effects of Contagion

It is common to think of dependencies among risks in a pool as stemming from the effects of **contagion.** This term likely originated from early models of infectious diseases. How can such important inter-dependencies arise?

Models of Infectious Disease. Actuarial models of contagion can be traced back to the work of Daniel Bernoulli in 1760. As described in Daw (1979), Bernoulli developed a model for the infectious disease smallpox. His work can be considered a precursor to the multi-decrement models that actuaries use to account for additional mortality effects of perils such as those due to infectious diseases such as smallpox, cholera, influenza, and so forth.

This development has led to the use of time-dependent Markov chain transition models in life and disability insurance (see, for example, Haberman and Pitacco (2018)). Such models enable analysts not only to examine additional effects but also to track the progression of infectious diseases as they spread among population segments. These models are essential for understanding and managing pandemic risks, including the AIDs epidemic of the 1980s and the more recent COVID-19 pandemic. In insurance, the term *pandemic risk* refers to the potential financial losses caused by a widespread disease outbreak. Insurers use transition models to predict the spread of diseases and their potential impact on claims.

Cyber Risks. Additionally, the term *contagion* is used within insurance to model dependencies among **cyber risks**. Like other perils, firms purchase insurance to shield themselves from potential financial losses resulting from digital security breaches, known as a cyber risk. These incidents can disrupt business operations, damage data, and expose sensitive information, leading to a various costs for businesses. Much like infectious diseases that affect people, cyber mishaps can spread contagiously through networks. See Eling and Schnell (2016) for a discussion of models and strategies relevant to understanding cyber risk for cyber insurance.

Catastrophes and Climate Change. Models of contagion are also used to quantify the impact on risks arising from *catastrophes and climate change*, as discussed in Toumi and Restell (2014). Natural disasters such as hurricanes, floods, and earthquakes can exhibit contagion effects, where a single disastrous event can simultaneously affect many different people, firms, and their properties within a geographic region. Due to contractual relationships, insurers naturally view different individuals and firms as separate risks. Additionally, different perils or coverages are often treated separately, partly due to historical reasons and partly due to differing risk distributions. For instance, as discussed in preceding chapters, a firm may have property, business interruption, motor vehicle coverage, and more. A single event, such as an earthquake, may simultaneously affect all coverages for a firm and many comparable firms within close proximity. Thus, models of contagion need to consider dependence both among individual risks and among different types of coverages.

In addition, one event can trigger or exacerbate others. That is, a disaster, such as an earthquake, can lead to secondary events, such as fires or tsunamis, affecting multiple policies. For example, the 1906 San Francisco earthquake led to widespread fires. Climate change may have even broader impacts that extend well beyond a specific geographic region. For instance, climate events can lead to widespread crop failures, critical for agriculture insurance.

The effects of contagion can be so extreme that private market insurance is not viable. For instance, if everyone in a village is subject to the same flood risk, who has resources to help rebuild in the event of a flood? With an extreme lack of diversification, risk-sharing among individual risks fails, and government intervention is required. Thus, it is common to have government-sponsored programs to provide insurance for earthquakes and floods. We are also witnessing the precarious marketplace for homeowners' insurance in some hurricane-prone regions such as the state of Florida.

Financial Contagion and Systemic Risk. Markets may not exist or, when they do exist, exhibit frailty because economic agents do not operate in isolation but are connected by various economic relationships. This interconnectedness allows shocks to propagate throughout a financial system.

This potential for the propagation of shocks is referred to as *financial contagion.* Simply put, financial contagion describes a situation where the failure of one financial institution, such as a bank, could lead to a cascade of failures across the financial system. For instance, with the global financial crisis of 2008, the concept of financial contagion became central as the collapse of major financial institutions and the housing market affected global financial systems. Note, however, that the phrase "financial contagion" can mean different things to different people; for example, Pericoli and Sbracia (2003) provide five different definitions of "contagion."

Closely related to financial contagion is the notion of *systemic risk.* According to Cummins and Weiss (2014):

> Systemic risk is the risk that an event will trigger a loss of economic value or confidence in a substantial segment of the financial system that is serious enough to have significant adverse effects on the real economy with a high probability.

Implementing this definition, Cummins and Weiss did not consider events such as the U.S. liability crisis of the 1980s and Hurricane Andrew in 1992 to be systemic events, although the financial crisis that gripped world markets in 2007-2010 was considered systemic. The criterion they applied was whether or not the event instigated sufficient adverse effects on real economic activity.

In other words, systemic risk refers to the risk of collapse of an entire financial system or market, as opposed to financial contagion, which may describe the risk associated with any one individual entity, group, or component. It is the risk that a shock will trigger severe instability or collapse in the financial system, leading to a significant economic downturn.

Models of financial contagion and systemic risks differ from the first three categories discussed: models of infectious disease, cyber risks, and catastrophes and climate change. They are similar in that they utilize models of contagions and interdependencies. However, models of financial contagion and systemic risks are different in that they typically concern relationships among financial institutions, including banks, insurers, and investment firms. In contrast, the first three categories are primarily focused on individual risks, such as individuals and firms, entities that could be considered to be policyholders of insurance companies.

12.4.2 Simulation Models of Contagion

Because of their economic impact, it is not surprising that deep and complex models of contagion have developed for each of the categories described in the previous subsection: (i) models of infectious disease, (ii) cyber risks, and (iii) catastrophes and climate change, and (iv) financial contagion and systemic risks. These models are based on knowledge of biology and demography, climatology and geology, as well as economic rules governing behavior of agents. These models are intricate, require substantial training and time to implement, and yet are capable of not only predicting the prevalence of contagious effects but also forecasting their propagation through a system. Government agencies, insurers, and other financial institutions rightfully invest substantial resources into developing such models of contagion due to their economic importance.

Linear Latent Variable Modeling. Nonetheless, there can be instances when an analyst wishes to incorporate effects of dependence but does not yet wish to commit substantial time and resources in doing so. In this scenario, it is common to use **latent effect** variables to induce dependencies using simulation techniques. Essentially, a latent variable is an *unobserved* variable that, in this instance, is common to several risks. Because it is common to several risks, it induces a (positive) dependence. For example, suppose that X_1^*, X_2^*

represent two (unobserved) risks and that the latent α represents a third variable that is also unobserved. Then, one might observe the outcomes:

$$\begin{aligned} X_1 &= X_1^* + \alpha \\ X_2 &= X_2^* + \alpha \end{aligned} . \tag{12.7}$$

Even when the underlying variables X_1^*, X_2^*, and α are independent, the observed versions X_1 and X_2 are dependent due to the common latent variable α.

An important advantage of the structure in Display (12.7) is that it is easy to generate using simulation techniques. One simply generates (typically independently) X_1^*, X_2^*, α and then combines them according to Display (12.7) to get dependent outcomes X_1 and X_2. Historically, this method has been used with normally distributed outcomes, so that if $\{X_1^*, X_2^*, \alpha\}$ are jointly normal, then so are $\{X_1, X_2\}$. However, for our purposes, other distributions are needed and this preservation of marginal distributions is not maintained. For example, if one starts with a Pareto distributed risk X_1^*, then the resulting X_1 will not have a Pareto distribution when perturbed by a random latent variable α. Therefore, another approach is more suitable for our purposes.

Latent Variable Modeling of Dependencies Approach. For a specific example, suppose that we wish to generate dependent variables $\{X_1, X_2\}$, where X_1 has a medium-tail gamma distribution and X_2 has a long-tail Pareto distribution, and they are joined by a Gaussian copula. As seen in Section 7.1, for each replicate $r = 1, \ldots, R$:

1. Create a realization of dependent bivariate uniform random variates $\{V_{1r}, V_{2r}\}$.
2. Create a realization of dependent bivariate losses $\{X_{1r}, X_{2r}\}$, for example by using the inverse of each distribution function, $X_{jr} = F_j^{-1}(V_{jr})$, $j = 1, 2$.

Let us focus on the first step, creating dependent bivariate uniform random variates. In practice, this can be done readily via, for example, the `R` package `copula` using the function `rCopula`, Hofert et al. (2018). To see how to do this using a latent variable modeling of dependencies, start with independent uniform random variates $\{V_{1r}^*, V_{2r}^*\}$. Next, convert them to independent standard normal random variables through the transform $\epsilon_{jr}^* = \Phi^{-1}(V_{jr}^*)$ for $j = 1, 2$. Now, let α_r be normal random variable independent of $\{\epsilon_{1r}^*, \epsilon_{2r}^*\}$, and define

$$V_{jr} = \Phi\left(\frac{\epsilon_{jr}^* + \alpha_r}{\sqrt{1 + \sigma_\alpha^2}}\right), \quad j = 1, 2.$$

That is, assuming that α has mean zero and variance σ_α^2, then $y_{jr}^* = \epsilon_{jr}^* + \alpha_r$ is a normally distributed random variable with mean zero and variance $1 + \sigma_\alpha^2$. Thus, the quantity inside the parens is simply a standard normal random variable and the standard normal distribution function $\Phi(\cdot)$ converts it to a uniform random variable. The variables $\{V_{1r}, V_{2r}\}$ are dependent because they share the common latent variable α. One can show that the amount of dependence is controlled by the uncertainty of the latent variable through the parameter σ_α^2.

Dependence Modeling for Pools. We now extend this approach to pools. The portfolios discussed in this book have focused on collections of different types of risks such as building and contents, cyber risks, liability, and so forth. Another perspective, particularly relevant for insurance companies and insurance pools, views a portfolio as consisting of several entities being insured, whether they are different insureds, policyholders, or fund members. Combining these two perspectives, each entity may have several risk types. For example, we have already seen the Wisconsin Property Fund, essentially an insurance pool, where we have data on $N = 1,098$ fund members each of which may have up to $p = 6$ different risk types.

Using a latent variable approach, we are able to quantify the impact of contagion across pool members. Motivated by the Wisconsin Property Fund, consider a pool with N entities having p risks, and suppose the ith entity has $p_i \leq p$ of these risks. Our main interest is in the effects of dependency, so assume we understand the marginal distributions of each risk. As before, we can apply these marginal distributions to the risks to get the corresponding uniform outcomes and apply the inverse normal distribution so that $\mathbf{y}_i^*$ represents a $p_i \times 1$ vector of risks that is normally distributed. Think of these transformed outcomes as being composed of two sources,

$$\mathbf{y}_i^* = \mathbf{D}_i^* \boldsymbol{\alpha} + \boldsymbol{\epsilon}_i,$$

where $\boldsymbol{\alpha}$ is a random vector that is common to all entities and $\boldsymbol{\epsilon}_i$ is specific to the ith entity. Here, $\mathbf{D}_i^*$ is a known (nonstochastic) $p_i \times p$ matrix of zeros and ones that indicate which risks the ith entity faces. Assuming that $\boldsymbol{\alpha}$ and $\boldsymbol{\epsilon}_i$ are independent, we can write

$$\begin{aligned} \mathrm{Var}(\mathbf{y}_i^*) &= \mathbf{D}_i^* \ \mathrm{Var}(\boldsymbol{\alpha}) \ \mathbf{D}_i^{*'} + \mathrm{Var}(\boldsymbol{\epsilon}_i) \\ &= \mathbf{D}_i^* \mathbf{A} \mathbf{D}_i^{*'} + \boldsymbol{\Omega}_i^* = \boldsymbol{\Sigma}_i^*, \end{aligned}$$

where $\mathbf{A} = \mathrm{Var}(\boldsymbol{\alpha})$ and $\mathrm{Var}(\boldsymbol{\epsilon}_i) = \boldsymbol{\Omega}_i^*$. To convert $\mathbf{y}_i^*$ to a vector of uniformly distributed variables, we rescale the variance to be one. To this end, define $\mathbf{y}_i = [diag(\boldsymbol{\Sigma}_i^*)]^{-1/2} \ \mathbf{y}_i^*$. Thus, we can compute dependent uniform random variates as $V_{ij} = \Phi(y_{ij})$, $j = 1, \ldots, p_i$. With this, we create realizations of dependent multivariate losses using $X_{ij} = F_j^{-1}(V_{ij})$, $j = 1, \ldots, p_i$ and $i = 1, \ldots, N$.

From this procedure, we see that dependency among entities is controlled by $\mathbf{A} = \mathrm{Var}(\boldsymbol{\alpha})$. If $\mathbf{A}$ is a zero matrix, then there is no dependency among entities. Conversely, as it grows large (in some sense), it can dwarf the effect of the dependencies within a risk, effectively driving each Ω_i^* to a matrix of zeros. That is, when $\mathbf{A}$ is "large," then dependencies among entities dominates and one can effectively ignore dependencies among risk types.

Example 12.7. Effects of Contagion Among Pool Members. This is a continuation of Examples 12.2 and 12.4. The dependency among risk types, denoted by $\boldsymbol{\Omega}_i^*$, has been given in Table 12.1. For dependency among pool members, we take $\boldsymbol{\alpha}$ to be a vector of independent normal random variables with mean zero and standard deviation σ_α so that $\mathbf{A} = \sigma_\alpha^2 \ \mathbf{I}$. This example uses $R = 2,000$ simulations, as with prior examples.

Table 12.12 summarizes six simulated distributions of the predicted pool claims that vary only by σ_α. The benchmark distribution uses $\sigma_\alpha = 0$ that corresponds to independence among pool members. Table 12.12 shows that $VaR = 29.30$ for $\alpha = 0.98$, as in Example 12.2. The table shows that each uncertainty measure increases with σ_α. In contrast, the table also shows that the mean of forecast claims is relatively constant over different levels of σ_α. This is by design; to isolate the effects of contagion, the latent variables are introduced in a way that affects only the uncertainty of the distribution, not the mean.

Figure 12.2 gives further detail of the six simulated distributions using density plots. The benchmark distribution corresponding to $\sigma_\alpha = 0$, the case of independence among pool members, is the most peaked and has the thinnest tails. As one increases the value of σ_α, the distribution becomes flatter and longer-tailed. We observed earlier in Example 12.2 that the normal distribution provided a sensible approximation to the simulated distribution when $\sigma_\alpha = 0$. Figure 12.2 demonstrates that this is no longer the case for large values of σ_α.

TABLE 12.12
Pool Uncertainty Measures (in Millions) by Latent Variable Standard Deviation.

	Latent Variable Standard Deviation (σ_α)					
	0	**0.05**	**0.10**	**0.25**	**0.50**	**0.75**
Mean	23.58	23.59	23.59	23.61	23.62	23.61
VaR - 0.80	25.66	25.91	26.78	29.68	33.25	34.84
Expected Shortfall - 0.80	27.42	27.93	29.34	35.68	47.07	56.71
VaR - 0.98	29.30	30.28	32.44	42.24	63.74	83.78
Expected Shortfall - 0.98	30.79	31.43	34.22	47.97	77.42	107.83

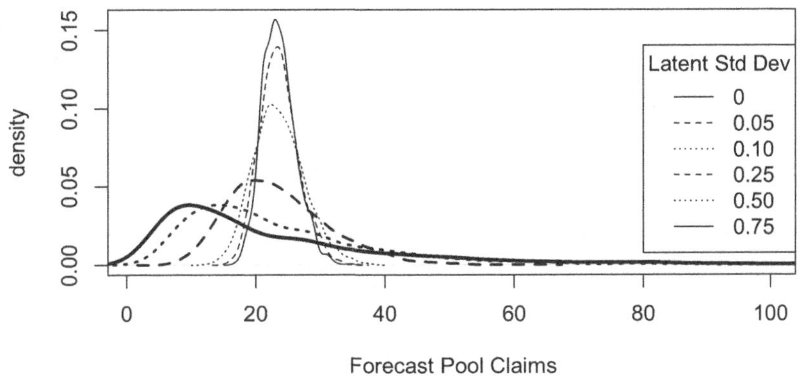

FIGURE 12.2
Density Plots of Simulated Portfolio Distribution by Latent Variable Standard Deviation.

The structure underpinning Example 12.7 can be readily extended to encompass more realistic scenarios. For instance, the specification $\mathbf{A} = \sigma_\alpha^2 \mathbf{I}$ implies that dependencies among entities are the same for each risk. However, one may wish to allow for these dependencies to vary by risk type so that $\mathbf{A}$ is still diagonal but with different values of $\mathrm{Var}(\alpha_j)$. Furthermore, for an application such as the Wisconsin Property Fund, one may wish to allow dependencies among entities of the same type such as schools or counties. The motivation is that outcomes are subject to common (latent) administrative regulations that restrict claim amounts. In addition, it is natural to introduce latent variables to account for spatial proximity due to common events such as weather occurrences (e.g. hail) that affect entities that are close to one another.

12.4.3 Pool Sensitivity to Contagion

One can get further insights into contagion among pool entities by applying the general structure developed for portfolio expectation dependence sensitivities in Section 12.2.3 to pools. Let us begin by rescaling the latent variable structure from the prior section using $[diag(\mathbf{\Sigma}_i^*)]^{-1/2}$ and defining:

- $\mathbf{D}_i = [diag(\mathbf{\Sigma}_i^*)]^{-1/2}\, \mathbf{D}_i^*$,
- $\mathbf{y}_i = [diag(\mathbf{\Sigma}_i^*)]^{-1/2}\, \mathbf{y}_i^*$,

- $\mathbf{\Omega}_i = [diag(\mathbf{\Sigma}_i^*)]^{-1/2}\, \mathbf{\Omega}_i^*[diag(\mathbf{\Sigma}_i^*)]^{-1/2}$, and
- $\mathbf{\Sigma}_i = [diag(\mathbf{\Sigma}_i^*)]^{-1/2}\, \mathbf{\Sigma}_i^*[diag(\mathbf{\Sigma}_i^*)]^{-1/2}$.

With this rescaling and equation (12.2), the weights for the pool sensitivity to contagion may be expressed as

$$W_\sigma(\mathbf{X}) \quad = \left[\mathbf{ns}(\mathbf{X})'\mathbf{\Sigma}^{-1}\mathbf{D}\right](\partial_\sigma \mathbf{A})\left[\mathbf{D}'\mathbf{\Sigma}^{-1}\mathbf{ns}(\mathbf{X})\right],$$

with the notation $\mathbf{D} = (\mathbf{D}_1', \mathbf{D}_2', \ldots, \mathbf{D}_N')'$ and $\mathbf{\Omega} = blkdiag(\mathbf{\Omega}_1, \mathbf{\Omega}_2, \ldots, \mathbf{\Omega}_N)$. The weights can be computed using

$$\mathbf{ns}(\mathbf{X})'\mathbf{\Sigma}^{-1}\mathbf{D} \quad = \left\{\sum_{i=1}^N \mathbf{ns}(\mathbf{X})_i'\mathbf{\Omega}_i^{-1}\mathbf{D}_i\right\}\left\{\mathbf{I} - \left[\mathbf{I} + \mathbf{A}\mathbf{D}'\mathbf{\Omega}^{-1}\mathbf{D}\right]^{-1}\mathbf{A}\mathbf{D}'\mathbf{\Omega}^{-1}\mathbf{D}\right\}$$

and $\mathbf{D}'\mathbf{\Omega}^{-1}\mathbf{D} = \sum_{i=1}^N \mathbf{D}_i'\mathbf{\Omega}_i^{-1}\mathbf{D}_i$. Although complex in appearance, these expressions are at most dimensions $p \times p$ and thus easy to calculate even when the number of entities N is in the thousands or more.

ⓘ *Under the Hood.* See the *Online Supplement* for confirmation of this result.

With these weights, one can calculate general portfolio expectation dependence sensitivities as in equation (12.1) and use these to compute pool dependence sensitivities for risk measures value-at-risk and expected shortfall, similar to Section 12.3.2. This is illustrated in the following example.

Example 12.8. Expected Shortfall Sensitivities for Contagion in an Insurance Pool. This is a continuation of Examples 12.4 and 12.7. This example uses $\mathbf{A} = \mathbf{0}$ so that we can utilize simulated values and make meaningful comparisons with prior examples. In addition, the weights are relatively easy to compute because the weighted sum of normal scores reduces to $\mathbf{ns}(\mathbf{X})'\mathbf{\Sigma}^{-1}\mathbf{D} = \sum_{i=1}^N \mathbf{ns}(\mathbf{X})_i'\ \mathbf{\Omega}_i^{-1}\mathbf{D}_i$. In addition, when $\mathbf{A} = \mathbf{0}$, all diagonal elements of $\mathbf{\Sigma}_i^*$ are one meaning that no rescaling is required.

With these weights, one may compute the expected shortfall sensitivities using the result in Example 12.5. Illustrative calculations are in Table 12.13. For this table, the confidence level is $\alpha = 0.80$ and the expected shortfall is expressed in millions of dollars. The sensitivity corresponding to `BC` means that a single latent variable, common to all entities, with dispersion parameter $\sigma_{1,\alpha}^2$ was used for the first risk. As this dispersion parameter increases by one unit, the table indicates one can anticipate $ES_{0.80}$ to increase by 268.219. Similar interpretations hold for the other risks.

For the `All Risks` category, now there is a single latent vector, common to all entities, with common dispersion parameter σ_α^2. This is the structure from the earlier Example 12.7. For interpretation, as this dispersion parameter increases by 0.05^2 units, the table indicates one can anticipate $ES_{0.80}$ to increase by $0.05^2\times$ 323.705, which is 0.81. This is comparable to Table 12.12, where we see the increase is $27.93 - 27.42 = 0.51$.

TABLE 12.13
Expected Shortfall Contagion Sensitivities.

Risk	*ES* Sensitivity	Risk	*ES* Sensitivity
BC	268.219	PO	-76.709
IM	6.494	CN	-0.002
PN	-1.394	CO	127.098
All Risks	323.705		

12.4.4 Insurance Companies as Risk Pools

Conceptually, an insurance company writes a contract in which it agrees to take on the uncertain risk of a consumer in exchange for a certain price. An insurance company is essentially a collection of such contracts. An insurer exists to spread the risks of many policyholders; insurance systems that can include many insurers are predicated on the pooling of contracts.

Insurers pool risks to enjoy the benefits of diversification. However, these benefits depend on the dependence among risks. Some risks are negatively associated and so provide a natural hedge for one another, offering diversification benefits. For example, consider the mortality risk in life insurance and annuities. If a policyholder lives longer than anticipated, benefit payments in life insurance are further in the future and thus lower in terms of the present value of today's money. Conversely, for a life annuity, the insurer pays while the policyholder is alive meaning a longer lifespan results in more payments. In sum, insurers pay less in life insurance but more in annuities, demonstrating a natural negative association.

In contrast, insurance risks are more commonly positively associated, such as the risk of flood to homes located close to one another. There may be few diversification benefits to pooling such positively associated risks.

The Chapter 6 risk controls represent actionable strategies that insurers can use to establish the composition of a pool. Selection of risks can be based on coverages, perils, or policyholders, or indirectly through ratemaking. Reinsurance treaties influence the insurer's retained risk, helping to determine the pool composition. This raises the question of how an insurer can measure the benefits achieved through a risk control strategy. And this is precisely why an insurer may wish to utilize the tools introduced in this book.

12.5 Summary and Concluding Remarks

At a fundamental level, dependence is a key concept underpinning risk sharing. This chapter provides approaches for quantifying the impact of dependence on risk sharing, and risk transfer, arrangements. These approaches are applied at both global and local levels, for collections of risks that are smaller (portfolios) and larger (pools). The portfolio expectation dependence sensitivity is highlighted as a useful tool for assessing local dependencies.

For managing pools, Section 12.4.1 emphasizes the importance of field-specific models of dependencies that offer insights into the effects and propagation of dependencies within a system. Additionally, the simulation models of contagion presented in this chapter can serve as a useful starting point for analyzing the effects of dependencies in large pools.

To emphasize the relevance of the tools and approaches advocated here, it is useful to acknowledge their limitations, as follows.

Absence of Dynamic Dependencies. One of main limitations of this chapter (and indeed, the entire book) has been the lack of development to address relationships over time, known as dynamic or temporal dependencies. This is significant because many dependencies occur over time and many insurance contracts are long-term, meaning they are in effect for more than one financial period. Indeed, the entire life insurance industry is based on long-term contracts. It is also a limitation from an academic perspective in that there is a host of dynamic models developed for the sister financial investments portfolio problems; presumably, these models could be leveraged to provide reasonable starting points for the insurable risks portfolios considered.

Despite these motivating factors, the approach of this chapter and book has been to emphasize one-period (sometimes known as cross-sectional) models of dependence. The goal has been to provide a careful analysis in a manner accessible to a broad audience, allowing for risks whose distributions have a mixture of discrete and continuous components, risks that are long-tail, risk retention rules that are nonlinear, and complex multivariate collections of retained risks. The aim is to provide a practically useful approach that also serves as a springboard for future investigations that may accommodate dynamic relationships.

On the Use of Elliptical Copulas. Another potential limitation is the emphasis on elliptical copulas for modeling dependencies. For example, Section 12.2.2 demonstrated how to compute sensitivity dependencies by considering elliptical copula dependence models such as the Gaussian copula. (Mathematically, this was achieved by essentially changing the basis of the distribution.) Some may find this notion of dependence to be restrictive. However, especially when combined with longitudinal and panel data models that incorporate mean effects and trends, we find that such copula models account for many situations of interest in actuarial and risk management applications, cf. Frees et al. (2016).

Note, however, for many applications, that the limitations of elliptical copulas have been well documented, cf. Frahm et al. (2003). Naturally, there are many non-elliptical copulas and other models of dependence, such as latent effects models, and this work may provide a platform for additional investigations.

12.6 Supplemental Materials

12.6.1 Further Resources and Reading

Naturally, effects of contagion have long been known to insurance researchers. As one example, an investigation of contagion due to a common interest environment for life and annuity risks was provided in earlier work by the author, Frees (1990).

For managing pools, some insights from the literature on optimal insurance are developed in the paper Bernard et al. (2020) that refers to pools as the case of "multiple policyholders."

Mathematically, the topic of *capital allocation* is closely related to the optimal risk retention problem. In insurance, an important practical problem is to allocate a firm's total pool of capital to individual business units. As emphasized in Dhaene et al. (2012), many of the principles for doing so can be expressed as a constrained optimization problem. There is a substantial actuarial literature on this problem; see, for example, Guo et al. (2021) for a recent review. Despite the parallels, this book does not take up this problem because the focus here has been on developing optimal *retention* policies for general risk-owning firms that may not include insurers. However, it is hoped that future researchers will be able to exploit potential synergies between these different problems.

This chapter (and book) has loosely described *diversification benefits* without providing formal definitions. As noted in Mainik and Embrechts (2013), a "mathematical basis of portfolio diversification was given by Markowitz (1952) for multivariate Gaussian models" but does not work well for risks typically faced in insurance modeling. As one step, their work proceeds to provide more formal definitions for representations known as "multivariate regular variation" models. The literature continues to develop. For example, see the recent work of Han et al. (2023) that are directed to the risk measures VaR and ES utilized here. It is hoped that the work in this book provides an application area that future researchers

TABLE 12.14
Independent Marginals, Simulation-Based Reinsurance Expected Payments by Upper Limit u and Retention per Limit (d/u).

Limit (u)	Retention per Limit (d/u)				
	0.0	0.25	0.50	0.75	0.95
10,000	19077	11561	6915	3192	606
100,000	37821	14420	7341	3059	546
500,000	52674	13139	6367	2594	458
1,000,000	58484	12214	5884	2385	419

will be able to use to motivate an agreed-upon formal definition of "diversification" for managing insurable risks.

12.6.2 Exercises

See the *Online Supplement* for selected exercise solutions.

Section 12.2 Exercises

Exercise 12.1. Reinsurance Expected Payments. Under the assumption of independence, determine simulation - based reinsurance expected payments by upper limits retention per limit. Your results should reproduce Table 12.14; these findings are consistent with Table 12.3 that was produced under a model that allows for dependence.

Exercise 12.2. Two Risk Portfolio. As seen in Chapter 5, when there are only $p = 2$ risks, distributions can be evaluated using double integration that can be readily computed without resorting to simulation techniques. It is helpful to go through this approach for two reasons. First, it provides some independent corroboration of the simulation based techniques. Second, one can quickly see the complexity of this approach and how this increases as the dimension p increases. Integration-based calculations certainly have their place, they can be quicker and are more accurate than simulation-based methods. However, they are limited in that they do not readily scale up to problems of interesting complexity.

a. **Portfolio Distribution.** When $p = 2$, let $\rho = \sigma$ be the dependence parameter of interest. The portfolio ($S = X_1 + X_2$) distribution function is

$$F(x; \rho) = \Pr(X_1 + X_2 \leq x) = \int_0^x C_2[F_1(x-z), F_2(z)]\ f_2(z)\ dz,$$

where recall the partial derivative of the copula $C_2(v,w) = \partial_w C(v,w)$ (see Appendix Chapter 14).

Show that one can express the differentials as

$$F_x(x;\rho) = \partial_x F(x;\rho) = \int_0^x c[F_1(x-z), F_2(z)] f_1(x-z) f_2(z) dz$$

and

$$F_\rho(x;\rho) = \partial_\rho F(x;\rho) = \int_0^x \partial_\rho\ C_2[F_1(x-z), F_2(z)] f_2(z) dz.$$

Fortunately, functions such as $\partial_\rho\ C_2(\cdot)$ are available in R package `VineCopula`.

b. **Special Case. Gaussian Copula.** To see how this works in a special case, consider uniform marginals and the Gaussian (normal) copula with partial/conditional distribution function. Derive computable versions of the results in part (a).

Exercise 12.3. Portfolio Distribution with Gamma and Pareto Risks. Returning to a familiar set up introduced in Example 4.6, assume:

- That X_1 has a gamma distribution with mean 10,000 and scale 5,000. With these parameters, the 95th percentile of X_1 is 23,719.
- That X_2 has a Pareto distribution with mean 1000 and scale 2,000. With these parameters, the 95th percentile of X_2 is 3,429.
- The dependence between risks X_1, X_2 is based on a Gaussian copula.

Using numerical integration, produce code that yields results comparable to Tables 12.15 and 12.16 for confidence levels α 0.95 and 0.90, respectively.

What do we learn from this exercise? Comparing the results of Tables 12.15 and 12.16 are interesting.

- In terms of value at risk, moving the level of confidence from 0.95 to 0.90 is more important than the dependence, except perhaps at the extremes.
- When $\alpha = 0.90$, we see that the dependence sensitivity becomes smaller for large values of ρ.

TABLE 12.15
Dependence Sensitivity Summary, $\alpha = 0.95$.

	Quantile	Density	Deriv:Rho Sum	Dependence Sensitivity (in percentages)
Rho= -0.9	23843	0.00000930	-0.0031	3.38
Rho= -0.3	24434	0.00000835	-0.0142	16.99
Rho= 0	25005	0.00000812	-0.0168	20.71
Rho= 0.3	25645	0.00000767	-0.0167	21.74
Rho= 0.9	26940	0.00000655	-0.0137	20.99

TABLE 12.16
Dependence Sensitivity Summary, $\alpha = 0.90$.

	Quantile	Density	Deriv:Rho Sum	Dependence Sensitivity (in percentages)
Rho= -0.9	19617	0.00001585	-0.0083	5.24
Rho= -0.3	20235	0.00001620	-0.0231	14.23
Rho= 0	20664	0.00001566	-0.0218	13.91
Rho= 0.3	21062	0.00001488	-0.0184	12.39
Rho= 0.9	21681	0.00001321	-0.0107	8.08

TABLE 12.17
Dependence Sensitivity Summary, Computed via Simulation.

	Quantile	Density	Deriv:Rho Sum	Dependence Sensitivity (in percentages)
Rho= -0.9	23858	0.00000828	-0.0016	1.96
Rho= -0.3	24461	0.00000833	-0.0136	16.32
Rho= 0	24992	0.00000817	-0.0168	20.53
Rho= 0.3	25643	0.00000765	-0.0167	21.83
Rho= 0.9	26949	0.00000661	-0.0117	17.69

Exercise 12.4. Portfolio Distribution - Simulation Based Computations. The results from Exercise 12.3 are based on numerical integration. This exercise investigates whether one can replicate these findings using a simulation approach. Specifically, consider five million simulated replicates using level of confidence $\alpha = 0.95$.

For this exercise, produce code that yields results comparable to Table 12.17. Note that the results from this table are largely consistent with those from Table 12.15. This suggests that we can be comfortable when calculating dependence sensitivities using simulation.

Section 12.3 Exercises

Exercise 12.5. Simulation. To supplement the portfolio expectation dependence sensitivity, this exercise develops an approach to computing $\partial_\sigma\ h(\mathbf{X}_r)$. This is a partial derivative with respect to a single simulated value of a function of risks $h(\mathbf{X})$.

For simplicity, assume that dependence among risks are captured by a Gaussian copula with dependence matrix $\mathbf{\Sigma}$. Further define f_j and F_j to be the density and distribution function of the jth marginal distribution and $h_j(x_1, \ldots, x_p) = \partial_{x_j} h(x_1, \ldots, x_p)$.

Starting with a vector of independent standard normal random vectors, $\mathbf{N}^{ind}$, show that one could compute the partial derivative using the following terms:

- $V_j = \Phi(\mathbf{1}_j' \mathbf{\Sigma}^{1/2} \mathbf{N}^{ind})$,
- $\partial_\sigma X_j = \frac{1}{f_j[F_j^{-1}(V_j)]}\ \phi\left(\mathbf{1}_j' \mathbf{\Sigma}^{1/2} \mathbf{N}^{ind}\right)\ \mathbf{1}_j'\ (\partial_\sigma \mathbf{\Sigma}^{1/2})\ \mathbf{N}^{ind}$, and
- $\partial_a\ h(X_1, \ldots, X_p) = \sum_{j=1}^{p}\ h_j[X_1, \ldots, X_p]\,(\partial_a X_j)$.

Note here that $\partial_\sigma \mathbf{\Sigma}^{1/2}$ can be determined via numerical differentiation.

Exercise 12.6. Simulation with Elliptical Copulas. Extend the work of Exercise 12.5 to include general elliptical copulas, beginning with t-copulas. *Hint*: Refer to Hofert et al. (2018) for discussions of simulating outcomes with elliptical copulas. In particular, see their Algorithm 3.1.8 for an approach for simulating t-copulas.

13

Appendix: Conventions for Notation

13.1 Risk Retention Conventions

Here is a set of symbols used in this book. The symbols are common although their use is not universal.

Symbol	Description
n, m	n risks with potentially m constraints
$\boldsymbol{\theta}$	a p - dimensional vector of risk retention parameters, such as d, c, u
d	deductible, a risk retention parameter
c	coinsurance, a risk retention parameter
u	upper limit, a risk retention parameter
$g(\cdot)$	a generic risk retention function
$S(u_1, \ldots, u_p)$	sum limited by $u_1, \ldots, u_p$, a type of risk retention function
$\mathbf{X}$	a n - dimensional vector of random risks, $X_1, \ldots, X_n$
RM	a generic risk measure, such as VaR, ES
VaR	value at risk
ES	expected shortfall
$RVaR$	range value at risk
α	confidence level for risk measures
LA	Lagrangian
RTC	risk transfer cost, the risk owner's cost of offloading a risk
RTC_{max}	maximal risk transfer cost, a fixed constant
LME	Lagrange multiplier for equality constraints
LMI	Lagrange multiplier for inequality constraints
CON_{eq}, CON_{in}	sets of equality, inequality constraints
$\mathbf{z}$	the decision variables, a p_z - dimensional vector
f_0	an objective function to be minimized
f_{con}	a generic constraint function
λ_{SL}, λ	security loading, generic penalty parameter
$U(\cdot)$	utility function
∂_x	short-hand notation for a partial derivative with respect to x, that is, $\frac{\partial}{\partial x}$
$\wedge$	wedge operator, take the minimum of two quantities
$(\cdot)_+$	take the positive part, e.g., $(-4)_+ = 0$
$\boldsymbol{\Sigma}$	matrix of association or dependence parameters

DOI: 10.1201/9781003469568-13

13.2 General Conventions

- Random variables are denoted by upper-case italicized Roman letters, with X or Y denoting a claim size variable, N a claim count variable, and S an aggregate loss variable. Realizations of random variables are denoted by corresponding lower-case italicized Roman letters, with x or y for claim sizes, n for a claim count, and s for an aggregate loss.
- Probability events are denoted by upper-case Roman letters, such as Pr(A) for the probability that an outcome in the event A occurs.
- Cumulative probability functions are denoted by $F(z)$ and probability density functions by the associated lower-case Roman letter: $f(z)$.
- For distributions, parameters are denoted by lower-case Greek letters. A caret or "hat" indicates a sample estimate of the corresponding population parameter. For example, $\hat{\beta}$ is an estimate of β .
- Upper-case boldface Roman letters denote a matrix other than a vector. Lower-case boldface Roman letters denote a (column) vector. Use a superscript prime "′" for transpose. For example, $\mathbf{x}'\mathbf{A}\mathbf{x}$ is a quadratic form.

13.2.1 Common Statistical Symbols and Operators

Here is a list of commonly used statistical symbols and operators.

$I(\cdot)$	binary indicator function. For example, $I(A)$ is one if an outcome in event A occurs and is 0 otherwise.
$\mathrm{Pr}(\cdot)$	probability function.
$F(\cdot)$	distribution function of the random variable X. For example, $F(4) = \mathrm{Pr}(X \leq 4)$.
$F_{(\cdot)}(\cdot)$	distribution function of a complex random variable or vector. For example, $F_{g(X)}(4) = \mathrm{Pr}[g(X) \leq 4]$.
F_{α}^{-1}	quantile function of the random variable X at confidence level α. For example, $F_{0.95}^{-1}$ is the 95th percentile of the distribution F. Also denoted as $F^{-1}(0.95)$.
$F_{(\cdot)}^{-1}$	quantile function of a complex random variable or vector . For example, $F_{g(X)}^{-1}(0.95)$ is the 95th percentile of the. distribution function for the random variable $g(X)$.
$\mathrm{E}(\cdot)$	expectation operator. For example, $\mathrm{E}(X)$ is the expected value of the random variable X, commonly denoted by μ.
$\mathrm{Var}(\cdot)$	variance operator. For example, $\mathrm{Var}(X)$ is the variance of the random variable X, commonly denoted by σ^2.
$\mathrm{Cov}(\cdot,\cdot)$	covariance operator. For example, $\mathrm{Cov}(X,Y) = \mathrm{E}\{(X - \mathrm{E}\ X)(Y - \mathrm{E}\ Y)\} = \mathrm{E}(XY) - (\mathrm{E}\ X)(\mathrm{E}\ Y)$ is the covariance between random variables X and Y.
$\mathrm{E}(X\vert\cdot)$	conditional expectation operator. For example, $\mathrm{E}(X\vert Y = y)$ is the conditional expected value of a random variable X given that the random variable Y equals y.
$\Phi(\cdot)$	standard normal cumulative distribution function
$\phi(\cdot)$	standard normal probability density function
$\sim$	means 'is distributed as' . For example, $X \sim F$ means that the random variable X has distribution function F.

Notes on Notation

- Var(·) is the variance operator in contrast to VaR which stands for value at risk
- E(·) is the expectation operator in contrast to ES which stands for expected shortfall.

14

Appendix: Characteristics of Copulas

Copulas were introduced in Section 4.2. This appendix describes selected characteristics of copulas that are useful in our study of risk retention.

14.1 Copula Derivatives

We wish to calculate partial derivatives of the copula distribution function. Following custom, we first consider the bivariate case.

14.1.1 Bivariate Copulas

Let C be a bivariate copula distribution and c be the corresponding density. Taking partial derivatives, we have

$$C_1(v,w) = \partial_v \ C(v,w) = \partial_v \int_0^v \int_0^w c(z_1,z_2) \ dz_1 dz_2 = \int_0^w c(v,z_2) \ dz_2$$

and similarly for C_2. Because the copula is symmetric in its arguments (the usual assumption), we have

$$C_2(v,w) = \partial_w \ C(v,w) = \int_0^v c(z_1,w) \ dz_1 = \int_0^v c(w,z_1) \ dz_1 = C_1(w,v).$$

We also need the mixed partial derivatives $C_{11}(v,w) = \partial_v \ C_1(v,w)$ and $C_{12}(v,w) = \partial_w \ C_1(v,w) = c(v,w)$. To illustrate, for the case of independence, we have $C(v,w) = v \cdot w$, $C_1(v,w) = w$, $C_2(v,w) = v$, $C_{11}(v,w) = 0$, and $C_{12}(v,w) = 1$.

Frank's Copula

To see how this works in a simple case, consider Frank's copula. The copula distribution function is

$$C(v,w) = -\frac{1}{\theta} \ln \left(\frac{1-e^{-\theta} - (1-e^{-\theta v})(1-e^{-\theta w})}{1-e^{-\theta}} \right).$$

Here, θ is the dependence parameter. With this, we have

$$\begin{aligned} C_1(v,w) &= -\frac{1}{\theta} \ \partial_v \ \ln \left(1-e^{-\theta} - (1-e^{-\theta v})(1-e^{-\theta w})\right) \\ &= \frac{e^{-\theta v}(1-e^{-\theta w})}{1-e^{-\theta} - (1-e^{-\theta v})(1-e^{-\theta w})} \end{aligned}$$

DOI: 10.1201/9781003469568-14

and

$$C_{11}(v,w) = \partial_v\, C_1(v,w) = (1-e^{-\theta w})\, \partial_v\, \frac{e^{-\theta v}}{1-e^{-\theta}-(1-e^{-\theta v})(1-e^{-\theta w})}$$
$$= -\frac{\theta(1-e^{-\theta w})e^{-\theta v}(e^{-\theta w}-e^{-\theta})}{[1-e^{-\theta}-(1-e^{-\theta v})(1-e^{-\theta w})]^2}$$

In the same way, the copula density is

$$C_{12}(v,w) = c(v,w) = \partial_w C_1(v,w)$$
$$= \frac{\theta(1-e^{-\theta})e^{-\theta(v+w)}}{[1-e^{-\theta}-(1-e^{-\theta v})(1-e^{-\theta w})]^2}.$$

See Joe (2014) (p. 165) for the distribution and density function. See also Schepsmeier and Stöber (2012), Schepsmeier and Stöber (2014).

Gaussian Copula

We will use the normal, or Gaussian, copula, so let us collect results for it in the simple bivariate case. The distribution function is:

$$C(v_1,v_2) = \Phi_2\left(\Phi^{-1}(v_1), \Phi^{-1}(v_2)\right) = \Phi_2\left(z_1, z_2\right),$$

where $z_j = \Phi^{-1}(v_j), j = 1, 2$ are normal scores and Φ_2 is a bivariate normal distribution function.

Taking partial derivatives, we have

$$C_2(v_1,v_2) = \Phi\left(\frac{z_1-\rho z_2}{\sqrt{1-\rho^2}}\right),$$
$$C_{12}(v_1,v_2) = c(v_1,v_2) = \frac{1}{\sqrt{1-\rho^2}}\frac{1}{\phi(z_1)}\phi\left(\frac{z_1-\rho z_2}{\sqrt{1-\rho^2}}\right), \quad \text{and}$$
$$C_{22}(v_1,v_2) = \frac{-\rho}{\sqrt{1-\rho^2}}\frac{1}{\phi(z_2)}\phi\left(\frac{z_1-\rho z_2}{\sqrt{1-\rho^2}}\right).$$

The density $C_{12}(\cdot,\cdot)$ and distribution function $C(\cdot,\cdot)$ are available from the `R` package `copula` (`dCopula` and `pCopula`, respectively). The first partial derivative $C_2(\cdot,\cdot)$ is available from the package `VineCopula` (the function `BiCopHfunc2`). We provide code for the function $C_{22}(\cdot,\cdot)$.

14.1.2 Multivariate Copulas

Now consider $p \geq 2$ risks, $X_1, \ldots, X_p$, with copula C that is indexed by a matrix of association parameters $\mathbf{\Sigma}$. We wish to calculate partial derivatives of the copula distribution function. For Gaussian, and later extensions to elliptical, copulas, it is natural to relate them to conditional copulas as follows.

$$
\begin{aligned}
&\partial_{v_{p-k+1}\cdots v_p} C(v_1, \dots v_p) \\
&= \frac{\partial^k}{\partial v_{p-k+1} \cdots \partial v_p} C(v_1, \dots v_p) \\
&= \int^{v_1} \cdots \int^{v_{p-k}} c\left(x_1, \dots, x_{p-k}, v_{p-k+1}, \dots, v_p\right) dx_1 \cdots dx_{p-k} \\
&= \int^{v_1} \cdots \int^{v_{p-k}} c\left(x_1, \dots, x_{p-k} | v_{p-k+1}, \dots, v_p\right) c\left(v_{p-k+1}, \dots, v_p\right) dx_1 \cdots dx_{p-k} \\
&= C\left(v_1, \dots, v_{p-k} | v_{p-k+1}, \dots, v_p\right) c\left(v_{p-k+1}, \dots, v_p\right).
\end{aligned}
$$

To illustrate, for $k = 1$, we have

$$\partial_{v_p} \ C(v_1, \dots, v_p) = C(v_1, \dots, v_{p-1} | v_p).$$

It is convenient to partition the association matrix as

$$\boldsymbol{\Sigma} = \begin{pmatrix} \boldsymbol{\Sigma}_{1:p-k,1:p-k} & \boldsymbol{\Sigma}_{1:p-k,p-k+1:p} \\ \boldsymbol{\Sigma}'_{1:p-k,p-k+1:p} & \boldsymbol{\Sigma}_{p-k+1:p,p-k+1:p} \end{pmatrix},$$

so that $\boldsymbol{\Sigma}_{1:p-k,1:p-k}$ is the submatrix for the first $p-k$ elements and similarly for the other entries.

14.1.3 Gaussian Copula

Although properties of the multivariate Gaussian copula has been developed extensively, it is worthwhile to collect facts needed for this book. This section is largely drawn from Joe (2014).

Consider a p dimensional multivariate normal distribution with variance-covariance matrix $\boldsymbol{\Sigma}$. As we will use this as a basis for defining copulas, consider the mean to be zero and variance to be 1 so that the diagonal elements of $\boldsymbol{\Sigma}$ equal 1. Let $\Phi_p(\cdot; \boldsymbol{\Sigma})$ be the corresponding distribution function. With this, the Gaussian copula can be expressed as

$$C\left(v_1, \dots, v_p\right) = \Phi_p\left(z_1, \dots, z_p; \boldsymbol{\Sigma}\right).$$

In this expression, we use the normal scores defined as $z_j = \Phi^{-1}(v_j), j = 1, \dots, d$.

For the Gaussian copula, we can utilize classic multivariate normal distribution theory to develop conditional distributions. With this, for the Gaussian copula, we have

$$
\begin{aligned}
&C\left(v_1, \dots, v_{p-k} | v_{p-k+1}, \dots, v_p\right) \\
&= \Pr\left(\Phi(N_1) \le v_1, \dots, \Phi(N_{p-k}) \le v_{p-k} | \right. \\
&\qquad\qquad \left. \Phi(N_{p-k+1}) = v_{p-k+1}, \dots, \Phi(N_p) = v_p\right) \\
&= \Pr\left(N_1 \le \Phi^{-1}(v_1), \dots, N_{p-k} \le \Phi^{-1}(v_{p-k}) | \right. \\
&\qquad\qquad \left. N_{p-k+1} = \Phi^{-1}(v_{p-k+1}), \dots, N_p = \Phi^{-1}(v_p)\right) \\
&= \Phi_{p-k}\left(z_1 - \mu_{1\cdot 2,1}, \dots, z_{p-k} - \mu_{1\cdot 2,p-k}; \boldsymbol{\Sigma}_{11\cdot 2}\right).
\end{aligned}
$$

Here, $\mu_{1\cdot 2,j}$ is the jth component of

$$\boldsymbol{\mu}_{1\cdot 2} = \boldsymbol{\Sigma}_{1:p-k,p-k+1:p}\boldsymbol{\Sigma}^{-1}_{p-k+1:p,p-k+1:p}\begin{pmatrix} z_{p-k+1} \\ \vdots \\ z_p \end{pmatrix}.$$

Further, $\Phi_{p-k}(\cdot;\boldsymbol{\Sigma}_{11\cdot 2})$ is a $p-k$ dimensional multivariate normal distribution function with mean zero and variance-covariance matrix

$$\boldsymbol{\Sigma}_{11\cdot 2} = \boldsymbol{\Sigma}_{1:p-k,1:p-k} - \boldsymbol{\Sigma}_{1:p-k,p-k+1:p}\boldsymbol{\Sigma}^{-1}_{p-k+1:p,p-k+1:p}\boldsymbol{\Sigma}'_{1:p-k,p-k+1:p}.$$

In the following, we sometimes require the correlation matrix, denoted as $\boldsymbol{\Sigma}$. Further, following standard notation, the subscripts in $\boldsymbol{\Sigma}_{ij}$ refer to the ith row and jth column of the matrix $\boldsymbol{\Sigma}$.

14.1.4 Derivatives for the Conditional Distributions

For the conditional distribution, we need the case where $k = 1$

$$\begin{aligned} C_p\left(v_1,\ldots,v_p\right) &= \partial_{v_p} C(v_1,\ldots,v_p) = C\left(v_1,\ldots,v_{p-1}|v_p\right) \\ &= \Phi_{p-1}\left(z_1-\mu_{\{1,\ldots,p-1\}\cdot p,1},\ldots,z_{p-1}-\mu_{\{1,\ldots,p-1\}\cdot p,p-1};\boldsymbol{\Sigma}_{\{1,\ldots,p-1\}\cdot p}\right), \end{aligned}$$

where

$$\boldsymbol{\mu}_{\{1,\ldots,p-1\}\cdot p} = \boldsymbol{\Sigma}_{1:p-1,p:p}\ z_p$$

and

$$\boldsymbol{\Sigma}_{\{1,\ldots,p-1\}\cdot p} = \boldsymbol{\Sigma}_{1:p-1,1:p-1} - \boldsymbol{\Sigma}_{1:p-1,p:p}\boldsymbol{\Sigma}'_{1:p-1,p:p}.$$

In the same way, for $k = 2$, we have

$$\begin{aligned} C_{p-1,p}\left(v_1,\ldots,v_p\right) &= \frac{\partial^2}{\partial v_{p-1}\partial v_p}C(v_1,\ldots,v_p) \\ &= C\left(v_1,\ldots,v_{p-2}|v_{p-1},v_p\right)c(v_{p-1},v_p) \\ &= \Phi_{p-2}\left(z_1-\mu_{\{1,\ldots,p-2\}\cdot\{p-1,p\},1},\ldots,z_{p-2}-\right. \\ &\quad \left.\mu_{\{1,\ldots,p-2\}\cdot\{p-1,p\},p-2};\boldsymbol{\Sigma}_{\{1,\ldots,p-2\}\cdot\{p-1,p\}}\right)c(v_{p-1},v_p), \end{aligned}$$

where

$$\boldsymbol{\mu}_{\{1,\ldots,p-2\}\cdot\{p-1,p\}} = \boldsymbol{\Sigma}_{1:p-2,p-1:p}\boldsymbol{\Sigma}^{-1}_{p-1:p,p-1:p}\begin{pmatrix} z_{p-1} \\ z_p \end{pmatrix}$$

and

$$\boldsymbol{\Sigma}_{\{1,\ldots,p-2\}\cdot\{p-1,p\}} = \boldsymbol{\Sigma}_{1:p-2,1:p-2} - \boldsymbol{\Sigma}_{1:p-2,p-1:p}\boldsymbol{\Sigma}^{-1}_{p-1:p,p-1:p}\boldsymbol{\Sigma}'_{1:p-2,p-1:p}.$$

14.1.5 Gaussian Dependency Sensitivity

For derivatives with respect to ρ, we start with the distribution function and note a result that, according to Plackett (1954), was well known even in the mid-1950's,

$$\frac{\partial}{\partial\rho}\Phi_2(z_1,z_2) = \phi_2(z_1,z_2).$$

For Gaussian copulas, this immediately yields

$$\frac{\partial}{\partial\rho}C(v_1,v_2) = \phi_2(z_1,z_2).$$

For other details in the bivariate case, we use the work of Schepsmeier and Stöber (2012), Schepsmeier and Stöber (2014). For example, they calculate the partial derivative

$$\frac{\partial C_2(v_1, v_2)}{\partial \rho} = \phi\left(\frac{\Phi^{-1}(v_1) - \rho\Phi^{-1}(v_2)}{\sqrt{1-\rho^2}}\right) \cdot \frac{\rho\Phi^{-1}(v_1) - \Phi^{-1}(v_2)}{(1-\rho^2)^{3/2}}.$$

We now consider derivatives with respect to dependency parameters. We work in the case of general elliptical copulas.

For a general multivariate approach, we cite a result due to Plackett (1954). Partition the correlation matrix as

$$\mathbf{\Sigma} = \left(\begin{array}{cc} \mathbf{\Sigma}_{1:2,1:2} & \mathbf{\Sigma}_{1:2,3:p} \\ \mathbf{\Sigma}'_{1:2,3:p} & \mathbf{\Sigma}_{3:p,3:p} \end{array}\right) \qquad \mathbf{\Sigma}_{1:2,1:2} = \left(\begin{array}{cc} 1 & \mathbf{\Sigma}_{12} \\ \mathbf{\Sigma}_{12} & 1 \end{array}\right),$$

so that $\mathbf{\Sigma}_{1:2,1:2}$ is the submatrix for the first two elements and $\mathbf{\Sigma}_{12}$ is the corresponding correlation coefficient. Then, from Plackett (1954), we have

$$\tfrac{\partial}{\partial \mathbf{\Sigma}_{12}} \Phi_p(x_1, \ldots, x_p; \mathbf{\Sigma}) = \phi_2\left(x_1, x_2; \mathbf{\Sigma}_{1:2,1:2}\right) \Phi_{p-2}(\mathbf{x}^*; \mathbf{\Sigma}_{\{3:p,3:p\}\cdot\{1:2\}}),$$

where

$$\mathbf{x}^* = \left(\begin{array}{c} x_3 \\ \vdots \\ x_p \end{array}\right) - \mathbf{\Sigma}'_{1:2,3:p}\mathbf{\Sigma}^{-1}_{1:2,1:2}\left(\begin{array}{c} x_1 \\ x_2 \end{array}\right) \quad \text{and}$$

$$\mathbf{\Sigma}_{\{3:p,3:p\}\cdot\{1:2\}} = \mathbf{\Sigma}_{3:p,3:p} - \mathbf{\Sigma}'_{1:2,3:p}\mathbf{\Sigma}^{-1}_{1:2,1:2}\mathbf{\Sigma}_{1:2,3:p}.$$

See also Gassmann (2003) for a more recent commentary on this interesting result.

14.2 Elliptical Dependence Sensitivity

14.2.1 Elliptical Distributions

Assume that a random p-dimensional vector $\mathbf{Y}_p$ has a multivariate elliptical distribution with density function

$$f_{\mathbf{Y}_p}(\mathbf{y}; \mathbf{\Sigma}, p) = \tfrac{c_p}{\sqrt{\det(\mathbf{\Sigma})}} \, f_p\left(\tfrac{1}{2}\mathbf{y}'\mathbf{\Sigma}^{-1}\mathbf{y}\right). \tag{14.1}$$

The function $f_p(\cdot)$ should satisfy $\int_0^\infty y^{p/2-1} f_p(y) dy < \infty$ and is known as the density generator function. As in Landsman and Valdez (2003), special cases of interest in insurance work include:

Multivariate Distribution	generator $f_p(y)$	derivative $f'_p(y)$	*NRatio* $-\frac{f'_p(y)}{f_p(y)}$
Normal distribution	e^{-y}	$-e^{-y}$	1
$t-$distribution with r degrees of freedom	$(1+2y/r)^{-(p+r)/2}$	$-\frac{p+r}{r}(1+2y/r)^{-(p+r+2)/2}$	$\frac{p+r}{2y+r}$
Cauchy	$(1+2y)^{-(p+1)/2}$	$-(p+1)(1+2y)^{-(p+3)/2}$	$\frac{p+1}{1+2y}$
Logistic	$\frac{e^{-y}}{(1+e^{-y})^2}$	$-\frac{e^{-y}(1-e^{-y})}{(1+e^{-y})^3}$	$\frac{1-e^{-y}}{1+e^{-y}}$
Exponential power	$\exp(-ry^s)$	$-rs\exp(-ry^s)y^{s-1}$	$rs\ y^{s-1}$

For dependence sensitivities, we first look to derivatives of the elliptical density. From equation (14.1), we have

$$
\begin{aligned}
\frac{1}{c_p}\partial_\sigma f_{\mathbf{Y}_p}(\mathbf{y};\boldsymbol{\Sigma},p) &= \partial_\sigma \left[[\det(\boldsymbol{\Sigma})]^{-1/2}\right] f_p\left(\frac{1}{2}\mathbf{y}'\boldsymbol{\Sigma}^{-1}\mathbf{y}\right) + [\det(\boldsymbol{\Sigma})]^{-1/2}\ \partial_\sigma f_p\left(\frac{1}{2}\mathbf{y}'\boldsymbol{\Sigma}^{-1}\mathbf{y}\right) \\
&= \left[-\frac{1}{2}[\det(\boldsymbol{\Sigma})]^{-3/2}\partial_\sigma \det(\boldsymbol{\Sigma})\right] f_p\left(\frac{1}{2}\mathbf{y}'\boldsymbol{\Sigma}^{-1}\mathbf{y}\right) \\
&\quad + [\det(\boldsymbol{\Sigma})]^{-1/2}\ f_p'\left(\frac{1}{2}\mathbf{y}'\boldsymbol{\Sigma}^{-1}\mathbf{y}\right)\frac{1}{2}\mathbf{y}'\partial_\sigma\boldsymbol{\Sigma}^{-1}\mathbf{y} \\
&= \left[-\frac{1}{2}[\det(\boldsymbol{\Sigma})]^{-3/2}\det(\boldsymbol{\Sigma})\ \mathrm{tr}(\boldsymbol{\Sigma}^{-1}\partial_\sigma\boldsymbol{\Sigma})\right] f_p\left(\frac{1}{2}\mathbf{y}'\boldsymbol{\Sigma}^{-1}\mathbf{y}\right) \\
&\quad - \frac{1}{2}[\det(\boldsymbol{\Sigma})]^{-1/2}\ f_p'\left(\frac{1}{2}\mathbf{y}'\boldsymbol{\Sigma}^{-1}\mathbf{y}\right)\mathbf{y}'\boldsymbol{\Sigma}^{-1}(\partial_\sigma\boldsymbol{\Sigma})\boldsymbol{\Sigma}^{-1}\mathbf{y} \\
&= \frac{1}{2}[\det(\boldsymbol{\Sigma})]^{-1/2} f_p\left(\frac{1}{2}\mathbf{y}'\boldsymbol{\Sigma}^{-1}\mathbf{y}\right) \\
&\quad \times\left\{-\frac{f_p'\left(\frac{1}{2}\mathbf{y}'\boldsymbol{\Sigma}^{-1}\mathbf{y}\right)}{f_p\left(\frac{1}{2}\mathbf{y}'\boldsymbol{\Sigma}^{-1}\mathbf{y}\right)}\mathbf{y}'\boldsymbol{\Sigma}^{-1}(\partial_\sigma\boldsymbol{\Sigma})\boldsymbol{\Sigma}^{-1}\mathbf{y} - \mathrm{tr}(\boldsymbol{\Sigma}^{-1}\partial_\sigma\boldsymbol{\Sigma})\right\}.
\end{aligned}
$$

Summarizing, we have

$$
\partial_\sigma f_{\mathbf{Y}_p}(\mathbf{y};\boldsymbol{\Sigma},p) = \frac{1}{2} f_{\mathbf{Y}_p}(\mathbf{y};\boldsymbol{\Sigma},p)\left\{NRatio\left(\frac{1}{2}\mathbf{y}'\boldsymbol{\Sigma}^{-1}\mathbf{y}\right)\mathbf{y}'\boldsymbol{\Sigma}^{-1}(\partial_\sigma\boldsymbol{\Sigma})\boldsymbol{\Sigma}^{-1}\mathbf{y} - \mathrm{tr}(\boldsymbol{\Sigma}^{-1}\partial_\sigma\boldsymbol{\Sigma})\right\}, \tag{14.2}
$$

where $NRatio(y) = -\frac{f_p'(y)}{f_p(y)}$.

14.2.2 Elliptical Copula Dependence Sensitivity

Because copulas are concerned with relationships, we may restrict our considerations to the case where diagonal elements of $\boldsymbol{\Sigma}$ are 1. For elliptical copulas, the marginal distributions are identical so that $F_{Y_j} = F_Y$ with density $f_Y(y) = c_1 f_1(y^2/2)$. With this, the copula distribution function is $C(v_1,\ldots,v_p) = F_{\mathbf{Y}_p}[F_Y^{-1}(v_1),\ldots,F_Y^{-1}(v_p)]$ and the corresponding probability density function

$$
\begin{aligned}
c(v_1,\ldots,v_p) &= f_{\mathbf{Y}_p}[F_Y^{-1}(v_1),\ldots,F_Y^{-1}(v_p);\boldsymbol{\Sigma},p]\prod_{j=1}^{p}\frac{1}{f_Y[F_Y^{-1}(v_j)]} \\
&= f_{\mathbf{Y}_p}[y_1,\ldots,y_p;\boldsymbol{\Sigma},p]\prod_{j=1}^{p}\frac{1}{f_Y(y_j)},
\end{aligned}
$$

where $y_j = F_Y^{-1}(v_j)$. Here, F_Y is the distribution function association with the copula, for example, the normal distribution function for the normal copula or a t-distribution function (with a specified degree of freedom) for the t-copula.

With equation (14.2), we have

$$
\begin{aligned}
\partial_\sigma c(v_1,\ldots,v_p) &= \{\partial_\sigma f_{\mathbf{Y}_p}(y_1,\ldots,y_p;\boldsymbol{\Sigma},p)\}\prod_{j=1}^{p}\frac{1}{f_Y(y_j)}\\
&= \left\{\frac{1}{2}f_{\mathbf{Y}_p}(y_1,\ldots,y_p;\boldsymbol{\Sigma},p)\left[NRatio\left(\frac{1}{2}\mathbf{y}'\boldsymbol{\Sigma}^{-1}\mathbf{y}\right)\mathbf{y}'\boldsymbol{\Sigma}^{-1}(\partial_\sigma\boldsymbol{\Sigma})\boldsymbol{\Sigma}^{-1}\mathbf{y}\right.\right.\\
&\quad \left.\left. - \operatorname{tr}(\boldsymbol{\Sigma}^{-1}\partial_\sigma\boldsymbol{\Sigma})\right]\right\}\prod_{j=1}^{p}\frac{1}{f_Y(y_j)}\\
&= \frac{1}{2}c(v_1,\ldots,v_p)\left[NRatio\left(\frac{1}{2}\mathbf{y}'\boldsymbol{\Sigma}^{-1}\mathbf{y}\right)\mathbf{y}'\boldsymbol{\Sigma}^{-1}(\partial_\sigma\boldsymbol{\Sigma})\boldsymbol{\Sigma}^{-1}\mathbf{y} - \operatorname{tr}(\boldsymbol{\Sigma}^{-1}\partial_\sigma\boldsymbol{\Sigma})\right].
\end{aligned}
\tag{14.3}
$$

This gives an expression for the elliptical copula dependence sensitivity.

14.2.3 Portfolio Expectation Dependence Sensitivity using Elliptical Copulas

We are interested in quantities of the form:

$$
\begin{aligned}
\mathrm{E}[h(X_1,\ldots,X_p)] &= \mathrm{E}\{h[F_1^{-1}(V_1),\ldots,F_p^{-1}(V_p)]\}\\
&= \int\cdots\int h[F_1^{-1}(v_1),\ldots,F_p^{-1}(v_p)]\; c(v_1,\ldots,v_p)\; dv_1\cdots dv_p.
\end{aligned}
$$

Here, the first equality is because $X_j = F_j^{-1}(U)$ in distribution (this is true for continuous, discrete, and hybrid/mixed distributions). Taking a derivative and using equation (14.3) yields the *portfolio expectation dependence sensitivity*

$$
\begin{aligned}
&\partial_\sigma \mathrm{E}[h(X_1,\ldots,X_p)]\\
&= \int\cdots\int h[F_1^{-1}(v_1),\ldots,F_p^{-1}(v_p)]\;\partial_\sigma c(v_1,\ldots,v_p)\; dv_1\cdots dv_p\\
&= \frac{1}{2}\int\cdots\int h[F_1^{-1}(v_1),\ldots,F_p^{-1}(v_p)]\left[\,NRatio\left(\frac{1}{2}\mathbf{y}'\boldsymbol{\Sigma}^{-1}\mathbf{y}\right)\mathbf{y}'\boldsymbol{\Sigma}^{-1}(\partial_\sigma\boldsymbol{\Sigma})\boldsymbol{\Sigma}^{-1}\mathbf{y}\right.\\
&\qquad\qquad \left. - \operatorname{tr}(\boldsymbol{\Sigma}^{-1}\partial_\sigma\boldsymbol{\Sigma})\right] c(v_1,\ldots,v_p)\; dv_1\cdots dv_p\\
&= \frac{1}{2}\,\mathrm{E}\left\{h[F_1^{-1}(V_1),\ldots,F_p^{-1}(V_p)]\; NRatio\left[\frac{1}{2}\mathbf{y}(U)'\boldsymbol{\Sigma}^{-1}\mathbf{y}(U)\right]\mathbf{y}(U)'\boldsymbol{\Sigma}^{-1}(\partial_\sigma\boldsymbol{\Sigma})\boldsymbol{\Sigma}^{-1}\mathbf{y}(U)\right\}\\
&\qquad - \frac{1}{2}\operatorname{tr}(\boldsymbol{\Sigma}^{-1}\partial_\sigma\boldsymbol{\Sigma})\;\mathrm{E}\; h[F_1^{-1}(V_1),\ldots,F_p^{-1}(V_p)],
\end{aligned}
$$

where $\mathbf{y}(\mathbf{V}) = [F_Y^{-1}(V_1),\ldots,F_Y^{-1}(V_p)]'$. The variables $V_1,\ldots,V_p$ have marginal uniform distributions with the same copula as $X_1,\ldots,X_p$.

Using matrix notation, we summarize this as

$$
\begin{aligned}
\partial_\sigma \mathrm{E}[h(\mathbf{X})] &= \frac{1}{2}\,\mathrm{E}\left\{h[\mathbf{q}(\mathbf{V})]\; NRatio\left[\frac{1}{2}\mathbf{y}(\mathbf{V})'\boldsymbol{\Sigma}^{-1}\mathbf{y}(\mathbf{V})\right]\mathbf{y}(\mathbf{V})'\boldsymbol{\Sigma}^{-1}(\partial_\sigma\boldsymbol{\Sigma})\boldsymbol{\Sigma}^{-1}\mathbf{y}(\mathbf{V})\right\}\\
&\qquad - \frac{1}{2}\operatorname{tr}(\boldsymbol{\Sigma}^{-1}\partial_\sigma\boldsymbol{\Sigma})\;\mathrm{E}\{h[\mathbf{q}(\mathbf{V})]\}
\end{aligned}
$$

where $\mathbf{X} = (X_1, \ldots, X_p)$ and $\mathbf{q}(\mathbf{V})' = [F_1^{-1}(V_1), \ldots, F_p^{-1}(V_p)]$. To promote interpretation, we re-label $\mathbf{ns}(\mathbf{V}) = \mathbf{y}(\mathbf{V})$ that, in the Gaussian case, are "normal scores" and define the weights $W_\sigma(\mathbf{V}) = \mathbf{ns}(\mathbf{V})'\, \mathbf{\Sigma}^{-1}(\partial_\sigma \mathbf{\Sigma})\mathbf{\Sigma}^{-1}\, \mathbf{ns}(\mathbf{V})$. With this notation, we have

$$\partial_\sigma \mathrm{E}[h(\mathbf{X})] = \frac{1}{2}\, \mathrm{E}\left\{ h[\mathbf{q}(\mathbf{V})]\ NRatio\left[\frac{1}{2}\mathbf{ns}(\mathbf{V})'\mathbf{\Sigma}^{-1}\mathbf{ns}(\mathbf{V})\right] W_\sigma(\mathbf{V}) \right\}$$
$$- \frac{1}{2}\, \mathrm{tr}(\mathbf{\Sigma}^{-1}\partial_\sigma \mathbf{\Sigma})\ \mathrm{E}\{h[\mathbf{q}(\mathbf{V})]\}.$$

Now, choosing $h(\mathbf{X}) = 1$, we have

$$\mathrm{tr}(\mathbf{\Sigma}^{-1}\partial_\sigma \mathbf{\Sigma}) = \mathrm{E}\left\{ NRatio\left[\frac{1}{2}\mathbf{ns}(\mathbf{V})'\mathbf{\Sigma}^{-1}\mathbf{ns}(\mathbf{V})\right] W_\sigma(\mathbf{V}) \right\}.$$

Thus,

$$\partial_\sigma \mathrm{E}[h(\mathbf{X})] = \frac{1}{2}\, \mathrm{Cov}\left\{ h[\mathbf{q}(\mathbf{V})],\ NRatio\left[\frac{1}{2}\mathbf{ns}(\mathbf{V})'\mathbf{\Sigma}^{-1}\mathbf{ns}(\mathbf{V})\right] W_\sigma(\mathbf{V}) \right\}. \tag{14.4}$$

For example, with a Gaussian copula, we can write equation (14.4) as

$$\partial_\sigma \mathrm{E}[h(\mathbf{X})] = \frac{1}{2}\, \mathrm{Cov}\left\{h[\mathbf{q}(\mathbf{V})],\ W_\sigma(\mathbf{V})\right\} \tag{14.5}$$

where $\mathbf{ns}(\mathbf{V})' = [\Phi^{-1}(V_1), \ldots, \Phi^{-1}(V_p)]$.

As another special case, if the marginal distributions $F_1, \ldots, F_p$ are continuous, then $F_j(X_j)$ has a uniform distribution and we may write

$$\partial_\sigma \mathrm{E}[h(\mathbf{X})] = \frac{1}{2}\, \mathrm{Cov}\left\{ h(\mathbf{X}),\ NRatio\left[\frac{1}{2}\mathbf{ns}(\mathbf{X})'\mathbf{\Sigma}^{-1}\mathbf{ns}(\mathbf{X})\right] W_\sigma(\mathbf{X}) \right\}, \tag{14.6}$$

where $\mathbf{ns}(\mathbf{X}) = (F_Y^{-1}[F_1(X_1)], \ldots, F_Y^{-1}[F_p(X_p)])'$ and $W_\sigma(\mathbf{X}) = \mathbf{ns}(\mathbf{X})'\, \mathbf{\Sigma}^{-1}(\partial_\sigma \mathbf{\Sigma})\mathbf{\Sigma}^{-1}\, \mathbf{ns}(\mathbf{X})$.

Bibliography

Abdikerimova, Samal and Runhuan Feng (2022). "Peer-to-peer multi-risk insurance and mutual aid," *European Journal of Operational Research*, Vol. 299, pp. 735–749, URL: https://doi.org/10.1016/j.ejor.2021.09.017.

Actuarial Community (2020). *Loss Data Analytics*, URL: https://openacttexts.github.io/Loss-Data-Analytics/index.html.

Albrecher, Hansjörg, Jan Beirlant, and Jozef L Teugels (2017). *Reinsurance: Actuarial and Statistical Aspects*. John Wiley & Sons.

Albrecht, Peter and Markus Huggenberger (2017). "The fundamental theorem of mutual insurance," *Insurance: Mathematics and Economics*, Vol. 75, pp. 180–188.

Alexander, Gordon J (2013). "From Markowitz to modern risk management," in *Asset Management and International Capital Markets*. Routledge, pp. 5–15.

ANU (2018). "ANU Self-Insurance Reserve Guidelines," URL: https://services.anu.edu.au/financial-management/insurance.

——— (2020). "ANU Annual Report 2020," URL: https://www.anu.edu.au/about/strategic-planning/annual-report-2020.

——— (2022a). "ANU Insurance Policy Statement," URL: https://policies.anu.edu.au/ppl/document/ANUP_000422.

——— (2022b). "ANU Risk Management Policy Statement," URL: https://policies.anu.edu.au/ppl/document/ANUP_000462.

——— (2023). "Workers' Compensation at ANU," URL: https://services.anu.edu.au/human-resources/health-safety/workers-compensation-at-anu.

Aon Risk Solutions (2019). "Global Risk Management Survey," URL: https://www.aon.com/2019-top-global-risks-management-economics-geopolitics-brand-damage-insights/index.html.

Arrow, Kenneth J (1963). "Uncertainty and the welfare economics of medical care," *American Economic Review*, Vol. 53, pp. 941–973.

Asimit, Alexandru V, Tao Gao, Junlei Hu, and Eun-Seok Kim (2018). "Optimal risk transfer: a numerical optimization approach," *North American Actuarial Journal*, Vol. 22, pp. 341–364, URL: https://doi.org/10.1080/10920277.2017.1421472.

Assa, Hirbod (2015). "On optimal reinsurance policy with distortion risk measures and premiums," *Insurance: Mathematics and Economics*, Vol. 61, pp. 70–75, URL: https://doi.org/10.1016/j.insmatheco.2014.11.007.

Ban, Gah-Yi, Noureddine El Karoui, and Andrew EB Lim (2018). "Machine learning and portfolio optimization," *Management Science*, Vol. 64, pp. 1136–1154.

Beard, Robert, Teivo Pentikäinen, and Erkki Pesonen (1984). *Risk Theory: The Stochastic Basis of Insurance, Third Edition.* Chapman and Hall.

Beketov, Mikhail, Kevin Lehmann, and Manuel Wittke (2018). "Robo Advisors: quantitative methods inside the robots," *Journal of Asset Management*, Vol. 19, pp. 363–370.

Belles-Sampera, Jaume, Montserrat Guillén, and Miguel Santolino (2014). "Beyond value-at-risk: GlueVaR distortion risk measures," *Risk Analysis*, Vol. 34, pp. 121–134, URL: https://doi.org/10.1111/risa.12080.

Benamraoui, Abdelhafid and Aljandali Abdulkader (2020). "FinTech Innovations: A review of the recent developments and prospects," *Bancaria*, Vol. 12, pp. 85–95.

Bernard, Carole, Fangda Liu, and Steven Vanduffel (2020). "Optimal insurance in the presence of multiple policyholders," *Journal of Economic Behavior & Organization*, Vol. 180, pp. 638–656.

Best, Michael J and Robert R Grauer (1991). "On the sensitivity of mean-variance-efficient portfolios to changes in asset means: some analytical and computational results," *The Review of Financial Studies*, Vol. 4, pp. 315–342.

Bierlaire, Michel (2018). *Optimization: Principles and Algorithms*, Lausanne. EPFL Press, 2nd edition, URL: http://optimizationprinciplesalgorithms.com/.

Billingsley, Patrick (2013). *Convergence of Probability Measures.* John Wiley & Sons.

Blanchet, Jose, Henry Lam, Qihe Tang, and Zhongyi Yuan (2019). "Robust actuarial risk analysis," *North American Actuarial Journal*, Vol. 23, pp. 33–63.

Boonen, Tim J and Mario Ghossoub (2023). "Bowley vs. Pareto optima in reinsurance contracting," *European Journal of Operational Research*, Vol. 307, pp. 382–391.

Borch, Karl H (1962). "Equilibrium in a reinsurance market," *Econometrica: Journal of the Econometric Society*, Vol. 30, pp. 424–444.

——— (1974). "The Mathematical Theory of Insurance: an annotated selections of papers on insurance published, 1960-1972."

Boyd, Stephen P and Lieven Vandenberghe (2004). *Convex Optimization.* Cambridge University Press.

Brockett, Patrick L, Samuel H Jr Cox, and Robert C Witt (1986). "Insurance versus self-insurance: A risk management perspective," *Journal of Risk and Insurance*, Vol. 53, pp. 242–257.

Bühlmann, Hans (1970). *Mathematical Methods in Risk Theory*, Vol. 172. Springer.

Cai, Jun and Yichun Chi (2020). "Optimal reinsurance designs based on risk measures: a review," *Statistical Theory and Related Fields*, pp. 1–13, URL: https://doi.org/10.1080/24754269.2020.1758500.

Cai, Jun, Haiyan Liu, and Ruodu Wang (2017). "Pareto-optimal reinsurance arrangements under general model settings," *Insurance: Mathematics and Economics*, Vol. 77, pp. 24–37.

Cai, Jun and Ken Seng Tan (2007). "Optimal retention for a stop-loss reinsurance under the VaR and CTE risk measures," *Astin Bulletin*, Vol. 37, pp. 93–112, URL: https://doi.org/10.1017/S0515036100014756.

Cai, Jun, Ken Seng Tan, Chengguo Weng, and Yi Zhang (2008). "Optimal reinsurance under VaR and CTE risk measures," *Insurance: Mathematics and Economics*, Vol. 43, pp. 185–196, URL: https://doi.org/10.1016/j.insmatheco.2008.05.011.

Candelon, Bertrand, Christophe Hurlin, and Sessi Tokpavi (2012). "Sampling error and double shrinkage estimation of minimum variance portfolios," *Journal of Empirical Finance*, Vol. 19, pp. 511–527.

Chalabi, Yohan and Diethelm Würtz (2015). "Portfolio allocation," in Arthur Charpentier ed. *Computational Actuarial Science with R*, pp. 447–470. CRC Press Taylor & Francis Group.

Cheung, Ka Chun, Wing Fung Chong, and Ambrose Lo (2019). "Budget-constrained optimal reinsurance design under coherent risk measures," *Scandinavian Actuarial Journal*, Vol. 2019, pp. 729–751, URL: https://doi.org/10.1080/03461238.2019.1598891.

Chopra, Vijay K and William T Ziemba (1993). "The effect of errors in means, variances, and covariances on optimal portfolio choice," *Journal of Portfolio Management;*, Vol. 19, pp. 6–11.

Cortis, Dominic, Jeremy Debattista, Johann Debono, and Mark Farrell (2019). "InsurTech," in *Disrupting Finance*. Palgrave Pivot, Cham, pp. 71–84.

Cummins, J David and Mary A Weiss (2014). "Systemic risk and the US insurance sector," *Journal of Risk and insurance*, Vol. 81, pp. 489–528.

D'Acunto, Francesco, Nagpurnanand Prabhala, and Alberto G Rossi (2019). "The promises and pitfalls of robo-advising," *The Review of Financial Studies*, Vol. 32, pp. 1983–2020.

Daw, RH (1979). "Smallpox and the double decrement table: a piece of actuarial prehistory," *Journal of the Institute of Actuaries*, Vol. 106, pp. 299–318.

Daykin, Chris D, Teivo Pentikäinen, and Martti Pesonen (1994). *Practical Risk Theory for Actuaries*. CRC Press.

De Finetti, Bruno (1940). "Il problema dei pieni," *Giorn. Ist. Ital. Attuari*, Vol. 11, pp. 1–88.

De Prado, Marcos Lopez (2016). "Building diversified portfolios that outperform out of sample," *The Journal of Portfolio Management*, Vol. 42, pp. 59–69.

DeMiguel, Victor, Lorenzo Garlappi, Francisco J Nogales, and Raman Uppal (2009). "A generalized approach to portfolio optimization: Improving performance by constraining portfolio norms," *Management science*, Vol. 55, pp. 798–812.

Denuit, Michel (2020). "Investing in your own and peers' risks: The simple analytics of P2P insurance," *European Actuarial Journal*, Vol. 10, pp. 335–359.

Denuit, Michel and Jan Dhaene (2012). "Convex order and comonotonic conditional mean risk sharing," *Insurance: Mathematics and Economics*, Vol. 51, pp. 265–270.

Denuit, Michel, Jan Dhaene, Marc Goovaerts, and Rob Kaas (2006). *Actuarial Theory for Dependent Risks: Measures, Orders and Models*. John Wiley & Sons.

Denuit, Michel, Jan Dhaene, and Christian Y Robert (2022). "Risk-sharing rules and their properties, with applications to peer-to-peer insurance," *Journal of Risk and Insurance*, Vol. 89, pp. 615–667.

Denuit, Michel and Christian Y Robert (2021a). "From risk sharing to pure premium for a large number of heterogeneous losses," *Insurance: Mathematics and Economics*, Vol. 96, pp. 116–126.

——— (2021b). "Risk sharing under the dominant peer-to-peer property and casualty insurance business models," *Risk Management and Insurance Review*, Vol. 24, pp. 181–205, URL: https://doi.org/10.1111/rmir.12180.

Dhaene, Jan, Andreas Tsanakas, Emiliano A Valdez, and Steven Vanduffel (2012). "Optimal capital allocation principles," *Journal of Risk and Insurance*, Vol. 79, pp. 1–28.

Dong, Yumo, Edward W Frees, and Fei Huang (2022). "Deductible costs for bundled insurance contracts," *Available at https://ssrn.com/abstract=4020299.*

Eling, Martin and Werner Schnell (2016). "What do we know about cyber risk and cyber risk insurance?" *The Journal of Risk Finance*, Vol. 17, pp. 474–491.

Embrechts, Paul, Rüdiger Frey, and Alexander McNeil (2005). *Quantitative Risk Management.* Princeton Series in Finance, Princeton.

Embrechts, Paul, Haiyan Liu, and Ruodu Wang (2018). "Quantile-based risk sharing," *Operations Research*, Vol. 66, pp. 936–949, URL: https://doi.org/10.1287/opre.2017.1716.

Feng, Runhuan (2023). *Decentralized Insurance: Technical Foundation of Business Models.* Springer Nature.

Feng, Runhuan, Chongda Liu, and Stephen Taylor (2023). "Peer-to-peer risk sharing with an application to flood risk pooling," *Annals of Operations Research*, Vol. 321, p. 813–842, URL: https://doi.org/10.1007/s10479-022-04841-x.

Fiacco, Anthony V (1976). "Sensitivity analysis for nonlinear programming using penalty methods," *Mathematical Programming*, Vol. 10, pp. 287–311.

Fiacco, Anthony V ed. (1983). *Introduction to Sensitivity and Stability Analysis in Nonlinear Programming*, Mathematics in Science and Engineering. Elsevier.

Fiacco, Anthony V and Garth P McCormick (1990). *Nonlinear Programming: Sequential Unconstrained Minimization Techniques.* SIAM.

Frahm, Gabriel, Markus Junker, and Alexander Szimayer (2003). "Elliptical copulas: applicability and limitations," *Statistics & Probability Letters*, Vol. 63, pp. 275–286.

Frank, Maurice J (1979). "On the simultaneous associativity of F(x, y) and x+y-F(x, y)," *Aequationes Mathematicae*, Vol. 19, pp. 194–226, URL: http://eudml.org/doc/136825.

Frees, Edward W (1990). "Stochastic life contingencies with solvency considerations," *Transactions of the Society of Actuaries*, Vol. 42, pp. 91–148, URL: https://www.soa.org/globalassets/assets/library/research/transactions-of-society-of-actuaries/1990-95/1990/january/tsa90v427.pdf.

——— (2009). *Regression Modeling with Actuarial and Financial Applications.* Cambridge University Press.

——— (2014). "Frequency and severity models," in Edward W Frees, Glenn Meyers, and Richard Derrig eds. *Predictive Modeling Applications in Actuarial Science*, Vol. 1, pp. 138–164. Cambridge University Press Cambridge.

——— (2017). "Insurance portfolio risk retention," *North American Actuarial Journal*, Vol. 21, pp. 526–551, URL: https://doi.org/10.1080/10920277.2017.1317272.

——— (2025a). *Constructing Insurable Risk Portfolios.* Taylor & Francis, CRC, URL: https://users.ssc.wisc.edu/~jfrees/InsurableRisks/.

——— (2025b). *Online Supplement to: Constructing Insurable Risk Portfolios.* Taylor & Francis, CRC, URL: https://users.ssc.wisc.edu/~jfrees/InsurableRisks/InsureRisksOnlineSupplement.pdf.

Frees, Edward W and Adam Butt (2022). "ANU Insurable Risks," URL: https://doi.org/10.25911/0SE7-N746.

Frees, Edward W and Lisa Gao (2020). "Predictive analytics and medical malpractice," *North American Actuarial Journal*, Vol. 24, pp. 211–227, DOI: http://dx.doi.org/10.1080/10920277.2019.1634597.

Frees, Edward W and Gee Lee (2015). "Rating endorsements using generalized linear models," *Variance*, Vol. 10, pp. 51–74, URL: https://www.casact.org/sites/default/files/2021-07/Rating-Endorsements-Frees-Lee.pdf.

Frees, Edward W, Gee Lee, and Lu Yang (2016). "Multivariate frequency severity regression models in insurance," *Risks*, Vol. 4, p. 4, URL: https://doi.org/10.3390/risks4010004.

——— (2024). "Data and Code to Support 'Multivariate frequency severity regression models in insurance'," URL: https://sites.google.com/a/wisc.edu/jed-frees/home.

Frees, Edward W, Glenn Meyers, and A David Cummings (2010). "Dependent multi-peril ratemaking models," *ASTIN Bulletin: The Journal of the IAA*, Vol. 40, pp. 699–726.

Frees, Edward W and Peng Shi (2024). "Insurable risk portfolios and data uncertainty."

Frees, Edward W and Emiliano A. Valdez (1998). "Understanding relationships using copulas," *North American Actuarial Journal*, Vol. 2, pp. 1–25, URL: https://doi.org/10.1080/10920277.1998.10595667.

Fu, Michael C (2008). "What you should know about simulation and derivatives," *Naval Research Logistics*, Vol. 55, pp. 723–736, URL: https://doi.org/10.1002/nav.20313.

Fu, Michael C ed. (2015a). *Handbook of Simulation Optimization.* Springer.

Fu, Michael C (2015b). "Stochastic gradient estimation," in Michael C Fu ed. *Handbook of Simulation Optimization*, pp. 105–148. Springer.

Gassmann, H. I. (2003). "Multivariate normal probabilities: implementing an old idea of Plackett's," *Journal of Computational and Graphical Statistics*, Vol. 12, pp. 731–752.

Genest, Christian and Josh Mackay (1986). "The joy of copulas: Bivariate distributions with uniform marginals," *The American Statistician*, Vol. 40, pp. 280–283, URL: https://doi.org/10.1080/00031305.1986.10475414.

Gerber, Hans and Gérard Pafumi (1998). "Utility functions: from risk theory to finance," *North American Actuarial Journal*, Vol. 2, pp. 74–91, URL: https://doi.org/10.1080/10920277.1998.10595728.

Gerber, Hans U (1979). *An Introduction to Mathematical Risk Theory.* S.S. Huebner Foundation for Insurance Education.

Givens, Geof H and Jennifer Hoeting (2013). *Computational Statistics, Second Edition*, Vol. 596. Wiley Online Library.

Glineur, François and Jean-François Walhin (2006). "de Finetti's retention problem for proportional reinsurance revisited," *Blätter der DGVFM*, Vol. 27, p. 451 462.

Gollier, Christian (2013). "The economics of optimal insurance design," in Georges Dionne ed. *Handbook of Insurance*. Springer, pp. 107–122, URL: https://doi.org/10.1007/978-1-4614-0155-1_4.

Gollier, Christian and Harris Schlesinger (1995). "Second-best insurance contract design in an incomplete market," *The Scandinavian Journal of Economics*, pp. 123–135, URL: https://doi.org/10.2307/3440833.

Gray, Roger J. and Susan M. Pitts (2012). *Risk Modelling in General Insurance: From Principles to Practice*. Cambridge University Press.

Grealish, Adam and Petter N Kolm (2021). "Robo-Advisory: From Investing Principles and Algorithms to Future Developments."

Guo, Qiheng, Daniel Bauer, and George H Zanjani (2021). "Capital allocation techniques: Review and comparison," *Variance*, Vol. 14, URL: https://variancejournal.org/article/29684-capital-allocation-techniques-review-and-comparison.

Gutterman, Sam (2017). "IAA Risk Book Chapter 17—Risk and Uncertainty," URL: https://www.actuaries.org/IAA/Documents/Publications/RiskBook/Ch17_Risk_and_Uncertainty_2017-06-06.pdf.

Haberman, Steven and Ermanno Pitacco (2018). *Actuarial Models for Disability Insurance*. Routledge.

Han, Xia, Liyuan Lin, and Ruodu Wang (2023). "Diversification quotients based on VaR and ES," *Insurance: Mathematics and Economics*, Vol. 113, pp. 185–197.

Härdle, Wolfgang (1990). *Applied Nonparametric Regression*, No. 19. Cambridge University Press.

Harrington, Scott E and Greg Niehaus (1999). *Risk Management and Insurance*. McGraw-Hill/Irwin.

Hastie, Trevor, Robert Tibshirani, and Jerome H Friedman (2009). *The Elements of Statistical Learning: Data Mining, Inference, and Prediction*, Vol. 2. Springer.

He, Guangliang and Robert Litterman (2002). "The intuition behind Black-Litterman model portfolios," *SSRN*, URL: https://ssrn.com/abstract=334304.

Henrion, René (2024). "Chance-constrained programming," URL: https://www.stoprog.org/what-stochastic-programming.

Hofert, Marius, Ivan Kojadinovic, Martin Mächler, and Jun Yan (2018). *Elements of Copula Modeling with R*. Springer.

Hong, L. Jeff (2009). "Estimating quantile sensitivities," *Operations Research*, Vol. 57, pp. 118–130, URL: https://doi.org/10.1287/opre.1080.0531.

Huebner, Solomon S, Kenneth Black, and Bernard Web (1996). *Property and Liability Insurance*. Prentice Hall.

Ingersoll, Jonathan E (1987). *Theory of Financial Decision Making*, Vol. 3. Rowman & Littlefield.

Insights, Deloitte (2019). "Global Risk Management Survey, 11th edition," URL: https://www2.deloitte.com/content/dam/insights/us/articles/4222_Global-risk-management-survey/DI_global-risk-management-survey.pdf.

Insurance Information Institute (2023a). "Finite Risk Reinsurance," URL: https://www.iii.org/article/finite-risk-reinsurance.

——— (2023b). "Small Business Insurance Basics," URL: https://www.iii.org/publications/insurance-handbook/insurance-basics/small-business-insurance-basics.

Jiang, Guangxin and Michael C Fu (2015). "Technical note. On estimating quantile sensitivities via infinitesimal perturbation analysis," *Operations Research*, Vol. 63, pp. 435–441, URL: https://doi.org/10.1287/opre.2015.1356.

Jiang, Wenjun, Hanping Hong, and Jiandong Ren (2021). "Pareto-optimal reinsurance policies with maximal synergy," *Insurance: Mathematics and Economics*, Vol. 96, pp. 185–198.

Joe, Harry (2014). *Dependence Modeling with Copulas.* CRC Press.

Kim, Sujin, Raghu Pasupathy, and Shane G Henderson (2015). "A guide to sample average approximation," *Handbook of Simulation Optimization*, pp. 207–243.

Klugman, Stuart A., Harry H. Panjer, and Gordon E. Willmot (2012). *Loss Models: From Data to Decisions.* John Wiley & Sons.

Knight, Frank H (1921). *Risk, Uncertainty, and Profit*, Boston/New York. Houghton Mifflin.

Koenker, Roger (2005). *Quantile Regression*, Vol. 38. Cambridge University Press.

Landsman, Zinoviy M and Emiliano A Valdez (2003). "Tail conditional expectations for elliptical distributions," *North American Actuarial Journal*, Vol. 7, pp. 55–71.

Ledoit, Olivier and Michael Wolf (2003). "Improved estimation of the covariance matrix of stock returns with an application to portfolio selection," *Journal of Empirical Finance*, Vol. 10, pp. 603–621.

Lee, Gee Yul (2017). *Multivariate insurance loss models with applications in risk retention.* University of Wisconsin-Madison, URL: https://na02.alma.exlibrisgroup.com/view/uresolver/01UWI_MAD/openurl?u.ignore_date_coverage=true&portfolio_pid=53842654580002122&Force_direct=true&rfr_id=info:sid/primo.exlibrisgroup.com, [Retrieved on January 1, 2023].

——— (2023). "Multivariate insurance portfolio risk retention using the method of multipliers," *North American Actuarial Journal*, URL: https://doi.org/10.1080/10920277.2022.2161578, [Retrieved on January 1, 2023].

Lemaire, Jean (1991). "Borch's theorem: A historical survey of applications," in *Risk, Information and Insurance.* Springer, pp. 15–37, URL: https://doi.org/10.1007/978-94-009-2183-2_2.

Levy, Paul S and Stanley Lemeshow (2013). *Sampling of Populations: Methods and Applications.* John Wiley & Sons.

Liu, Guangwu and Liu Jeff Hong (2009). "Kernel estimation of quantile sensitivities," *Naval Research Logistics*, Vol. 56, pp. 511–525.

Lo, Ambrose (2017). "A unifying approach to risk-measure-based optimal reinsurance problems with practical constraints," *Scandinavian Actuarial Journal*, Vol. 2017, pp. 584–605, URL: https://doi.org/10.1080/03461238.2016.1193558.

de Lourdes Centeno, Maria and Onofre Simões (2009). "Optimal reinsurance," *RACSAM-Revista de la Real Academia de Ciencias Exactas, Fisicas y Naturales. Serie A. Matematicas*, Vol. 103, pp. 387–404, URL: https://rac.es/ficheros/doc/00787.pdf.

Luenberger, David G, Yinyu Ye et al. (2016). *Linear and Nonlinear Programming.* Springer.

Machina, Mark J (2013). "Non-Expected Utility and the Robustness of the Classical Insurance Paradigm," in *Handbook of Insurance, Second Edition.* Springer, pp. 59–106.

Mainik, Georg and Paul Embrechts (2013). "Diversification in heavy-tailed portfolios: properties and pitfalls," *Annals of Actuarial Science*, Vol. 7, pp. 26–45.

Markowitz, Harry (1952). "The utility of wealth," *Journal of Political Economy*, Vol. 60, pp. 151–158.

Mayers, David and Clifford W Smith (2000). "Organizational forms within the insurance industry: Theory and evidence," *Handbook of Insurance*, pp. 689–707.

Metropolis, Nicholas and Stanislaw Ulam (1949). "The monte carlo method," *Journal of the American Statistical Association*, Vol. 44, pp. 335–341.

Metz, Jason and Ashlee Valentine (2023). "Business Owners Policy (BOP): Coverage and Costs," URL: https://www.forbes.com/advisor/business-insurance/business-owners-policy/.

Mildenhall, Stephen A and John Major (2022). *Pricing Insurance Risk: Theory and Practice.* John Wiley & Sons.

Morningstar Report (2022). "2022 Robo-Advisor Landscape," URL: https://www.morningstar.com/lp/robo-advisor-landscape.

Mossin, Jan (1968). "Optimal multiperiod portfolio policies," *The Journal of Business*, Vol. 41, pp. 215–229.

NAIC (2020). "Captive Insurance companies," URL: https://content.naic.org/cipr_topics/topic_captive_insurance_companies.htm.

——— (2023). "Business Interruption/Businessowner's Policies (BOP)," URL: https://content.naic.org/cipr-topics/business-interruptionbusinessowners-policies-bop.

National Association of Insurance Commissioners (2023a). "Peer-to-Peer (P2P) Insurance," URL: https://content.naic.org/cipr-topics/peer-peer-p2p-insurance.

——— (2023b). "Risk Retention Groups," URL: https://content.naic.org/cipr_topics/topic_risk_retention_groups.htm.

National Association of Risk Retention (2023). "NARR," URL: https://www.riskretention.org/.

Net, ABC (2020). "Hail storm sweeps through Canberra, damaging countless cars and smashing windows," *ABC Net*, URL: https://www.abc.net.au/news/2020-01-20/11882472.

Nocedal, Jorge and Stephen Wright (2006). *Numerical Optimization*. Springer Science & Business Media.

OECD (2020). "Insurance Statistics Yearbook," URL: https://stats.oecd.org/Index.aspx?DatasetCode=INSIND.

——— (2022). *OECD Insurance Statistics 2021*, pp.182, URL: https://www.oecd-ilibrary.org/content/publication/841fa619-en, DOI: http://dx.doi.org/https://doi.org/10.1787/841fa619-en.

Osborne, Martin J (2021). "Mathematical methods for economic theory: a tutorial," *University of Toronto, Toronto*, URL: https://mjo.osborne.economics.utoronto.ca/index.php/tutorial/index/1/toc.

Panjer, Harry H, Phelim P Boyle, Samuel H Cox, D Dufresne, HU Gerber, HH Mueller, HW Pedersen, SR Pliska, M Sherris, E Shiu et al. (1998). *Financial Economics: With Applications to Investments, Insurance, and Pensions*. Actuarial Foundation Schaumburg, Ill.

Pericoli, Marcello and Massimo Sbracia (2003). "A primer on financial contagion," *Journal of Economic Surveys*, Vol. 17, pp. 571–608.

Plackett, Robin L. (1954). "A reduction formula for normal multivariate integrals," *Biometrika*, Vol. 41, pp. 351–360.

Pressacco, Flavio, Paolo Serafini, and Laura Ziani (2011). "Mean variance efficient strategies in proportional reinsurance under group correlation in a gaussian framework," *European Actuarial Journal*, Vol. 1, p. 433 454.

Puccetti, Giovanni and Ruodu Wang (2015). "Extremal dependence concepts," *Statistical Science*, Vol. 30, pp. 485–517.

Rego, Margarida Lima and Joana Campos Carvalho (2020). *Insurance in Today's Sharing Economy: New Challenges Ahead or a Return to the Origins of Insurance?*, pp. 27–47. Springer International Publishing, URL: https://doi.org/10.1007/978-3-030-27386-6_2, DOI: http://dx.doi.org/10.1007/978-3-030-27386-6_2.

Robinson, Stephen M (1974). "Perturbed Kuhn-Tucker points and rates of convergence for a class of nonlinear-programming algorithms," *Mathematical programming*, Vol. 7, pp. 1–16.

Rockafellar, R Tyrrell and Stanislav Uryasev (2002). "Conditional value-at-risk for general loss distributions," *Journal of Banking & Finance*, Vol. 26, pp. 1443–1471, URL: https://doi.org/10.1016/S0378-4266(02)00271-6.

Rosset, Saharon and Ji Zhu (2007). "Piecewise linear regularized solution paths," *The Annals of Statistics*, pp. 1012–1030.

Schäfer, Juliane and Korbinian Strimmer (2005). "A shrinkage approach to large-scale covariance matrix estimation and implications for functional genomics," *Statistical Applications in Genetics and Molecular Biology*, Vol. 4.

Schepsmeier, Ulf and Jakob Stöber (2012). "Web supplement: Derivatives and Fisher information of bivariate copulas,"Technical report, Tech. rep. TU München.

——— (2014). "Derivatives and Fisher information of bivariate copulas," *Statistical Papers*, Vol. 55, pp. 525–542.

Schlesinger, Harris (2013). "The theory of insurance demand," in *Handbook of Insurance*. Springer New York, pp. 167–184, URL: https://doi.org/10.1007/978-1-4614-0155-1_7.

Seal, Hilary L (1969). *Stochastic Theory of a Risk Business*. Wiley.

Serfling, Robert J. (1980). *Approximation Theorems of Mathematical Statistics*, Vol. 162. John Wiley & Sons.

Serpa, Juan Camilo and Harish Krishnan (2017). "The strategic role of business insurance," *Management Science*, Vol. 63, pp. 384–404.

Shapiro, Alexander (2013). "Sample average approximation," *Encyclopedia of Operations Research and Management Science*, Vol. 3, pp. 1350–1355.

Simon, Carl P and Lawrence Blume (1994). *Mathematics for Economists*. W.W. Norton and Company, New York.

Skipper, Harold D. and W. Jean Kwon (2007). *Risk Management and Insurance: Perspectives in a Global Economy*. Blackwell.

Sklar, M (1959). "Fonctions de repartition a N dimensions et leurs marges," *Publ. inst. statist. univ. Paris*, Vol. 8, pp. 229–231.

State of Vermont (2020). "Captive basics," URL: https://www.vermontcaptive.com/captive/.

Stout, William F (1974). "Almost Sure Convergence."

Sun, Haoze, Chengguo Weng, and Yi Zhang (2017). "Optimal multivariate quota-share reinsurance: A nonparametric mean-CVaR framework," *Insurance: Mathematics and Economics*, Vol. 72, pp. 197–214.

Swiss.Re (2013). "The essential guide to reinsurance," URL: https://www.swissre.com/Library/the-essential-guide-to-reinsurance.html, [Retrieved on June 10, 2020].

Tan, Ken Seng and Chengguo Weng (2014). "Empirical approach for optimal reinsurance design," *North American Actuarial Journal*, Vol. 18, pp. 315–342, URL: https://doi.org/10.1080/10920277.2014.888008.

Toumi, Ralf and Lauren Restell (2014). "Catastrophe Modelling and Climate Change," URL: https://assets.lloyds.com/assets/pdf-modelling-and-climate-change-cc-and-modelling-template-v6/2/pdf-modelling-and-climate-change-CC-and-modelling-template-V6.pdf.

Tse, Yiu-Kuen (2009). *Nonlife Actuarial Models: Theory, Methods and Evaluation*. Cambridge University Press.

Vajda, Stefan (1962). "Minimum variance reinsurance," *ASTIN Bulletin: The Journal of the IAA*, Vol. 2, pp. 257–260, URL: https://doi.org/10.1017/S0515036100009995.

Verlaak, Robert and Jan Beirlant (2003). "Optimal reinsurance programs: an optimal combination of several reinsurance protections on a heterogeneous insurance portfolio," *Insurance: Mathematics and Economics*, Vol. 33, pp. 381–403, URL: https://doi.org/10.1016/j.insmatheco.2003.08.002.

Wang, Ruodu and Ričardas Zitikis (2021). "An axiomatic foundation for the Expected Shortfall," *Management Science*, Vol. 67, pp. 1413–1429, URL: https://pubsonline.informs.org/doi/pdf/10.1287/mnsc.2020.3617.

Willams, C Arthur, Michael L. Smith, and Peter C. Young (1995). *Risk Management and Insurance, Seventh Edition.* McGraw-Hill.

Xiao, Yugu and Emiliano A Valdez (2015). "A Black–Litterman asset allocation model under elliptical distributions," *Quantitative Finance*, Vol. 15, pp. 509–519.

Yates, Natalie (2023). "Reciprocal Insurance Exchange," URL: https://www.usnews.com/insurance/glossary/reciprocal-insurance-exchange.

Yin, Chenyang, Romain Perchet, and François Soupé (2021). "A practical guide to robust portfolio optimization," *Quantitative Finance*, Vol. 21, pp. 911–928.

Index